T5-CQC-541

INTRAOCULAR
DRUG
DELIVERY

INTRAOCULAR DRUG DELIVERY

edited by

Glenn J. Jaffe
Duke University
Durham, North Carolina, U.S.A.

Paul Ashton
Control Delivery Systems,
Watertown, Massachusetts, U.S.A.

P. Andrew Pearson
University of Kentucky,
Lexington, Kentucky, U.S.A.

Taylor & Francis
Taylor & Francis Group
New York   London

Published in 2006 by
Taylor & Francis Group
270 Madison Avenue
New York, NY 10016

© 2006 by Taylor & Francis Group, LLC

No claim to original U.S. Government works
Printed in the United States of America on acid-free paper
10 9 8 7 6 5 4 3 2 1

International Standard Book Number-10: 0-8247-2860-2 (Hardcover)
International Standard Book Number-13: 978-0-8247-2860-1 (Hardcover)
Library of Congress Card Number 2005046669

Library of Congress Cataloging-in-Publication Data

Intraocular drug delivery / edited by Glenn J. Jaffe, Paul Ashton, Andrew Pearson.
    p. ; cm.
  Includes bibliographical references and index.
  ISBN-13: 978-0-8247-2860-1 (alk. paper)
  ISBN-10: 0-8247-2860-2 (alk. paper)
  1. Ocular pharmacology. 2. Drug delivery systems. 3. Therapeutics, Opthalmological. I. Jaffe, Glenn J.
II. Ashton, Paul, 1960- III. Pearson, Andre, 1961-
  [DNLM: 1. Drug Delivery Systems. 2. Drug Administration Routes. 3. Eye Diseases--drug therapy. WB 340 I61 2005]

RE994.I62 2005
617.7'061--dc22
                                                     2005046669

informa
Taylor & Francis Group
is the Academic Division of Informa plc.

**Visit the Taylor & Francis Web site at
http://www.taylorandfrancis.com**

Preface

The development of drug treatments for diseases of the retina and back of the eye has been slow. Among the principal causes for this have been a failure of the pharmaceutical industry to appreciate the potential size of the market these diseases represent, a poor understanding of the disease processes themselves, and technical difficulty in delivering drugs to the back of the eye. There have been recent rapid advances in all three areas with many more changes likely to occur in the next decade.

Until the 1990s, very few drugs had ever been developed specifically for ophthalmology. Virtually all drugs used in ophthalmology had initially been developed for other applications and subsequently found to be useful in ophthalmology. One potential reason for this is economics. In 2001 it was estimated that it took over 12 years and cost over $800 million to develop and commercialize a new drug (1). For a company to undertake such an investment there must be a reasonable expectation that eventually sales of a new drug will, after allowing for development risk, at least recoupe its development costs. In 1996 the total world market for drugs for back-of-the-eye diseases was less than $500 million, providing little impetus to develop drugs for these conditions.

A major contributor to both the cost and the time it takes to develop a drug is the regulatory approval process. Following animal experiments, drugs enter limited clinical trials that often involve very few patients. These early studies, often called Phase I or Phase I/II trials, are generally designed to get a preliminary indication of safety and possibly efficacy while exposing as few subjects to the drug as possible. Once these studies have been successfully completed, a product can proceed to larger, Phase II trials. The goal of these larger trials, often involving 50 to 100 people, is to generate sufficient efficacy data to adequately power the next, Phase III, studies. It is these studies, sometimes called pivotal trials, that are designed to provide sufficient data to satisfy the regulatory agencies that a product is both safe and effective. Data collected in Phase II is generally used to ensure pivotal studies are appropriately designed and have sufficient statistical power to meet these objectives. These larger trials involve hundreds to thousands of patients. In clinical trials of an agent to treat a previously untreated disease it can be difficult to decide on the primary clinical trial endpoint to demonstrate drug efficacy. This is particularly true for diseases that are slowly progressing, where a clinically significant progression of the disease can take years. Any drug therapy designed to slow down the progression of such a disease is likely to require very long term clinical trials, increasing the time, the cost and the risk of developing a drug. Diseases in this group include diabetic retinopathy, neovascular and non-neovascular age-related macular degeneration, retinitis pigmentosa and others. For a company developing a drug to treat these conditions, while risks from competitors are always present, they become magnified in the face of very long-term and expensive clinical trials. As a trial progresses, science advances and a competitor may develop a better drug or a more creative way through the regulatory system.

The difficulty of the Food and Drug Administration's (FDA's) task in approving drugs, especially for previously untreated diseases, should not be underestimated. Considerable pressure is exerted on the FDA to both approve drugs quickly and to ensure drugs meet the appropriate standards of safety and efficacy. The FDA is in a difficult position. If after approval significant side effects are encountered, the FDA is likely deemed to be at fault. On the other hand, if a drug is not approved quickly, the FDA is likely deemed to be at fault. The voices decrying the "glacial" pace of drug approval are often the same ones decrying the "cavalier attitude" of the FDA should a drug be withdrawn. Despite these pressures, the FDA can move extremely quickly to approve new drug treatments. Although it takes an average of 12 years for a drug to be developed, Vitrasert®, a sustained release delivery device to treat AIDS associated cytomegalovirus retinitis, progressed from in vitro tests to FDA approval in eight years. The total development time for Rertisert®, which recently became the first drug treatment approved for uveitis, was seven years. Both of these products were supported initially by grants from the National Eye Institute and without such support, the industry has rarely funded the development of such high-risk programs. For major pharmaceutical companies the risks of developing drugs for well understood diseases are high enough. Add to these risks an unknown market size, unfamiliar regulatory approval process, new drug delivery requirements and novel pharmacological drug targets, and the process becomes truly daunting. "Big Pharma" has not perceived the opthalmic marketplace as large enough to support a fully-fledged development effort. Pharmaceutical development has instead been largely limited to smaller, so-called "specialty" pharmaceutical companies.

A turning point in ophthalmology came with the approval of Latanaprost, a prostaglandin analogue. This molecule was developed specifically for glaucoma and has been commercially extremely successful, generating over \$1 billion per year in sales in 2003 (2). This appears to have triggered the realization that ophthalmology has the potential to support billion dollar products and has lead to an increased focus on the area by the pharmaceutical industry.

In recent years there has been a dramatic increase in the understanding of the pathologies of ocular diseases and, perhaps not coincidentally, many new therapeutic candidates and pharmacological treatments. Unlike such mature fields as hypertension, there is as yet no clear consensus of the pharmacological targets best hit to generate an optimal therapeutic response. Not only are there now a large number of drugs under development but there are also a large number of different classes of drugs in development. Into the mix of increased commercial focus and rapidly advancing biology there is also the rapidly evolving field of drug delivery for the posterior segment of the eye. This state of high flux is exemplified by the three treatments for wet age-related macular degeneration that are either approved or awaiting approval. The first approved, Visudyne®, is an intravenous injection followed by an ocular laser to activate the drug in the eye. In 2005 it was followed by Macugen®, a vascular endothelial growth factor (VEGF) inhibitor, given by intravitreal injections every six weeks. Retaane™ is pending approval and is an angiostatic steroid given as a peri-ocular injection every six months. All three of these treatments have completely different modes of action and completely different means of administration.

This book is a snap shot in time. In it the contributors have attempted to describe some of the parameters influencing drug delivery and some of the attempts made, with varying degrees of success, to achieve therapeutic drug concentrations in the posterior of the eye. Also described are disease states of the back of the eye, some of which, like wet age-related macular degeneration, affect many people. Following the approval of Visudyne and Macugen, one could expect rapid changes in clinical management of these diseases. Other conditions, like retinitis pigmentosa, are very slowly progressing (making the design of clinical trials extremely difficult) or else affect only a small number of people, such as proliferative vitreoretinopathy (PVR). For these conditions there is as yet no precedent with the FDA for what constitutes an approvable drug. Progress in the management of such conditions is unfortunately likely to be much slower.

Glenn J. Jaffe
Paul Ashton
P. Andrew Pearson

REFERENCES

1. DiMasi JA, Hansen RW, Grabowski HG. The price of innovation. New estimates of drug development costs. J Health Econ 2003; 22:151–185.
2. Form 10-K. SEC. Pfizer Annual Report Year End December 31, 2003.

Contents

Contributors

Gustavo D. Aguirre   James A. Baker Institute for Animal Health, College of Veterinary Medicine, Cornell University, Ithaca, New York, U.S.A.

Michael M. Altaweel   Department of Ophthalmology and Visual Sciences, University of Wisconsin-Madison Medical School, Madison, Wisconsin, U.S.A.

Jayakrishna Ambati   Department of Ophthalmology and Visual Sciences and Physiology, University of Kentucky, Lexington, Kentucky, U.S.A.

David A. Antonetti   Departments of Cellular and Molecular Physiology and Ophthalmology, Penn State College of Medicine, Hershey, Pennsylvania, U.S.A.

Paul Ashton   Control Delivery Systems, Watertown, Massachusetts, U.S.A.

Sanjay Asrani   Duke University Eye Center, Durham, North Carolina, U.S.A.

Sophie J. Bakri   The Cole Eye Institute, Cleveland Clinic Foundation, Cleveland, Ohio, U.S.A.

Alistair J. Barber   Department of Ophthalmology, Penn State College of Medicine, Hershey, Pennsylvania, U.S.A.

Caroline R. Baumal   Department of Ophthalmology, Vitreoretinal Service, New England Eye Center, Tufts University School of Medicine, Boston, Massachusetts, U.S.A.

Jean Bennett   F.M. Kirby Center for Molecular Ophthalmology, Scheie Eye Institute, University of Pennsylvania, Philadelphia, Pennsylvania, U.S.A.

Mark T. Cahill   Duke University Eye Center, Durham, North Carolina, U.S.A.

Thomas W. Gardner   Departments of Cellular and Molecular Physiology and Ophthalmology, Penn State College of Medicine, Hershey, Pennsylvania, U.S.A.

Morton F. Goldberg   Wilmer Ophthalmological Institute, Johns Hopkins University, Baltimore, Maryland, U.S.A.

Lewis J. Gryziewicz   Regulatory Affairs, Allergan, Irvine, California, U.S.A.

Michael S. Ip   Department of Ophthalmology and Visual Sciences, University of Wisconsin-Madison Medical School, Madison, Wisconsin, U.S.A.

Glenn J. Jaffe   Duke University Eye Center, Durham, North Carolina, U.S.A.

Peter K. Kaiser   The Cole Eye Institute, Cleveland Clinic Foundation, Cleveland, Ohio, U.S.A.

Ivana K. Kim   Department of Ophthalmology, Harvard Medical School, Massachusetts Eye and Ear Infirmary, Boston, Massachusetts, U.S.A.

Hideya Kimura   Nagata Eye Clinic, Nara, Japan

Alan Laties   Department of Ophthalmology, University of Pennsylvania School of Medicine, Philadelphia, Pennsylvania, U.S.A.

Albert M. Maguire   F.M. Kirby Center for Molecular Ophthalmology, Scheie Eye Institute, University of Pennsylvania, Philadelphia, Pennsylvania, U.S.A.

Melissa J. Mahoney   Departments of Ophthalmology and Neurobiology, Duke University Medical Center, Durham, North Carolina, U.S.A.

Travis A. Meredith   Department of Ophthalmology, University of North Carolina, Chapel Hill, North Carolina, U.S.A.

Joan W. Miller   Department of Ophthalmology, Harvard Medical School, Massachusetts Eye and Ear Infirmary, Boston, Massachusetts, U.S.A.

Yuichiro Ogura   Ophthalmology and Visual Science, Nagoya City University Graduate School of Medical Science, Nagoya, Aichi, Japan

P. Andrew Pearson   Department of Ophthalmology and Visual Science, Kentucky Clinic, Lexington, Kentucky, U.S.A.

Ward M. Peterson   Department of Biology, Inspire Pharmaceuticals, Durham, North Carolina, U.S.A.

Stephen J. Phillips   Duke University Eye Center, Durham, North Carolina, U.S.A.

Zeshan A. Rana   Department of Ophthalmology and Visual Science, Kentucky Clinic, Lexington, Kentucky, U.S.A.

Kourous A. Rezaei   Department of Ophthalmology, Rush University Medical Center, University of Chicago, Chicago, Illinois, U.S.A.

Dennis W. Rickman   Departments of Ophthalmology and Neurobiology, Duke University Medical Center, Durham, North Carolina, U.S.A.

Weng Tao Neurotech USA, Lincoln, Rhode Island, U.S.A.

Rong Wen Department of Ophthalmology, University of Pennsylvania School of Medicine, Philadelphia, Pennsylvania, U.S.A.

Scott M. Whitcup Research and Development, Allergan, Irvine and Department of Ophthalmology, Jules Stein Eye Institute, David Geffen School of Medicine at UCLA, Los Angeles, California, U.S.A.

Ran Zeimer Wilmer Ophthalmological Institute, Johns Hopkins University, Baltimore, Maryland, U.S.A.

1

Retinal Drug Delivery

Paul Ashton
Control Delivery Systems, Watertown, Massachusetts, U.S.A.

In any drug treatment, the overall goal of drug delivery is to achieve and maintain therapeutic concentrations of the drug at its site of action for sufficient time to produce a beneficial effect. A secondary aim is to avoid exposing any other tissues to concentrations of the agent high enough to cause a deleterious effect. The efficacy of a compound is governed by its intrinsic effects on the target site (and any other sites with which it comes into contact), its distribution throughout and its elimination from the body. Alterations to the and elimination of a compound can thus radically alter its efficacy. For regions of the body with a significant barrier to drug permeation, such as the eye and brain, great care should be taken to deliver drugs appropriately.

In the design of a drug delivery system for the eye a balance must be struck between the limitations imposed by the physicochemical properties of the drugs, the limitations imposed by the anatomy and disease state of the eye, and the dosing requirements of the drug for that particular disease. This chapter gives an overview of some general concepts and tactics in drug delivery, barriers to getting drugs into the vitreous and retina, mechanisms by which drugs are cleared, and drug delivery for some specific ophthalmic problems.

BASIC PRINCIPLES OF DRUG DELIVERY

Delivery Rates

As a means to predict the properties of drug delivery systems, it is useful to briefly review some basic thermodynamic functions. The rate or speed of a reaction is given by

$$dc/dt$$

where dc is the change in concentration and dt is the time interval over which that change occurs.

In the simplest case, dc/dt is constant, i.e., the rate of change does not vary over time, this situation is termed zero order and can be expressed as

$$-dc/dt = k \qquad (1)$$

where k is a constant, also known as the rate constant.

Integrating this with respect to time gives

$$C_t - C_0 = -kt \tag{2}$$

where C_0 is the initial concentration and C_t is the concentration at time t. A graph of concentration (C_t) versus time would therefore have a constant gradient with k as the slope (Fig. 1).

In drug delivery, a more common situation is one where the rate of change of concentration is directly proportional to the concentration of drug present. This situation is termed first order and can be expressed as

$$-dc/dt = kc \tag{3}$$

Integrating this with respect to time gives

$$\ln C_t - \ln C_0 = -kt \tag{4}$$

Expressing (4) another way

$$C = C_0 e^{-kt} \tag{5}$$

A graph of C versus time is shown in Figure 2A and a graph of $\ln C$ versus time is shown in Figure 2B.

In the situation where C_t is half of C_0 (i.e., the concentration has decreased by 50%), and where t is the half-life, this equation becomes

$$\ln 2 = kt_{1/2} \tag{6}$$

or

$$\ln 2/k = t_{1/2} \tag{7}$$

i.e., larger the rate constant, shorter is the half-life.

Zero-order reactions describe processes such as output from a pump or, in some cases, diffusion from a suspension. Radioactive decay is an example of a first-order process.

Generally, a drug's potential to diffuse from one region to another is directly proportional to its chemical potential, which can usually be approximated to its concentration. The aforementioned equations can thus be readily applied to drug delivery.

Considering the setup in Figure 3, drug diffusion from chamber A to chamber B is driven by the difference in concentration (or chemical potential) between the two chambers. Assuming that chamber B is a perfect sink and that chamber A is well stirred, as the drug diffuses from A, the concentration in this chamber is decreased, and with it the driving force for diffusion. This results in a progressively slower release

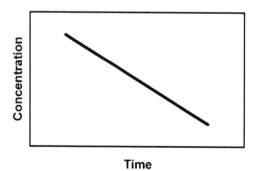

Figure 1 Zero-order or linear kinetics showing the decrease in drug concentration in a dosage form versus time.

(A)

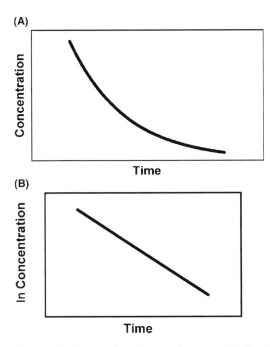

Figure 2 First order kinetics showing, (**A**) the decrease in drug concentration in a dosage form versus time, (**B**) the natural log(ln) of the same concentration data plotted against time.

rate. Drug diffusion from chamber A to chamber B in this system will follow first-order kinetics. Figure 3B describes a similar situation except that here chamber A contains a suspension of the drug. In this situation, drug delivery from A into B is again determined by the difference in chemical potential between A and B, but in this case as the drug diffuses from A, some of the solid drug in A dissolves so as to maintain the concentration in A. As long as the dissolution of the drug in A is able to keep

(**A**)

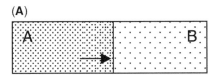

(**B**)

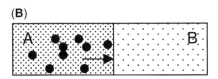

Figure 3 (**A**) Chamber A contains a diffusant that is fully dissolved in A and is at a higher concentration than in chamber B, which is a perfect sink. Diffusion from A to B is driven by the concentration difference between the two chambers. (**B**) Chamber A contains a suspension of a diffusant. As the diffusant moves to chamber B, the decrease in concentration can be off-set by the dissolution of the suspended particles, which acts to maintain the concentration gradient.

pace with the diffusion across the membrane, the concentration of dissolved drug will be maintained and consequently the diffusion across the membrane will be constant provided B remains a perfect sink. Release rate from A will therefore follow zero-order kinetics.

The aforesaid situations apply in special cases where diffusion through the material in chamber A is not important (A is well stirred) and where the dissolution rate of the drug particles in A is rapid. A more common situation arises when drug release is both a function of its concentration within a vehicle and its ability to diffusion through it. When placed into a release medium, the drug closest to the surface is released the fastest. Over a period of time, the drug must diffuse from further and further back within the bulk of the device, which progressively slows the release. Systems such as this can be described by solutions to Ficks's second law of diffusion (1).

$$\delta C/\delta t = D\delta^2 c/dx^2 \tag{8}$$

where C is the concentration in a reservoir, t the time, x the distance and D the diffusion coefficient the diffusant through the media. Partial derivatives (δ) are used because C is a function of both t and x.

In the 1960s, Higuchi (2) proposed that if diffusion through a vehicle is rate limiting, then the amount of drug released from a vehicle (in which the drug is fully dissolved) can be described by

$$Q = 2\ C_0(Dt/\pi)^{1/2} \tag{9}$$

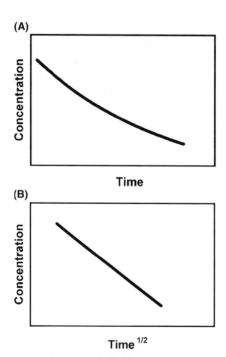

Figure 4 Square root time kinetics typical of release from a gel or ointment as described by Higuchi (2,3). (**A**) Decrease in concentration of drug versus time, (**B**) same data plotted against square root of time.

where Q is the quantity of drug released per unit area, t the time of application, and C_0 is the initial concentration of drug in the vehicle (2). It is assumed that the composition of the vehicle is initially homogeneous and that the receptor acts as a perfect sink (once out of the vehicle the drug is immediately removed from the surface).

Similarly, an equation was proposed to describe release from an emulsion or suspension in which diffusion through the vehicle is rate limiting (3)

$$Q = (2\, C_0\, Dt\, C_s)^{1/2} \tag{10}$$

where C_0 is the initial concentration of the diffusant in the vehicle and C_s the solubility limit of the diffusant in the vehicle.

These formulae (10) can be readily applied to drug release from ointments and gels typically used in eye drops. The unfortunate implication is that even if a gel or ointment system remains in the cul de sac of the eye for an extended period, any drug in such a vehicle will be released by square root time kinetics. Further, in each case the amount of drug released from the vehicle is proportional to D, the diffusion coefficient of the diffusant within the vehicle.

D has been described in classical physics by the Stokes–Einstein equation (1)

$$D = KT/6\pi\eta r \tag{11}$$

where K is the Boltzmann constant, T is the absolute temperature, r is the diffusion radius of the molecule, and η is the viscosity of the vehicle. Thus, from Eqs. (4) and (5), the release of the drug from a diffusion controlled vehicle is both inversely proportional to the viscosity of the vehicle and a function of the square root of time. To have a significant effect on release rates, a gel or ointment may have to be highly viscous (see discussion below).

Similar approaches based on Higuchi's work have been developed over the last 40 years describing many of the hypothetical situations that arise in drug delivery systems. Considerably more detailed mathematical analyses can be found in textbooks for those with an interest in such matters.

Gels, Creams, and Ointments

One of the simplest approaches to sustained drug delivery is the formulation of a drug in a viscous gel, cream, or ointment. Release from such a system is governed by both diffusion (discussed previously) and partitioning of the drug from the vehicle into the release medium. The more soluble the drug in the release medium (and the lower its chemical potential) and the less soluble in the vehicle (higher its chemical potential), the more readily the drug will partition into the release medium. If diffusion through the vehicle is very fast and the partitioning is so slow as to be rate determining, release is likely to approximate first-order kinetics (the concentration of drug in the vehicle at any time will be homogeneous). If, however, partitioning from the vehicle is relatively fast and diffusion through the vehicle slow (and rate limiting) square root time kinetics can be anticipated. An additional consideration is that release will be a function of the viscosity actually encountered by the diffusant. For hydrogels this is not normally the same as the perceived bulk viscosity. The bulk viscosity of a gel is the so-called "large-scale" viscosity. When a relatively large object moves through a gel it must either break or move aside gel fibers and this creates the viscosity. On the molecular scale, however, a small molecule may be able to diffuse through the fiber matrix between strands. For such a molecule, the microviscosity would be similar to that of the water between the strands. A small drug molecule such as pilocarpine may move very quickly through a highly viscous gel while a larger molecule, such as a protein, is more likely to become entangled in the strands of a gel and will thus diffuse more slowly. Hence, although some sustained release properties have been described for small molecules in hydrogels, these systems may have more promise in controlling the delivery of large molecules such as proteins or peptides (4).

Microparticulates and Liposomes

Many microencapsulated drug delivery systems, such as nanoparticles and microspheres, often composed of biocompatible, bioerodible polymers, have been developed. The polylactide/glycolide (PLGA) polymer system in particular has been extensively investigated and a microsphere system based on this polymer that releases leuprolide acetate is in clinical use (5). Release from such microspheres is a function of both diffusion of a compound from the particles and degradation of the particle. These systems and many other bioerodible devices typically exhibit complex release kinetics. There is generally a fast initial phase followed by a slower second phase typical of square root time kinetics as the drug molecules diffuse through the matrix. The second phase is often followed by a burst effect as the PLGA polymers undergo bulk erosion and the systems lose structural integrity.

Liposomes are small vesicles, typically ranging in size from 0.01 to 10 μm, composed of single or concentric phospholipid bilayers entrapping water in their center. They are formed by the dispersion of phospholipids in water (6,7) and have been investigated since the 1970s as a means to achieve controlled and targeted drug delivery. Hydrophilic drugs can be entrapped within the aqueous liposome core while lipophilic drugs can partition into the lipid bilayers, either process dramatically altering their biodistribution (6,8).

There are several mechanisms by which drugs can be released from liposomes. In the simplest case, the liposome acts as a sustained release reservoir slowly releasing drug into the surrounding fluid. Release, being a function of drug concentration within the liposome, can be expected to decrease exponentially with time (first order). Another way for drugs to be released from liposomes involves diffusion from degraded or destabilized liposomes; if this is the primary mechanism release can be expected to match the degradation rate of the liposomes. Other absorption mechanisms for microparticluate systems (microspheres and liposomes) involve direct interaction with cells. Small particles (less than 1 μm) can enter cells by phagocytosis; liposomes can also enter cells by membrane fusion. This mechanism offers the potential to actively target certain cell types or systems such as the Kupfer cell and the reticular endothelial system (9). Changes in the size and lipid composition of the liposomes (10) and their surface change (11) can be used to try to tailor their absorption. Greater specificity of interaction can be obtained by tagging antibodies onto the liposome surface (12,13).

As a result of work in the last two decades, at least two liposomal drug suspensions have been approved by the Food and Drug Administration (FDA), one a less toxic amphotericin B for systemic fungal infections (14) and the other a liposomal formulation of doxorubicin for Kaposi's sarcoma (15).

With their ability to achieve delivery to the cytoplasm, fusogenic liposomes, possibly containing polyethylene glycol (16) offer the potential for clinical application for cellular delivery of enzymes and DNA into cells (12,17–19).

Implantable and Injectable Devices

Another approach has been the use of solid, sustained release devices that are injected or surgically implanted and which slowly release the drug. These systems can, in general, be classified as either membrane-controlled or matrix-controlled.

In membrane-controlled devices, a reservoir of drug is contained within polymer coatings and is released by either permeation across a rate-limiting membrane or diffusion through an aperture of a fixed size. If constant (zero-order) release is desired, the permeability of the rate-limiting membrane or the size of the aperture should remain constant over the lifetime of the device, as should the chemical potential of the drug within the reservoir. This is readily achievable if, for example, the rate at which the drug diffuses out of the device is matched by the rate at which drug in the reservoir dissolves (see earlier discussion, Fig. 3). In such a situation, release will be constant but as the reservoir is depleted, release will slow down as the dissolution rate of drug within the device can no longer keep up with the diffusion rate from the device. To satisfy the requirement that the diffusion properties of the membrane or the size of the aperture do not change, it is often simpler to design the device as a nonerodible delivery system. Examples of membrane-controlled nonerodible systems include the Retisert™ and Vitrasert™ (20–22).

In nonerodible matrix systems, the drug is dispersed within a matrix of a polymer and is released as it diffuses through the polymer according to the kinetics described earlier. If polymers that erode before substantially releasing the entire drug are used, the kinetics is more complex. In many PLGA-based systems, drug is released by both diffusion through the polymer and as the matrix breaks up giving rise to the "s" kinetics described earlier. A notable exception is Gliadel™, approved by the FDA for the treatment of brain tumors (23,24). This is composed of a bioerodible hydrophobic A–B block copolymer (poly-[bis-9-carboxyphenoxy propane]-sebacic acid) containing the almost insoluble anticancer drug carmustine (BCNU). The matrix is in the form of a thin wafer and drug is released as the matrix undergoes slow surface dissolution. As BCNU has exceptionally low aqueous solubility, it is released primarily as the polymer breaks down and little drug is released by diffusion from the polymer. The thin, flat shape of these devices dictates that the surface area of the matrix, and hence the release rate, does not change substantially until a large fraction of the matrix has eroded. Because of its lipophilicity, the polymer used also has the unusual property of being resistant to bulk erosion. Thus, release follows zero-order kinetics for a substantial portion of the duration of the implant (23). However, if more hydrophilic drugs are used in these implants, release becomes first order as the drugs dissolve and leach out of the matrix independently of the polymer breakdown (25,26).

Many of the technologies described in the preceding text, although developed for systemic administration, have been investigated, with varying degrees of success, for ocular use and will be reviewed in this and other chapters.

Prodrugs

A frequently used technique in the optimization of drug delivery, regardless of the route selected, has been prodrug synthesis. In this approach, a drug is chemically modified to optimize its delivery properties. Normally, this means increasing either its stability or its absorption. Once absorbed, the prodrug is either chemically or enzymatically converted back to its original, active form (27). One of the earlier successful examples of prodrug formation is levodopa (L-dopa). Dopamine does not cross the blood–brain barrier after systemic administration and has no therapeutic effect on Parkinsonism. Systemically administered L-dopa, however, is transported into the brain where it is enzymatically decarboxylated to the active dopamine. A comprehensive review of the early work on L-dopa was provided by Barbeau and McDowell (28).

Another example of prodrug design is the antiherpetic drug acyclovir. Acyclovir, a synthetic purine nucleoside analog, is readily absorbed into many cell types but is phosphorylated to the active form, acyclovir monophosphate, almost entirely by viral thymidine kinase; its affinity for the viral form of thymidine kinase is 200 times greater than the mammalian form of the enzyme and phosphorylation in the uninfected cell is minimal. Within the virally infected cell the triphosphory-lated form of the drug both selectively inhibits viral DNA polymerase and is incorporated into elongating viral DNA causing termination of synthesis (29,30). In other experiments, lipophilic produgs of timolol were investigated as a means to increase corneal absorption after topical application. These prodrugs were formed by covalently linking timolol to inert alkyl chains. The resulting produgs, being more lipophilic than the parent drug, were more readily absorbed across the cornea and hence had reduced systemic absorption. Once absorbed, the produgs were hydro-lyzed to regenerate the active parent drug, timolol, and the inert promoiety, the alkyl chain (31,32). A more recent development of this approach has been the development of codrugs. Codrugs are formed by linking two active drug molecules together to form an inactive compound that is cleaved at the target site to regenerate the two active parent molecules. This has the potential for simultaneous delivery of synergistic drugs. Using this approach, Ingrams et al. (33,34) found that 5-FU/ triamcinolone acetonide codrugs significantly inhibited tracheal stenosis in the rabbit model.

The prodrug approach thus has the potential to increase bioavailability, opti-mize elimination kinetics, and affect biodistribution. The approach has found broad application in many areas of the pharmaceutical industry and a more comprehensive review can be found in the book *Prodrug Design and Synthesis* by the late Prof. Bungaard (27).

DRUG DELIVERY TO THE POSTERIOR SEGMENT OF THE EYE

Topical Drops

Topical application allows a drug to be placed in direct contact with the eye, but drops are quickly cleared. Although eye drops are typically 25–50 μL, the tear film volume is only 5–7 μL (35). After instillation of an eye drop, the tear volume rapidly returns to normal and the mean retention times of most drugs in the tear film is less than five minutes giving little time for significant ocular absorption. Compounding rapid clearance is the efficiency of the barrier function of the cornea, the principal entry route for most topically applied compounds. Noncorneal ocular absorption (across the conjunctiva and sclera) has been hypothesized and may be significant for delivery to the anterior chamber for some compounds. A review of posterior delivery via the transcleral route is provided in Chapter 13.

Topical Drops and Iontophoresis

Therapeutic intraocular drug levels are achievable by topical drops with the use of iontophoresis. In iontophoresis, a potential difference is applied across a membrane causing a current to be driven through the tissue. If the solutions bathing the membrane contain an ionized drug the current is, at least partially, carried by these ions thus increasing flux across the membrane. Iontophoresis has been explored as a means to increased ocular drug delivery for over 50 years (36). The technique has been reported to successfully increase both transcorneal (37,38) and transscleral delivery (39). Despite these promising findings, iontophoresis has not, as yet, become popular. This may be due partly to the large (up to 20-fold) variation reported in drug delivered to the eye by this route (40). Another potential concern is safety and although currents of up to 20 mA appear to be well tolerated for up to five minutes (41), burns over the area of current application are not uncommon (42). Renewed efforts are now underway to examine the feasibility of iontophoresis as a means to achieve therapeutic concentrations of antivirals in the posterior chamber, as a means to treat cytomegalovirus (CMV) retinitis (43,44).

Systemic Delivery

Systemic administration has the advantage of relatively uniform dosing, although this does not necessarily translate into uniform ocular bioavailability. Even after local administration, drug levels within ocular tissues are subject to intrapatient variability due to differences in permeability through the various tissues and membranes of the eye (such as the retinal pigmented epithelium), differences in clearance rates from the eye, and the disease state. Systemic administration compounds these complexities by adding intrapatient variability in bioavailability of the drug, metabolism of the drug, and patient compliance. Systemic administration also dictates that any agent will be widely dispersed throughout the body and thereby expose more tissues to the potentially damaging drug substances. Further, the blood–eye barrier greatly reduces drug penetration into the eye, and thus for most compounds, the concentration of drug in the eye will be lower than that in most other tissues. To achieve therapeutic concentrations in the eye, the rest of the body must be exposed to yet higher concentrations, a potential problem for drugs with narrow therapeutic indices. Despite these potential concerns, systemic administration has historically been the route of choice for drug delivery to the posterior segment.

Periocular Injections

Periocular injections, subconjunctival, subtenons, and retrobulbar injection of drugs have been frequently investigated as a means to increase ocular availability. Subtenon injections of steroids, such as triamcinolone acetonide, are frequently used to control inflammatory conditions of the posterior segment such as cystoid macular edema, although this delivery route carries a risk of inadvertent intraocular injection (45).

The ocular bioavailability from periocular injections is not well studied and the routes by which drugs penetrate the eye after such injections have never been satisfactorily elucidated. Levine and Aronson (46) used radiopaque media to demonstrate the diffusion of injected compounds from these sites in rabbits and found that subtenons and subconjunctival injections disperse and spread circumferentially around the eye but do not diffuse back to the orbit, while retrobulbar injections tend to disperse throughout the orbit but not into the anterior chamber. This work supports earlier findings of Hyndiuk (47), and Hammeshige and Potts (48).

Davis et al. (49) investigated topical versus subconjunctival versus intramuscular (IM) tobramycin in a rabbit model of *Pseudomonas* keratitis. Topical application (two drops every 30 minutes) was found to be the most effective followed by subconjunctival and IM administrations. Neither local application had a significant effect on disease in the contralateral eye. Similar findings were reported by Leibowitz et al. (50) who showed that 16 hourly topical antibiotic drops were more effective than large single intravenous (IV) or subconjunctival injections. Ocular drug levels were not reported in either of these studies. Subconjunctival injections of ciprofloxacin have been studied in the rabbit (51). It was found that although potentially therapeutic levels were detected in the aqueous, vitreous levels were consistently low.

A possible explanation for these findings was provided by Wine et al. (52) in an elegant experiment published in 1964. They investigated ocular absorption of radiolabeled hydrocortisone from subconjunctival injections administered either through the conjunctiva (providing a possible path to the tear film) or through the upper eye lid and threaded through to the limbus (avoiding puncture of the conjunctiva). Samples of tear fluid and whole eyes were assayed periodically and it was found that ocular bioavailability was 10 to 100 times greater in eyes where drainage from the injection site to the tear film was possible than in those with an intact conjunctiva. In conjunctiva punctured eyes, peak ocular levels of over 170 µg were achieved at 30 minutes and then rapidly declined to 40 µg after two hours. Further, in these eyes approximately twice as much steroid was detected in the tear film as in the entire eye. In eyes with intact conjunctiva, the peak ocular steroid recovery (less than 4 µg) was 30 minutes after administration suggesting that some steroid does diffuse directly into the eye. It should be emphasized that even with the conjunctiva puncture method ocular bioavailability was less than 3%. Thus, this work indicates that subconjunctival injections, which make a hole in the conjunctiva, act as a reservoir draining the drug into the tear film and hence across the cornea into the eye. The poor total availability (5–6% in tears, 2–3% intraocular) indicates that another form of drug loss, presumably systemic absorption, is high from this route and possibly higher if the conjunctiva is not punctured. Oakley et al. (53) studied the corneal distribution of antibiotics after subconjunctival injection using the conjunctiva sparing technique. For all antibiotics studied (penicillin G, gentamicin, and chloramphenicol) highest concentrations were found nearest the injection site. This indicates that these compounds at least can pass through the corneoscleral limbus and diffuse across the cornea; however, no estimate of the ocular bioavailability was given.

Barza et al. (54) later reported that after subconjunctival injection of gentamicin, higher drug concentrations were found in ocular tissues from normal eyes than from inflamed, infected (*Staphylococcus aureus* endophthalmitis) eyes, despite the presumed reduction in blood–eye barrier in the inflamed eye. This result was not due to altered drainage into the tear film but may have been caused by increased ocular and orbital vascularity or decreased half-life within the eye (55). Similar results have been reported by Levine and Aronson (46) who found that inflammation caused a twofold decrease in ocular absorption of radiolabeled cortisol after retrobulbar injection although no such difference was seen following subconjunctival injection. Peak ocular concentrations were observed five minutes after administration. These authors also speculated that the difference in ocular absorption after retrobulbar injection was probably due to more rapid steroid removal from the injection site in the inflamed eye. This finding suggests that systemic absorption after extraocular injection can be expected to be significant and that a systemic contribution to ocular bioavailability may be important. This is especially likely after peribulbar injections, particularly if a conjunctival sparing injection technique is used.

Retrobulbar injections of radiolabeled methyl prednisolone acetate (insoluble) were found to produce elevated concentrations of steroid in the uvea, vitreous, and lens compared to IM injections on the squirrel monkey (56). This finding was not reproduced by Barry et al. (57), who also used radiolabeled methylprednisolone acetate, but failed to show a significant difference in either vitreal or uveal concentrations between treated and contralateral, untreated eyes after retrobulbar steroid injections in the albino rabbit. This apparent difference may have been due to interspecies variation or may be artifactual due to differences in tissue handling.

Weijtens et al. (58) compared vitreous and serum concentrations after peribulbar injections of steroid in patients scheduled for vitrectomy. In this study, 61 eyes of 61 patients received a peribulbar injection, 5 mg of dexamethasone disodium phosphate, at various times before their vitrectomy. The steroid was injected transcutaneously avoiding puncture of the conjunctiva. Analysis of vitreous and serum samples taken at the same time showed that the mean peak serum concentration was more than four times that of the intravitreal concentration at all time points. These eyes were, for reasons not discussed, all scheduled for vitrectomies. This may have significantly affected the kinetics of the eye, possibly increasing intravitreal permeation. Thus, it is entirely possible that in healthy eyes the difference between systemic exposure and vitreal absorption could be higher. The effect of periocular steroid injections to the vitreous concentration in the contralateral eye was not investigated.

Intravitreal Systems

Direct intravitreal injection has the obvious advantages of being able to achieve immediate therapeutic concentrations in the eye while largely avoiding systemic exposure. However, following injection, drugs are rapidly eliminated from the vitreous, typically by a first-order diffusion process, with half-lives of 24 hours or less. If suspensions of relatively insoluble drugs, such as triamcinolone acetonide are injected into the vitreous, their apparent half-life can be increased, although the intrinsic half-life of the drug itself does not change. In this case dissolution of the suspended drug is rate limiting rather than clearance of the drug itself (59). Repeated injections are therefore required to maintain therapeutic concentrations in the eye and associated risks of endophthalmitis, cataract formation, and retinal detachment have reduced the application of this technique to a few, sight-threatening diseases such as endophthalmitis and cytomegalovirus retinitis.

There have been many attempts to achieve sustained intraocular drug levels after intravitreal injection. Moreira et al. (60,61) investigated the use of local gentamicin injections dissolved in sodium hyaluronate gel (Healon™, Pharmacia, Piscataway, New Jersey, U.S.A.) in an attempt to achieve more efficacious, prolonged delivery; however, there was no significant difference in either the half-life or the efficacy of gentamicin injected in saline or Healon. This result may be attributable to the composition of the gel. As described previously, although the bulk viscosity of Healon is high, the microviscosity is much lower and thus for relatively small molecules little sustained effect can be anticipated from their incorporation in such gels.

Intravitreal liposome injection as a means to achieve sustained intraocular levels has been studied extensively. After injection, liposomes appear to be eliminated from the vitreous via a diffusional process through the anterior chamber with small unilamellar vesicles (SUVs) having a half-life of 9 to 10 days and large unilamellar vesicles (LUVs) having a half-life of approximately 20 days (62). Although liposomes appear to be retained in the vitreous for a prolonged period, release of the drug from the liposomes into the vitreous does not follow the same kinetic profile. As described previously, release from liposomal systems is generally first order (i.e., dependent on the concentration of the drug in the liposome). This kinetics gives rise to high initial drug concentrations followed by progressively lower levels. For diseases that require long-term therapeutic drug levels this may represent a considerable problem, as would the potential for vitreous clouding. For endophthalmitis, however, these drawbacks may be acceptable. To date, results obtained with liposomes as a sustained release system have been mixed.

Liposome entrapment reduces the toxicity of intravitreally injected amphotericin B in both rabbits and monkeys (63,64). However, reduced toxicity does not necessarily correlate to increased efficacy. In 1989, Liu et al. (65) reported that liposome entrapment actually decreased the efficacy of amphotericin B in a rabbit model of fungal endophthalmitis (*Candida albicans*), presumably because the drug was entrapped within the liposome and not available to exert its pharmacological action. Potentially, this reduced efficacy could be offset by the administration of more amount of the drug. Liu et al. (65) found that 20 µg of liposome encapsulated amphotericin appeared as effective as 10 µg of free amphotericin after intravitreal injection. Unfortunately, the pharmaceutical advantage of liposomal delivery, an increase in the AUC (area under the curve) of free drug in the vitreous, did not translate into a more effective treatment in this model. The potential exists, however, to use liposomes either as a means to achieve intracellular targeting within the posterior segment, perhaps via cationic liposomes for gene transfer (66), or to achieve cellular targeting, perhaps to phagocytotic cells such as the retinal pigment epithelium (RPE) or macrophages (67,68). For these interactions the use of liposomes perse may not be necessary, similar effects are achievable by the use of microspheres. This was demonstrated by Kimura et al. (69) who showed intracellular release of a dye from polylactic acid microspheres following their phagocytosis by cultured RPE cells.

Other attempts have focused on intravitreal injections of a suspension of polylactic acid (PLA) microspheres. Moritera et al. (70) showed that in vitro PLA microspheres containing 1% adriamycin (doxorubicin) can give sustained, first-order release for approximately two weeks.

DRUG ELIMINATION MECHANISMS

Drugs are cleared from the vitreous primarily by two routes. The anterior route involves drainage into the anterior chamber and clearance with aqueous turnover via bulk flow. The posterior route involves either active or passive permeation across the retina and RPE and subsequent systemic dissipation (71,72). There is evidence that for some, highly lipophilic, compounds clearance via the retinal blood vessels may also play an important role (73).

Anterior Elimination

The anterior route has been found to be the primary elimination pathway for small, hydrophilic compounds such as gentamicin (74), fluorescein glucuronide

(75), and also for some relatively large compounds such as oligonucleotides (76). These compounds typically have half-lives of 24 to 30 hours and are among the slowest small molecules cleared from the vitreous. Some compounds appear to be transported both by the anterior route and the posterior route (77) and thus have shorter half-lives.

Posterior Elimination

The posterior route involves transport across the lipophilic retina and the RPE. Posterior elimination appears to occur via passive diffusion for lipophilic compounds although active transport mechanisms also exist for some endogenous compounds and also for some drugs. Forbes and Becker studied the vitreous elimination of radio-labeled iodopyracet in rabbits and found that while 300 µg labeled with 1 µCi was cleared rapidly with a half-life of three hours, 1.4 mg (also labeled with 1 µCi) was eliminated with an apparent half-life of 17 hours. This increase in half-life, they speculated, was due to saturation of an active transport mechanism (78). Weiner et al. (79) showed that systemic administration of probenecid, which acts by competitive inhibition of weak organic acid transport, increased the intravitreal half-life of iodopyracet to 17 hours. Similar findings were reported by Barza et al. (80,81) who demonstrated that systemic administration of probenecid increased the half-life of intravitreally injected carbenicillin and cefazolin in rabbits and monkeys.

The effects of agents that stimulate active elimination mechanisms have also been investigated. Carbonic anhydrase is present in many ocular tissues including the retina, the RPE, and ciliary processes and is important in the regulation of ion transport (82). Systemically administered acetazolamide can enhance subretinal fluid absorption across the RPE and has been investigated as a means to treat chronic macular edema (83–85). Acetazolamide also increases vitreal elimination of fluorescein in primates and patients with retinitis pigmentosa (86,87).

Effects of Disease States

Similarly, disease states can have a marked effect on drug clearance from the eye. The intraocular elimination of aminoglycosides such as gentamicin, which are normally eliminated via the anterior pathway, can be increased in inflamed eyes (55). Conversely, inflammation can reduce elimination of predominantly posterior, actively eliminated, beta-lactam antibiotics (81). Presumably, this is due to the opposing effects of increased ocular permeability versus reduced active transport.

The elimination of fluorescein, which is actively transported out of the healthy eye (75), is slowed in disease states such as retinitis pigmentosa (88), diabetes (89,90), and chronic cystoid macular edema (91). Administration of sodium iodate to disrupt the function of both the retina and RPE (92) was found to increase the half-life of fluorescein from 4 to 12 (93).

In experimental models of diabetes, there are many changes in both RPE and capillary permeability. Early in the course of diabetes, folding of the RPE surface has been documented with a consequent increase in surface area (94,95). This appears to be accompanied by an increase in permeability; both systemically administered fluorescein and horseradish peroxidase (HRP) have been reported to leak through the RPE into the retinal extracellular space (95–97). Other investigators, while confirming findings with fluorescein, have failed to reproduce the results with HRP (98). Still other groups have examined the integrity of the diabetic blood–eye barrier by determining the accumulation of fluorescein-labeled albumin in the vitreous after systemic administration. This was found to be mediated by intraretinal histamine levels but was independent of the development of diabetic retinopathy (99). Clinically, increased permeability to fluorescein is one of the earliest clinically detectable abnormalities in diabetic retinopathy (100,101).

Another effect of disease states such as diabetes and proliferative diabetic retinopathy is an increase in the vitreous protein concentration, which may be partly, but not entirely, due to vascular leakage (102,103). Thus, the decreased permeability barrier (systemic to vitreous) and increased potential for protein binding within the vitreous will tend to increase vitreous drug concentrations. The increased intravitreal drug concentration is opposed by the decreased permeability barrier (vitreous to serum) that may enhance vitreous elimination. The process is further complicated by the effects of ocular diseases on active transport mechanisms.

Effects of Ocular Surgery on Drug Elimination

The integrity of the eye can also greatly affect drug clearance. Vitrectomy and lensectomy can both greatly increase elimination of drugs regardless of their primary elimination route in the intact eye (104–106). Similarly, the use of gas to repair a retinal detachment can affect the vitreous levels and the duration of sustained release systems in the eye (107). Where possible, these factors should be considered and adjusted for when contemplating pharmacological treatments of the posterior segment.

POSTERIOR DELIVERY IN DISEASE STATES

Direct intravitreal administration of agents has the advantage of achieving high intraocular drug levels and of minimizing systemic exposure; however, this approach clearly has the potential for considerable trauma. The acceptability of such a direct approach is likely to be a function of the disease state of the eye, the likely outcome, and of the possibility to combine administration with any other procedures already being performed. The clinical acceptance of implanting a device for a relatively trivial disease would be considerably lower than for a blinding disease in which the eye is already open. It is thus appropriate to discuss the development of intraocular delivery systems on a disease basis. The following discussion illustrates the principles of intraocular drug delivery and the various delivery approaches in an ocular disease specific context. A more detailed discussion of each of these conditions, and the results of clinical studies using intraocular drug delivery systems to treat them, appears in subsequent chapters.

Endophthalmitis

Intravitreal injections of antibiotics (which do not easily penetrate the eye by any other route) are the cornerstone of the management of endophthalmitis although adjunctive treatments, such as IV antibiotics and early vitrectomy have often been employed. A recent multicenter study on postoperative endophthalmitis found that IV antibiotics had no significant effect on clinical outcomes while early vitrectomy was beneficial only in eyes with light perception or worse. In eyes with better vision, early vitrectomy offered no significant benefit (see Chapter 6). The role of IV antibiotics in other causes of endophthalmitis such as trauma was not established in this study (108). An ideal antibiotic for intravitreal use would have a long half-life in the vitreous, have a broad spectrum of action, and be nontoxic to the eye. Unfortunately, no agent currently possesses all these properties so combinations are typically used. Initially, a cephalosporin such as Cefazolin® was used in conjunction with an aminoglycoside such as gentamicin (109). Unfortunately, although aminoglycosides have a relatively long intravitreal half-life they may cause toxicity (110) and some investigators consider that their use be limited to proven gram-negative infections (111) or that they be replaced with third-generation cephalosporins such as ceftazidime (112).

The possible advantages of sustained delivery systems have been investigated by several groups although as these infections are often acute, the maintenance of high drug levels in the vitreous for periods of over one to two weeks is likely to be of little direct benefit. Intravitreal injections are a simple procedure compared to a vitrectomy but repeated injections are not considered attractive and control of vitreous concentrations from such a regimen is difficult. A more thorough review of pharmacological treatments in endophthalmitis is provided in Chapter 6.

Cytomegalovirus Retinitis

Cytomegalovirus (CMV) is relatively common in patients with acquired immunodeficiency syndrome (AIDS), especially when the immune system is seriously compromised (CD4 typically less than 50) (113). Although there are several drugs available to treat the disease, their systemic toxicity and poor ability to cross the blood–eye barrier complicates the clinical treatment. In the late 1980s, clinicians discovered that intravitreal injections of these drugs can maintain adequate control of CMV (114,115). The rapid elimination of ganciclovir and foscarnet necessitates administration once or twice each week, while intravitreal therapy with cidofovir (HPMPC), which has a very long intravitreal half-life, is fraught with difficulty due to the narrow therapeutic range of this compound (116–119).

Several groups have investigated intraocular antiviral sustained release systems for CMV retinitis. Intravitreal injections of liposomally entrapped ganciclovir have been used clinically but even in the liposomal form, the requirement for repeated injections make this approach unappealing (120). One of the first successful methods was a nonerodible intravitreal device composed of a solid core of drug coated with several rate-limiting membranes (a membrane controlled system). This relatively simple device gave zero-order release of ganciclovir (121) and initial studies showed that the device was effective in the control of CMV retinitis in patients intolerant of existing therapies (than IV ganciclovir or foscarnet) (122). Subsequent studies confirmed the efficacy of the device and after two pivotal phase III studies, one of which was conducted by the National Eye Institute, the device was approved by the FDA and is currently sold under the name Vitrasert™ (22). The device must be inserted through a 6 mm incision through the pars plana and should be replaced once it is depleted of drug (after 6 months), (123). Although this can be readily achieved, one patient has received 14 implants for bilateral CMV over the last four years, the situation is far from optimal (124). Fortunately, the development of highly active antiretroviral therapy (HAART) has dramatically reduced the incidence of this disease (see Chapter 6).

Proliferative Vitreoretinopathy

Proliferative vitreoretinopathy (PVR) is characterized by the proliferation of cells, thought to be mainly retinal pigment epithelial cells, macrophages, and fibroblasts on the retinal surface, undersurface, and within the vitreous. Despite the many advances made in surgical technique, the need for a means to pharmacologically inhibit cellular proliferation in the vitreous has long been identified (125).

Glucocorticosteroids were the first compounds that were used to pharmacologically reduce vitreoretinal scarring, and were initially selected on the basis of their ability to inhibit fibroblast growth in vitro and in vivo as indicated by delayed corneal wound healing after topical application (126). It is now accepted, however, that one of the major mechanisms of their action is reduction of ocular inflammation and moderation of blood–retina barrier breakdown. Corticosteroids have a bimodal effect on fibroblast proliferation in vitro and can stimulate proliferation at low doses and inhibit it at high doses (124). Intraocular penetration from topical drops is very low and no studies have demonstrated efficacy against PVR. Ocular penetration from systemic dosing (oral or IV) is also very low and high doses are required for an ocular effect (127). The systemic toxicities resulting from prolonged systemic steroid administration are well known and include hypertension, hyperglycemia, increased susceptibility to infection, peptic ulcers, aseptic necrosis of the femoral head, and hirsutism (128).

Corticosteroid suspensions were among the earliest (and simplest) sustained release injectable drug delivery systems investigated. In the early 1980s, Tano et al. (129,130) published their work on intravitreal injection of 1 mg of dexamethasone (alcohol) suspended in 100 μL saline into the mid-vitreous in a rabbit model of PVR. Once injected, particles of dexamethasone slowly dissolved over 7–14 days, thus maintaining levels of the freely dissolved, biologically active drug in the eye. In the model used, the first reported fibroblast injection PVR model, suspensions of dexamethasone appeared to inhibit PVR development although statistical significance was only achieved on day seven. Injection of the more soluble dexamethasone phosphate (1 mg in solution) was less effective due to its rapid clearance from the vitreous ($t_{1/2}$ of three hours). The same group achieved better results with suspensions of triamcinolone acetonide (kenalog). This compound is less soluble than dexamethasone alcohol and provided longer sustained release; drug particles were seen in the vitreous several months after injection (131). Pretreatment with intravitreal triamcinolone acetonide has been shown to inhibit the proliferation of injected fibroblasts in a rabbit model of PVR and hence reduce the occurrence of retinal detachment (132).

Injection of low-solubility drug suspensions, although technically simple, has many disadvantages including poor drug release kinetics. The concentration of drug freely dissolved in solution (and hence pharmacologically active) is a function of the suspension dissolution rate which, within the vitreous, is difficult to control. Within the vitreous, the concentration achieved is governed by the balance between the dissolution rate of the particles and the elimination of dissolved drug from the vitreous. High initial drug levels followed by progressively lower levels are therefore likely.

Optimal steroid therapy for PVR would maintain therapeutic levels in the vitreous for a prolonged period; at present this can only be achieved by prohibitively high systemic dosing or intravitreal injections (which do not maintain constant drug levels). A sustained release device that maintains constant therapeutic intravitreal levels with minimal systemic exposure may prove useful in the clinical management of PVR, although the potential for ocular effects such as elevation of intraocular pressure and cataract cannot be ignored.

The local use of cytotoxic chemothetic agents has also been investigated as a means to treat or prevent PVR. Intravitreal injections of daunorubicin have been used clinically to treat traumatic PVR with some success; however, concerns exist as to the safety of this agent (133). Daunorubicin, although an effective antiproliferative agent, is not cell-cycle specific and can be expected to interact with a variety of cells in the eye, causing toxicity. A cell-cycle specific agent, being preferentially toxic to actively dividing cells within the vitreous, may prove more effective.

Five–Fluorouracil (5-FU) has been widely investigated for possible use against PVR. Although 5-FU is effective, at least in vitro, in suppressing fibroblast proliferation, its half-life in the vitreous is less than eight hours (134). Early on, it was found that the retina is relatively sensitive to 5-FU with some evidence of toxicity apparent following intravitreal injections of 1 mg 5-FU (135). These findings indicated that high bolus intravitreal dosing was not likely to achieve prolonged therapeutic levels and stimulated the investigation of various delivery systems as a means to achieve sustained release.

The potential exists to use silicone oil as an intravitreal reservoir for drugs lipophilic enough to dissolve in this highly lipophilic fluid. The stability and release of the anticancer agent carmustine (or BCNU) from silicone oil was studied by Chung et al. (136) in the late 1980s. The same group then went on to examine silicone oil as a means to deliver the highly lipophilic compound retinoic acid. In the rabbit model, retinoic acid dissolved in silicone oil appeared to decrease the rate of retinal detachments due to PVR although this was not statistically significant (137). In an attempt to extend the number of drugs incorporated into the silicone, others investigated a series of highly lipophilic prodrugs of 5-FU (138). These were designed to dissolve in silicone oil and be slowly released into the residual vitreous. Once in solution, the prodrugs were rapidly hydrolyzed, regenerating 5-FU.

Despite some encouraging data, a major problem with this approach is the inability to control drug release from silicone oil. Release can once more be expected to follow either square root time kinetics (if diffusion through the oil is rate limiting) or first-order kinetics (if partitioning out of the oil into the vitreous is rate limiting). Thus, constant drug levels would be difficult to achieve using release from such a vehicle. The use of such a delivery system would, in any case, be limited to cases of severe PVR in which silicone oil is employed.

Liposomes have been investigated as an alternate delivery system; most works to date have used liposomes as drug reservoir systems rather than as means to achieve cellular or intracellular targeting. Joondeph et al. (139) reported that liposome encapsulation of 5-FU reduces the toxicity of intravitreal 5-FU injections allowing higher doses to be administered (1.6 mg of the liposomal form). This reduction in toxicity is, however, marginal and of limited practical use. Liposomal injections of 5-fluorouridine have been tested in monkeys but were accompanied by unacceptable toxicity, possibly due to the large doses of drug used (over 20 mg) (140). Five–Fluorouridine is approximately 100 times more potent than 5-FU (141). Maignen et al. (142) reported that liposomally entrapped mitoxantrorie was as effective as the free drug in the inhibition of rabbit subconjunctival fibroblasts in an ex-vivo model.

Degradable microspheres of PLA have been investigated as a means to sustain the release of antimetabolites for the treatment of intraocular proliferative disease. Moritera et al. (70) showed that in vitro PLA microspheres containing 1% adriamycin (doxorubicin) can give sustained, first-order release for approximately two weeks. This sustained release reduced the toxicity of adriamycin compared to injections of a free solution as measured by electroretinogram (ERG).

As PVR is generally treated by a surgical procedure in which the eye is opened, the barrier to inserting a drug delivery device is likely to be small, especially if such a device offers advantages in terms of improved drug delivery. An implantable device could be kept out of the visual axis and provide better drug release and distribution. Several investigators have prepared implantable delivery systems for PVR. The primary requirements of an ideal implant are that it should cause no inflammation when implanted in the eye, be easily inserted, cause no damage to ocular structures, and it should release its drug (or drugs) at the required rate for the required duration. An ideal system should not require removal at the completion of therapy, i.e., it should rapidly and safely erode when it has released its drug or at least be safe to remain in the eye indefinitely.

A bioerodible sustained release implant for 5-FU composed of a PLA matrix of 5-FU coated with PLA was found to significantly reduced the severity of PVR in a rabbit model (143). This device maintained levels of 5-FU between 1 and 13 µg/mL in the vitreous for 14 days. Importantly, no evidence of drug- or polymer-related toxicity was noted even though the device did not fully erode during the experiment (28–day duration). Although the model used did not progress to total retinal detachment, the finding that sustained release 5-FU can significantly inhibit the development of PVR is extremely important.

Another approach has been the use of codrug implants. In this case, the codrugs used formed an insoluble conjugate. Parent drugs were released as the conjugate dissolves and hydrolyzed in the vitreous. In this system, the release rate can be adjusted to give either first- or zero-order release and enables either simultaneous or asynchronous release of the two drugs. Another advantage is that no polymer is necessary to control release. Jaffe and coworkers reported that a codrug pellet providing constant, equimolar release of 5-FU and triamcinolone was effective in the prevention of PVR in a rabbit model (144). They subsequently found that a more advanced codrug that provided release of 5-FU over approximately two weeks, and a longer release of fluocinolone acetonide was even more effective (145). A more detailed discussion of drug delivery and PVR is provided in a later chapter.

PHOTODYNAMIC THERAPY

In recent years, photodynamic therapy (PDT) has received considerable attention as a new means to treat age-related macular degeneration. The basic concept is to deliver a pharmacologically inactive compound to a target tissue and then activate the compound by exposure to light. Ideally, the only region to be exposed to the active form of the compound would be those tissues exposed to the drug and to light. Thus, there is increased potential for achieving a localized effect. Despite the recent flurry of activity, the idea is an old one and can be traced back to the work of Tappeneir and Jesionek who, in 1903, reported the effects of topical administration of eosin and white light as a means to treat skin cancer.

In an extremely inventive approach, PDT with liposomes has been used to achieve localized delivery of highly toxic, short-lived, free radicals to neovascular tissue. A dye, photoporphorin, is encapsulated in liposomes which are injected intravenously. A laser fired into the eye simultaneously destabilizes the dye molecule causing it to form a free radical and destabilizes the liposome causing it to release the free radical. The free radical interacts with the first tissue it encounters, in this case the neovascular tissue. PDT is one of the only treatments currently available for age-related macular degeneration (see Chapter 6).

FUTURE OPPORTUNITIES AND CHALLENGES

Increased understanding of diseases such as retinitis pigmentosa, proliferative diabetic retinopathy, and age-related macular degeneration can be expected to produce new cellular targets and drug candidates. Our increasing ability to deliver these agents in a safe and effective way will offer new opportunities to treat currently blinding diseases. The technologies required to deliver agents specifically and effectively to the eye are rapidly evolving. These technologies will have the potential to radically alter the way many, especially vitreoretinal, diseases are treated. The next decade promises great strides in therapy for many poorly treated or untreatable ocular diseases.

REFERENCES

1. Alfred Martin. In Physical Pharmacy. Lee and Febiger, 1992:326.
2. Higuchi T. Physical chemical analysis of percutaneous absorption process from creams and ointments. J Soc Cosmet Chem 1960; 11:85–97.
3. Higuchi T. Release of medicaments from ointment bases containing drugs in suspension. J Pharm Sci 1961; 50:874–875.
4. Peppas NA. Controlling protein diffusion in hydrolgels. In: Lee VH-L, Hashida M, Misushima Y, eds. Trends and Future Perspectives in Peptide and Protein Delivery. Chur, Switzerland: Harwood Academic Publishers, 1995:23–38.
5. Ogawa Y, Okada H, Yamamoto M, Shimamoto T. In vivo release profiles of leuprolide acetate from microsphere prepared with polylactic acids or copoly(lactic/glycolic) acids and in vivo degradation of these polymers. Chem Pharm Bull 1988; 36:2576–2581.
6. Gregoriadis G. Liposome Technology. Boca Roton, FL: CRC Press, 1984.
7. Cornell BA, Hetcher GC, Middlehurst J, Separovic F. The lower limit of the size of small sonicated phospholipid vesicles. Biochim Biophys Acta 1982; 690:15–19.
8. Knight CG, ed. Liposomes: From Physical Structure to Therapeutic applications. Amsterdam: Elsevier/North Holland Biomedical Press, 1981.
9. Son K, Huang L. Liposomal DNA delivery. In: Lee VH-L, Hashida M, Misushima Y, eds. Trends and Future Perspectives in Peptide and Protein Delivery. Chur, Switzerland: Harwood Academic Publishers, 1995:321–336.
10. Rahman YE, Cerney EA, Patel KR, et al. Differential uptake of liposomes varying in size and lipid composition by parenchymal and Kupfer cells of mouse liver. Life Sci 1982; 31:2061–2071.
11. Farhood H, Gao X, Son K, et al. Cationic liposomes for direct gene transfer in therapy of cancer and other disease. Ann NY Acad Sci 1994; 716:23–35.
12. Barbet J, Machy P, Leserman LD. Monoclonal antibody covalently coupled to liposomes: specific targeting to cells. J Supramol Struct Cell Biochem 1981; 16:243–258.
13. Shen DF, Huang A, Huang L. An improved method for covalent attachment of antibody to liposomes. Biochim Biophys Acta 1982; 689:31–37.
14. Lopez-Bernstein G, Fainstein V, Hopfer MK, et al. Liposomal amphotericin B for the treatment of systemic fungal infections in patients with cancer. A preliminary study. J Infect Dis 1985; 151:704–710.
15. Sunalp M, Wiedemann P, Sorgente N. Effects of cytotoxic drugs on proliferative vitreoretinopathy in the rabbit cell injection. Curr Eye Res 1984; 3:619–623.
16. Szoka F, Magnussen KE Wojcieszn J, et al. Use of lectins and polyethylene glycol for fusion of glycolipid containing liposomes with eukaryotic cells. Proc Natl Acad Sci USA 1981; 78:1685–1689.
17. Weissmann G, Bloomgarden D, Kaplan R, et al. A general method for the introduction of enzymes, by means of immunoglobulin coated liposomes, into lysosomes of deficient cells. Proc Natl Acad Sci, USA 1975; 72:88.

18. Feigner PL, Gadek TR, Holm M, et al. Lipofectin: a highly efficient lipomediated DNA-transfection procedure. Proc Natl Acad Sci, USA 1987; 84:7413–7417.

19. Itani T, Ariga H, Yamaguchi N, Tadakuma T, Yasuda T. A simple and efficient liposome mediated transfection of DNA into mammalian cells grown in suspension. Gene 1987; 56:267–276.

20. Jaffe GJ, Ben-Nun J, Guo H, et al. Fluocinolone acetonide sustained drug delivery device to treat severe uveitis. Ophthalmology 2000; 107:2024–2033.

21. Jaffe GJ, Yang CH, Guo H, et al. Safety and pharmacokinetics of an intraocular fluocinolone acetonide sustained delivery device. Invest Ophthalmol Vis Sci 2000; 41: 3569–3575.

22. Musch DC, Martin DF, Gordon JF, et al. Treatment of cytomegalovirus retinitis with a sustained release ganciclovir implant. N Engl J Med 1997; 337:83–90.

23. Tamada J, Langer R. The development of polyanhydrides for drug delivery applications. J Biomater Sci Polym Ed 1992; 3:315–353.

24. Brem H, Ewend MG, Piantadosi S, Greenhoot J, Burger PC, Sisti M. The safety of interstitial chemotherapy with BCNU loaded polymer followed by radiation therapy in the treatment of newly diagnosed malignant gliomas: Phase I trial. J Neurooncol 1995; 26:111–123.

25. Lee DA, Leong KW, Panek WC, et al. The use of bioerodible polymers and 5-fluorouracil in glaucoma filtration surgery. Invest Ophthalmol Vis Sci 1988; 29:1692–1697.

26. Jampel HD, Leong KW, Dunkelburger GR, Quigley HA. Glaucoma filtration surgery in monkeys using 5-fluorouridine in polyanhydride disks. Arch Ophthalmol 1990; 108:430–435.

27. Bungaard H, ed. Design of Prodrugs. Amsterdam: Elsevier, 1985.

28. Barbeau A, McDowell FH, eds. L-Dopa and Parkinsonism. Philadelphia, PA: FA Davis Co, 1970.

29. McGuirt PV, Furman PA. Acyclovir inhibition of viral DNA chain elongation in herpes simplex virus infected cells. Am J Med 1982; 73(suppl):67–71.

30. Furman PA, St Clair MH, Spector T. Acyclovir triphosphate is a suicidal inactivator of the herpes simplex DNA polymerase. J Biol Chem 1984; 83:8333–8337.

31. Bungaard H, Buur A, Chang S-C, Lee VHL. Prodrugs of timolol for improved ocular delivery: synthesis hydrolysis kinetics and lipophilicity of various timolol esters. Int J Pharm 1986; 33:15–26.

32. Chang SC, Bungaard H, Buur A, Lee VHL. Improved corneal penetration of timolol by prodrugs as a means to reduce systemic load. Invest Ophthalmol Vis Sci 1987; 28: 487–491.

33. Ingrams DR, Sukin SW, Ashton P, Valonen HJ, Shapshay SM. Does slow release 5-fluorouracil and triamcinolone reduce subglotic stenosis? Otolaryngol Head Neck Surg 1998; 118:174–177.

34. Ingrams DR, Dhingra J, Shah R, Ashton P, Shapshay SM. Slow release 5-fluorouracil reduces subglottic stenosis in a rabbit model. Ann Otol Rhinol Laryngol 2000; 109:422–424.

35. White Glover, Buckner AB. Effect on blinking on tear elimination as evaluated by dacryoscintigraphy. Ophthalmol 1991; 98:367–369.

36. Von Sallmann L. Sulfadiazine iontophoresis in pyocyaneus infection of rabbit cornea. Am J Ophthalmol 1945; 34:195–201.

37. Hill JM, Park NH, Gamgarosa LP, et al. Iontophoresis of vidarabine monophosphate into rabbit eyes. Invest Ophthalmol Vis Sci 1978; 17:473–476.

38. Rootman DS, Hobden JA, Jantzen JA, et al. Iontophoresis of tobramycin for treatment of experimental *Pseudomonas* keratitis in the rabbit. Arch Ophthalmol 1988; 106:262–265.

39. Burnstein NL, Leopold IH, Bernacci DB. Transscleral iontophoresis of gentamicin. J Ocular Pharmacol 1985; 1:363–368.

40. Barza M, Peckman C, Baum J. Transscleral iontophoresis of cefazolin, tricarcillin and gentamicin in the rabbit. Ophthalmology 1986; 93:133–139.

41. Highes L, Maurice DM. A fresh look at iontophoresis. Arch Ophthalmol 1984; 102:1825–1829.
42. Barza M, Peckman C, Baum J. Transscleral iontophoresis of gentamicin in monkeys. Invest Ophthalmol Vis Sci 1987; 28:1033–1036.
43. Dessouki AL, Yoshizumi MO, Lee D, Lee G. Multiple applications of ocular iontophoresis of foscarnet. Invest Ophthalmol Vis Sci 1987; 38:S1–S117.
44. Gautier S, Kasner L, Behar-Cohen F. Transscleral coulomb controlled iontophoresis of ganciclovir in rabbits: Safety and pharmacokinetics. Invest Ophthalmol Vis Sci 1997; 38:S147.
45. Lincoff H, Zweifach P, Brodie S, et al. Intraocular injection of lidocaine. Ophthalmology 1985; 92:1587–1591.
46. Levine ND, Aronson SB. Orbital infusion of steroids in the rabbit. Arch Ophthalmol 1970; 83:599–607.
47. Hyndiuk RA. Subconjunctival radioactive depot corticosteroid penetration into monkey ocular tissue [abstract]. Invest Ophthalmol 1969; 8:352.
48. Hammeshige S, Potts AM. The penetration of cortisone and hydrocortisone into ocular structures. Am J Ophthalmol 1955; 40:3211–3215.
49. Davis AD, Sarff LD, Hyndiuk RA. Comparison of therapeutic routes in experimental *Pseudomonas* keratitis. Am J Ophthamol 1979; 87:710–716.
50. Leibowitz HM, Ryan WJ, Kupferman. Route of antibiotic administration in bacterial keratitis. Arch Ophthalmol 1981; 99:1420–1423.
51. Behrens-Baumann W, Martell J. Ciprofloxacin concentrations in the rabbit aqueous humor and vitreous following intravenous and subconjunctival administration. Infection 1988; 16:54–57.
52. Wine NA, Gornall AG, Basu PK. The ocular uptake of subconjunctivally injected C14hydrocortisone. Am J Ophthalmol 1964; 58:362–366.
53. Oakley DA, Weeks RD, Ellis PP. Corneal distribution of subconjunctival antibiotics. Am J Ophthalmol 1976; 81:307–312.
54. Barza, M, Kane A, Baum J. Excretion of gentamicin in rabbit tears after subconjunctival injection. Am J Ophthalmol 1979; 85:118–120.
55. Kane A, Barza M, Baum J. Intravitreal injection of gentamicon in rabbits: effect of inflammation and pigmentation on half-life. Invest Ophthalmol Vis Sci 1981; 20:593–597.
56. Hyniuk RA, Reagan MG. Radioactive depot-corticosteroid penetration into monkey ocular tissue. Arch Ophthalmol 1968; 80:499–503.
57. Barry A, Rousseau A, Babineau LM. The penetration of steroids into the rabbit's vitreous, choroid and retinal following retrobulbar injections. Can J Ophthalmol 1969; 4:395–399.
58. Weijtens OVD, Sluijs FA, Schoemaker RC, et al. Peribulbar corticosteroid injection: vitreal and serum concentrations after dexamethasone disodium phosphate injection. Am J Ophthalmol 1997; 123:358–363.
59. Beer PM, Bakri SJ, Singh RJ, et al. Intraocular concentration and pharmacokinetics of triamcinolone acetonide after a single intravitreal injection. Ophthalmology 2003; 110:681–688.
60. Moreira CA, Moreira AT, Armstrong DK, et al. In vitro and in vivo studies with sodium hyaluronate as a carrier for intraocular gentamicin. Acta Ophthalmol 1991; 69:50–56.
61. Moreira CA, Armstrong DK, Jelliffe RW, et al. Sodium hyaluronate as a carrier for intravitreal gentamicin. An experimental study. Acta Ophthalmol 1990; 68:133.
62. Barza M, Stuart M, Szoka F. Effect of size and lipid composition on the pharmacokinetics of intravitreal liposomes. Invest Ophthalmol Vis Sci 1987; 28:893–900.
63. Tremblay C, Barza M, Szoka F, et al. Reduced toxicity of liposome-associated amphoteracin B injected intravitreally in rabbits. Invest Ophthalmol Vis Sci 1985; 26:711–718.
64. Barza M, Baum J, Tremblay C, et al. Ocular toxicity of intravitreally injected liposomal amphotericin B in rhesus monkeys. Am J Ophthalmol 1985; 100:259–263.

65. Liu K-R, Peyman GA, Koobehi B. Efficacy of liposome bound amphteracin B for the treatment of experimental fungal endophthalmitis in rabbits. Invest Ophthamol Vis Sci 1989; 30:1527–1533.

66. Masuda I, Matsuo T, Yasuda T, Matsuo N. Gene transfer with liposomes to the intraocular tissues by different routes of administration. Invest Ophthalmol Vis Sci 1996; 37:1914–1920.

67. Urtti A, Polansky J, Lui GM, Szoka FC. Gene delivery and expression in human retinal pigmented epithelial cells: effects of synthetic carriers, serum, extracellular matrix and viral promoters. J Drug Target 2000; 7:413–421.

68. Abrahm NG, Da Silva JL, et al. Retinal pigmented epithelial cell based gene therapy against hemoglobin toxicity. Int J Mol Med 1998; 1:657–663.

69. Kimura H, Ogura Y, Honda Y, et al. Intracellular sustained release with biodegradable polymer microspheres in cultured retinal pigment epithelial cells. Invest Ophthalmol Vis Sci 1993; 34:1487.

70. Moritera T, Ogura Y, Yoshimura N, et al. Biodegradable microspheres containing adriamycin in the treatment of proliferative vitreoretinopathy. Invest Ophthalmol Vis Sci 1992; 33:3125–3130.

71. Pearson PA, Jaffe G, Ashton P. Letter to editor. Am J Ophthalmol 1993; 115:686–687.

72. Berthe P, Baudouin C, Garraffo R, et al. Toxicologic and pharmacokinetic analysis of intravitreal injections of foscarnet, either alone or in combination with ganciclovir. Invest Ophthalmol Vis Sci 1994; 35:1038–1045.

73. Pearson PA, Jaffe GJ, Martin DP, et al. Evaluation of a delivery system providing long term release of cyclosporine. Arch Ophthalmol 1996; 114:311–317.

74. Cobo LM, Forster RK. The clearance of intravitreal gentamicin. Am J Ophthalmol 1981; 92:59–62.

75. Seto C, Araie M, Takase M. Study of fluorescein glucoronide. Graefes Arch Clin Exp Ophthalmol 1986; 224:113–117.

76. Leeds JM, Kornburst D, Truong L, Henry S. Metabolism and pharmacokinetic analysis of a phosphothioate oligonucleotide after intravitreal injection (abstract). Pharm Res 1994; 11(suppl):S353.

77. Lam TT, Edward DP, Zhu X-A, Tso MOM. Transcleral inotophoresis of dexamethasone. Arch Ophthalmol 1989; 107:1368–1371.

78. Forbes M, Becker B. The transprot of organic anions by the rabbit eye. In vivo transport of iodopyrocet (diodrast). Am J Ophthalmol 1960; 50:867–873.

79. Weiner IM, Blanchard KC, Mudge GH. Factors influencing the renal excretion of foreign organic acids. Am J Physiol 1964; 207:953–963.

80. Barza M, Kane A, Baum J. Pharmacokinetics of intravitreal carbenicillin, cefazolin and gentamicin in rhesus monkeys. Invest Ophthalmol Vis Sci 1983; 24:1602–1606.

81. Barza M, Kane A, Baum J. The effects of infection and probenicid on the transport of carbenicillin from the rabbit vitreous humor. Invest Ophthalmol Vis Sci 1982; 22:720–726.

82. Lutjen-DrecollE, Lonnerholm G, Eichhorn M. Carbonic anhydrase distribution in the human and monkey eye by light microscopy. Graefes Arch Clin Exp Ophthalmol 1983; 220:285–291.

83. Marmor MF, Negi A. Pharmacologic modifications of subretinal fluid absorption in the rabbit eye. Arch Ophthalmol 1986; 104:1674–1677.

84. Marmor MF, Maack T. Enhancement of retinal adhesion and subretinal fluid resorption by acetazolamide. Invest Ophthalmol Vis Sci 1982; 23:121–124.

85. Cox NS, Hay E, Bird AC. Treatment of chronic macular edema. Arch Ophthalmol 1988; 106:1190–1195.

86. Tsuboi S, Pederson JE. Experimental retinal detachment X. Effect of acetazolamide on vitreous fluorescein disappearance. Arch Ophthalmol 1985; 103:1557–1558.

87. Moldow B, Sander B, Larsoen M, et al. The effect of acetazolamide on passive and active transport of fluorescein across the blood–retina barrier in retinitis pigmentosa complicated by macular edema. Graefes Arch Clin Exp Ophthalmol 1998; 236:881–889.

88. Mallick KS, Zeimer RC, Fishman GA, et al. Transport of fluorescein in the ocular posterior segment in retinitis pigmentosa. Arch Ophthalmol 1984; 102:691–696.

89. Krupin T, Waltman SR. Fluorophotometry in juvenile onset diabetes: long term follow-up. Jpn J Ophthalmol 1985; 29:139–145.

90. Krupin T, Waltman SR, Szewczyk P, et al. Fluorometric studies on the blood-retinal barrier in experimental animals. Arch Ophthalmol 1982; 100:631–634.

91. Miyake, K. Vitreous fluorophotometry in aphakic or pseudophakic eyes with persistent cystoid macular edema. Jpn J Ophthalmol 1985; 29:146–152.

92. Grignolo A, Orzalesi N, Calabria GA. Studies on the fine structure and the rhodopsin cycle of the rabbit retina in experimental degeneration induced by sodium iodate. Exp Eye Res 1966; 5:86–97.

93. Kitano S, Hori S, Nagataki S. Transport of fluorescein in the rabbit eye after treatment with sodium iodate. Exp Eye Res 1988; 46:863–870.

94. Grimes PA, Laties AM. Early morphological alteration of the pigmented epithelium in streptozocin-induced diabetes: increased surface area of the basal cell membrane. Exp Eye Res 1980; 30:631–639.

95. Blair NP, Tso MOM, Dodge JT. Pathologic studies of the blood-retinal barrier in the spontaneously diabetic BB rat. Invest Ophthalmol Vis Sci 1984; 25:302–311.

96. Tso MOM, Cunha-Vaz J, Shih CY. Clinicopathologic study of blood–retinal barrier in experimental diabetes mellitus. Arch Ophthalmol 1978; 98:725–728.

97. Wallow IHL, Engerman RL. Permeability and patency of the retinal blood vessel in experimental diabetes. Invest Ophthalmol Vis Sci 1983; 24:1259–1268.

98. Kirber WM, Nichols CVW, Grimes PA, et al. A permeability defect of the retinal pigmented epithelium. Occurrence in early streptozocin diabetes. Arch Ophthalmol 1980; 98:725–728.

99. Enea ME, Hollis TM, Kern JAK, Gardner TW. Histamine HI receptors mediate increased blood–retinal barrier permeability in experimental diabetes. Arch Ophthalmol 1989; 107:270–274.

100. Krupin T, Waltman SR, Oestrich C, et al. Vitreous fluorophotometry in juvenile-onset diabetes mellitus. Arch Ophthalmol 1978; 96:812–814.

101. Cunha-Vaz J, Faria de Abreu JR, Canmpos AJ, Figo GM. Early breakdown of the blood retinal barrier in diabetes. Br J Ophthalmol 1975; 59:649.

102. Shires TK, Faeth JA, Pulido JS. Protein levels in the vitreous of rat with streptozotocin-induced diabetes mellitus. Brain Res Bull 1993; 30:85–90.

103. Hawkins KN. Contribution of plasma proteins to the vitreous of the rat. Curr Eye Res 1986; 5:655–663.

104. Jarus G, Blumenkranz M, Hernandez E, Sosi N. Clearance of intravitreal fluorouracil. Normal and aphakic vitrectomized eyes. Ophthalmology 1995; 92:91–96.

105. Pearson PA, Hainsworth DP, Ashton P. Clearance and distribution of ciprofloxacin after intravitreal injection. Retina 1993; 13:326–330.

106. Wingard LB, Zuravleff JJ, Doft BH, et al. Intraocular distribution of intravitreally administered amphoteracin B in normal and vitrectomized eyes. Invest Ophthalmol Vis Sci 1989; 30:2184–2189.

107. Perkins SL, Gallemore RP, Yang CH, et al. Pharmacokinetics of the fluocinolone/5-fluorouracil codrug ion the gas filled eye. Retina 2000; 20:514–519.

108. Doft BH. The endophthalmitis vitrectomy study. Arch Ophthalmol 1991; 109:188–195.

109. Forster RK, Abbott RL, Gelender H. Management of endophthalmitis. Ophthalmology 1980; 87:313–319.

110. Campochiaro PA, Lim JL. Aminoglycoside toxicity in the treatment of endophthalmitis. Arch Ophthalmol 1994; 112:48–53.

111. Donahue AP, Kowalski RP, Eller AW, et al. Empiric treatment of endophthalmitis. Are aminoglycosides necessary? Arch Ophthalmol 1994; 112:45–47.

112. Aaberg TM, Flynn HW, Murray TG. Intraocular ceftazidine as an alternative to aminoglycosides in the treatment of endophthalmitis. Arch Ophthalmol 1994; 112:18–19.

113. Pauriah M, Ong EL. Retrospective study of CMV retinitis in patients with AIDS. Clin Microbiol Infect 2000; 6:14–18.
114. Henry K, Cantrill H, Fletcher C, et al. Use of intravitreal ganciclovir (dihydroxyprop-xymethyl guanine) for cytomegalovirus retinitis. Am J Ophthalmol 1987; 103:17–23.
115. Ussery FM, Gibson SR, Conklin RH, et al. Intravitreal ganciclovir in the treatment of AIDS-associated cytomegalovirus retinitis. Ophthalmology 1988; 95:640–648.
116. Diaz-llopis M, Chipont E, Sanchez S, et al. Intravitreal foscarnet for cytomegalovirus retinitis in a patient with acquired immunodeficiency syndrome. Am J Ophthalmol 1992; 14:742–747.
117. Hodge WG, Lalonde RG, Sampalis J, Deschenes J. Once weekly intraocular injections of ganciclovir for maintenance therapy of cytomegalovirus retinitis: clinical and ocular outcome. J Infect Dis 1996; 174:393–396.
118. Taskintuna I, Rahhal FM, Arevalo JR, et al. Low-dose intravitreal cidofovir (HPMPC) therapy of cytomegalovirus retinitis in patients with acquired immune deficiency syndrome. Ophthalmology 1997; 104:1049–1057.
119. Berthe P, Baudouin C, Garraffo R, et al. Toxicologic and pharmacokinetic analysis of intravitreal injections of foscarnet, either alone or incombination with ganciclovir. Invest Ophthalmol Vis Sci 1994; 35:1038–1045.
120. Akula SK, Ma PE, Peyman GA, et al. Treatment of cytomegalovirus retinitis with intra-vitreal injection of liposome encapsulated ganciclovir in a patient with AIDS. Br J Ophthalmol 1994; 78:677–688.
121. Smith TJ, Pearson AP, Blandford DL, et al. Intravitreal sustained-release ganciclovir. Arch Ophthalmol 1992; 110:255–258.
122. Sanborn GE, Anand R, Torti RE, et al. Sustained-release ganciclovir therapy for treat-ment of cytomegalovirus retinitis. Arch Ophthalmol 1992; 110:188–195.
123. Morley MG, Duker J, Ashton P, Robinson M. Replacing ganciclovir implants. Ophthalmology 1995; 102:388–394.
124. Duker JS, Ashton P, Davis JL, et al. Long-term successful maintenance of bilateral cytomegalovirus retinitis using exclusively local therapy. Arch Ophthalmol 1996; 14:881–882.
125. Ryan SJ. The pathophysiology of proliferative vitreoretinopathy in its management. Am J Ophthalmol 1985; 100:188–193.
126. Ruhmann AG, Berliner DL. Influence of steroids on fibrosis. The fibroblast as an assay system for topical antiinflammatory potency of corticosteroids. J Invest Dermatol 1967; 49:123.
127. Blumenkranz MS, Claflin A, Hajek AS. Selection of therapeutic agents for intraocular proliferative disease. Cell culture evaluation. Arch Ophthalmol 1984; 102:598–694.
128. Goodman AG, Rail TW, Nies AS, Taylor P. eds. Goodman and Gilman's Pharmaco-logical Basis of Therapeutics. New York: Pergamon Press, 1990:1431–1462.
129. Tano Y, Sugita G, Abrams C, Machemer R. Inhibition of intraocular proliferation with intravitreal corticosteroids. Am J Ophthalmol 1980; 89:131–136.
130. Tano Y, Chandler DB, McCuen BW, Machemer. Glucocorticosteroid inhibition of intraocular proliferation after injury. Am J Ophthalmol 1981; 91:184–189.
131. McCullen BW, Bessler M, Tano Y, et al. The lack of toxicity of intravitreally adminis-tered triamcinolone acetonide. Am J Ophthalmol 1981; 91:785–788.
132. Chandler RB, Hida T, Sheta S, et al. Improved efficacy of corticosteroid therapy in an animal model of proliferative retinopathy by pretreatment. Graefes Arch Clin Exp Ophthalmol 1987; 225:259–265.
133. Wiedemann P, Lemmen K, Schmiedl R, Heimann K. Intraocular daunorubicin for the treatment and prophylaxis of traumatic proliferative vitreoretinopathy. Am J Ophthal-mol 1987; 104:10–14.
134. Blumenkrantz MS, Ophir A, Claflin AJ, Hajek A. Fluorouracil for treatment of massive periretinal proliferation. Am J Ophthalmol 1982; 94:458–467.

135. Stern WH, Lewis GP, Erickson PA, et al. Fluorouracil therapy for proliferative vitreoretinopathy after vitrectomy. Am J Ophthalmol 1983; 96:33–42.

136. Chung H, Tolentino FI, Cajita VN, et al. BCNU in silicone oil in proliferative vitreoretinopathy. 1. Solubility, stability (in vitro and in vivo), and antiproliferative in vitro studies. Curr Eye Res 1988; 7:1199–1206.

137. Araiz JJ, Refojo MF, Arroyo HM, et al. Antiproliferative effect of retinoic acid in intravitreous silicone oil in an animal model of proliferative vitreoretinopathy. Invest Ophthalmol Vis Sci 1993; 34:522–530.

138. Steffansen B, Ashton P, Buur A. Intraocular drug delivery. In vitro release studies of 5-Fluorouracil from N-1 alkoxycarbonyl prodrugs in silicone oil. Int J Pharm 1996; 132: 243–250.

139. Joondeph BC, Peyman GA, Khoobehi B, Yue BY. Liposome-encapsulated 5-fluorouracil in the treatment of proliferative vitreoretinopathy. Ophthalmic Surg 1988; 19:252–256.

140. Skuta GL, Assil K, Parrish RK, et al. Filtering surgery in owl monkeys with the antimetabolite; 5-flourouridine S-monophosphate entrapped in muluvesicular liposomes. Am J Ophthalmol 1987; 103:714–716.

141. Blumenkranz MS, Hartzer MK, Hajek AS. Selection of therapeutic agents for intraocular proliferative disease, H: Differing antiproliferative activity of the fluoropyrimidines. Arch Ophthalmol 1987; 105:396–399.

142. Maignen, F, Tilleul P, Billardon C, et al. Antiproliferative activity of a liposomal delivery system of mitoxantrone on rabbit subconjunctival fibroblasts in an ex-vivo model. J Ocul Pharmacol Ther 1996; 12:289–298.

143. Rubsamen PE, Davis PA, Hernandez E, et al. Prevention of experimental proliferative vitreoretinopathy with a biodegradable intravitreal implant for the sustained release of fluorouracil. Arch Ophthalmol 1994; 112:407–413.

144. Berger AS, Cheng CK, Pearson PA, et al. Intravitreal sustained release corticosteroid 5-fluorouracil conjugate in the treatment of experimental proliferative vitreoretinopathy. Invest Ophthalmol Vis Sci 1996; 37:2318–2325.

145. Yang CS, Khawley JA, Hainsworth DP, et al. An intravitreal sustained release triamcinolone 5-FU codrug in the treatment of experimental proliferative vitreoretinopathy. Arch Ophthalmol 1998; 116:69–77.

2

Blood–Retinal Barrier

David A. Antonetti and Thomas W. Gardner
Departments of Cellular and Molecular Physiology and Ophthalmology,
Penn State College of Medicine, Hershey, Pennsylvania, U.S.A.

Alistair J. Barber
Department of Ophthalmology, Penn State College of Medicine,
Hershey, Pennsylvania, U.S.A.

INTRODUCTION

The blood–retinal barrier controls the flux of fluid and blood-borne elements into the neural parenchyma, helping to establish the unique neural environment necessary for proper neural function. Loss of the blood–retinal barrier characterizes a number of the leading causes of blindness including diabetic retinopathy and age-related macular degeneration. In this chapter, the structure of the tight junctions that constitute the blood–retinal barrier will be examined with specific emphasis on the transmembrane tight junction proteins occludin and claudin, which form the seal between adjacent endothelial cells. In addition, alterations that occur to the tight junction proteins in diseases such as diabetic retinopathy will be addressed. Finally, the use of glucocorticoids to restore barrier properties and the effect of this hormone on tight junctions will be discussed.

FUNCTION OF THE BLOOD–RETINAL BARRIER

The blood vessels of the retina, like those of the brain, develop a barrier that partitions the neural parenchyma from the circulating blood. Together with the retinal pigmented epithelium, the blood vessels of the retina create the blood–retinal barrier. This unique barrier is composed of the junctional complex that includes the tight junctions, originally called the zonula occludens (ZO), the adherens junctions, and desmosomes. The unique barrier properties of the blood vessels in neural tissues are the result of well-developed tight junctions. The initial ultrastructural characterization of this barrier was achieved by electron microscopy. Most notably, horseradish peroxidase, used as a tracer in electron microscopy, diffuses only up to the tight junction in brain cortical capillaries: in other tissues without tight junctions, this marker diffuses out of the vascular lumen (1). Similar studies in the retina with tracers reveal that tight junctions mediate the blood–retinal barrier, preventing solute flux into the retinal parenchyma (2,3).

This tight control of blood elements into the retinal parenchyma is necessary for a number of reasons related to neural function. First, the neural tissue maintains constant exchange of metabolites between glia and neurons. For example, glucose is metabolized by glia and provided to the neurons as lactate for oxidation and energy production. Thus, the neural tissue requires a defined and controlled environment. Second, the ionic environment must be tightly controlled to allow neurons to establish and control membrane potentials and depolarization in neuronal signaling. Third, the blood contains amino acids used as protein building blocks as well as intermediate metabolites. These amino acids are used by the neural tissue as signaling molecules; for example, glutamate and aspartate. The blood typically maintains relatively high concentrations of these excitatory amino acids. Their entry into the neural parenchyma must be tightly controlled to maintain proper neural signaling. Thus, the blood–retinal barrier protects neural tissue by regulating flow of essential metabolites into the tissue to control the composition of the extracellular environment.

FORMATION OF THE BLOOD–NEURAL BARRIER

The formation of the tight junction complex and the blood–neural barrier depends on the close association of glia with the endothelial cells in the capillaries and arterioles traversing the neural tissue. Evidence for glial induction of endothelial barrier properties comes from a variety of experimental approaches. First, on a morphologic level, astrocytes make close contact with the endothelial cells of both arterioles and capillaries in the retina. Figure 1 depicts whole mount immunostaining for a specific tight junction protein, occludin in panel A and in panel B, the same section of retina stained for glial fibrillary acid protein is shown. This close association between astrocytes and endothelia is also observed in brain blood vessels, suggesting a role for glia in endothelial barrier induction. In the capillary plexus of the retinal outer plexiform layer, the Müller cells may provide the glial support supplied by the astrocytes in the

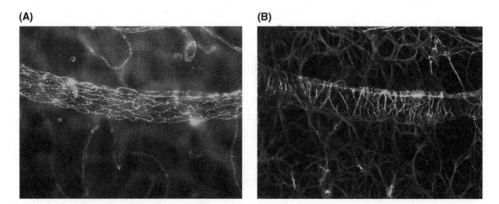

(A)   **(B)**

Figure 1 Astrocytes make close contact with endothelial cells within the retina. (**A**) Immunostaining for the tight junction protein occludin reveals a high degree of well-organized tight junctions in the arterioles and capillaries of the retina. (**B**) Glial fibrillary acid protein staining demonstrates that astrocytes make close contact with the endothelial cells within the retina.

capillary plexus of the ganglion cell layer. Further support is obtained by coculture experiments that demonstrate that close contact of astrocytes or brain slices can confer increased barrier properties to endothelial cells (4–6). In addition, astrocyte-conditioned media supplemented with agents that increase cAMP can dramatically increase barrier properties of endothelial cell culture, suggesting a soluble component may confer barrier properties (7). Finally, introduction of astrocytes (8) or Müller cells adjacent to normally leaky blood vessels increases barrier properties (9). The ability of glia to induce endothelial barrier properties suggests that loss of the blood–retinal barrier in eye disease could be related to changes in glial function or association with the retinal endothelium.

OCULAR DISEASE AND LOSS OF THE BLOOD–RETINAL BARRIER

While normal retinal function requires the blood–retinal barrier, loss of this barrier characterizes a wide array of retinal complications and precedes neovascularization. Increased vascular permeability, observed as macular edema, is a common characteristic of diabetic retinopathy, with a prevalence of 20.1% and 25.4% of type 1 and type 2 diabetic patients, respectively (10,11). Furthermore, 27% of patients in the secondary intervention arm of the diabetes control and complications trial developed macular edema within nine years (12). Indeed, loss of the blood–retinal barrier in diabetic retinopathy is still one of the earliest detectable events in diabetic retinopathy and macular edema is the clinical feature most closely associated with loss of vision (13). Loss of the blood–retinal barrier includes increased permeability in both the blood vessels and retinal pigmented epithelium but altered vascular permeability appears to precede changes in the pigmented epithelium in diabetes (14). In addition, retinal vein occlusion results in blood–retinal barrier breakdown as seen upon vascular reperfusion, as does uveoretinitis and age-related macular degeneration. Changes in the pigmented epithelium likely dominate in the latter. Thus, loss of the normal blood–retinal barrier is a common feature to many retinal degenerative diseases that are the leading causes of vision loss in Western society, making development of therapies to prevent loss of barrier properties or restore barrier properties a high priority in vision research.

Increased growth factor production from the neural retina and cytokine production from inflammation both contribute to the loss of the blood–retinal barrier in diabetic retinopathy. Changes in ocular growth factors and their receptors include insulin-like growth factor 1 and its binding proteins, platelet-derived growth factor, fibroblast growth factor, and vascular endothelial growth factor (VEGF) (15–18). Immunohistochemistry and in situ hybridization studies demonstrate that the expression of VEGF and its receptors increase by six months of experimentally induced diabetes within the retinal parenchyma (19–21); in Goto–Kakizaki rats, a model of type 2 diabetes, the level of hormone is significantly elevated over control by 28-weeks. In addition, measurements of VEGF content in patients with proliferative diabetic retinopathy reveal that many, but not all patients, have increased hormone in the vitreous fluid (22,23) and in epiretinal membranes (24). VEGF expression in the retina occurs before the onset of proliferative retinopathy, suggesting a role for this growth factor specifically in vascular permeability (25,26).

In addition to neural production of VEGF, inflammation contributes to vascular permeability as well. Leukostasis increases in the capillaries of the retina in animals made diabetic by streptozotocin. Inhibition of leukostasis with antibodies to adhesion molecule intracellular adhesion molecule (ICAM), which block the leukocyte-endothelial interaction, also reduce retinal vascular permeability (27). The contribution of various cytokines and chemokines to vascular permeability in diabetic retinopathy are now under intense investigation and a functional role for these cytokines in permeability has already been demonstrated (28). Furthermore, oxygen free-radicals may cause disruption of the blood–retinal barrier. In vitro studies of the retinal-pigmented epithelium (29) and endothelial cells (30,31) suggest that hydrogen peroxide may disrupt barrier properties. Oxygen free-radical production may be due to an inflammatory response, ischemia reperfusion, or, in the case of diabetes, from dysregulation of metabolism. Thus, the contribution of free-radical production on barrier properties in disease states is an area in need of further study. These studies demonstrate that multiple insults alter the blood–retinal barrier in diabetic retinopathy. Understanding how diabetes changes the molecules that constitute this barrier may provide a means to prevent or reverse the loss of the barrier regardless of the insult.

MOLECULAR ARCHITECTURE OF THE BLOOD–RETINAL BARRIER

Tight junctions confer the barrier properties to the endothelial cells within the retinal vasculature creating the blood–retinal barrier. The tight junctions are composed of two transmembrane proteins, occludin and claudin, known to provide barrier properties. These proteins are linked through adaptor proteins, such as the ZO family members, to the cell actin cytoskeleton. Occludin and claudin share a common structural motif; specifically, both proteins span the membrane four times, creating two extracellular loops that dimerize with proteins in the tight junction of adjacent endothelial cells, helping to create the paracellular seal. However, occludin and claudin contribute unique functionality to the tight junction. This chapter will focus on how these transmembrane proteins are involved in barrier formation. Additional junction-specific proteins may provide important differences to the composition and function of the junctional complex between endothelial and epithelial cells. For example, cingulin is an epithelial restricted tight junction protein (32,33) and junction-enriched and associated protein (JEAP) is an exocrine specific protein (34). However, the differences between endothelial cell and retinal pigmented epithelial cell junctional proteins have not yet been characterized.

CLAUDINS

The claudins are made of at least 24 separate gene products whose expression helps to determine barrier properties of the tight junctions (35–38). Claudin family members exhibit distinct tissue expression patterns (39–41). Claudin 5 expression is largely restricted to the endothelium (42) but in some cases is expressed in retinal vasculature as well (43). The brain endothelium also expresses claudin 1 (44); however, little has been done to examine additional claudin expression in the retinal vasculature. Expression of claudins in cell lines that normally lack tight junctions has helped in proposing important principles. First, claudin expression in cells that do not express additional junctional components shows that these cells are capable of forming limited strands that mimic tight junctions in vivo (45). In contrast, occludin forms a punctate staining pattern with much less extended tight junction-like strands (45). However, cotransfection of occludin with claudins results in occludin integration into the tight junction strands. Expression studies have also demonstrated that claudins can form homomeric and heteromeric complexes with specific restrictions. For example, coculture studies with cells expressing claudin 1, 2, or 3 indicate that claudin 3 interacts with claudin 1 and claudin 2 on adjacent cells; however, claudin 1 and claudin 2 do not interact (46). Finally, gene deletion studies have demonstrated the role of claudins in barrier formation. Claudin 1-deficient mice die within one day of birth due to transepidermal water loss (47). Specifically relevant to the blood–neural barrier, claudin 5-deficient mice demonstrate increased permeability across the blood–retinal barrier, specifically to molecules of less than 800 Da (48). These studies reveal that claudins help to create the barriers that comprise the tight junctions.

Specific expression patterns of claudins provide the character of tight junctions, particularly in relation to electrical resistance. Transfection experiments demonstrate that expression of claudin isotypes can directly affect ion selectivity and conductance (49). The effect of charge selectivity was most dramatically shown when three amino acids in the first extracellular loop of claudin 15 were mutated from a negative charge to a positive charge. This mutation changed the tight junction from allowing Na^+ flux and preventing Cl^- flux to becoming permissive for Cl^- flux and inhibitory of Na^+ flux. Thus, claudins can form barriers to specific ions and create conductance channels for other ions. To date, little is known regarding the nature of the tight junctions in the retina in relation to ionic selectivity.

OCCLUDIN

Occludin is encoded by a single gene but may be alternatively spliced or initiated from an alternative promoter, yielding novel variants (50,51). The expression of occludin correlates well with the degree of barrier properties in various tissues. For example, arterial endothelial cells express 18-fold more occludin protein than venous endothelial cells and form a tighter solute barrier (52). Similarly, occludin is highly expressed in brain endothelium coincident with the formation of the blood–brain barrier and is expressed at much lower levels in endothelial cells of non-neuronal tissue, which have less barrier properties (53). In the retina, the endothelium of the arteries, arterioles, and capillaries express a relatively high degree of occludin that is well organized at the cell border. In contrast, the venules and veins express a lower amount of occludin and localization to the cell border is minimal (43,54).

A number of experiments, performed mostly in epithelial cells, demonstrate that occludin contributes to the barrier function of tight junctions. Antisense oligonucleotide experiments demonstrate a decrease in barrier properties associated with a reduction of occludin content (52). Expression of chicken occludin in Madin-Darby Canine Kidney Epithelial (MDCK) cells under the control of an inducible promoter substantially increased transcellular electrical resistance (TER) and increased the number of tight junction strands compared to untreated cells (55). In contrast, synthetic peptides targeting the second extracellular loop of occludin (OCC2) significantly decreased the TER and increased the flux of several paracellular tracers in confluent monolayers of a *Xenopus* kidney epithelial cell line (56). Furthermore, the OCC2 peptide promotes the degradation of occludin by competitively inhibiting occludin-mediated cell-cell adhesion. In a similar study, synthetic peptides homologous to regions of the first extracellular loop of occludin prevented junction resealing after calcium depletion and readdition, as measured by TER (57). These studies support a role for occludin in barrier formation of tight junctions.

Gene-deletion experiments have demonstrated a more complex role for occludin in tight junction barrier formations. Embryonic stem cells from occludin null mice formed cystic embryoid body structures with an outermost layer of epithelial cells, similar to wild-type embryonic cells (58). Ultrastructural analysis revealed no changes in the tight junctions; the tight junction protein ZO-1 exhibited normal localization at apical junctional regions in the outermost layer of epithelial cells and no change in barrier properties was observed in the occludin null cells. However, the adult occludin homozygous null mice, although viable, possessed a host of abnormalities (59). Occludin-deficient mice exhibited postnatal growth retardation, male knockout mice were infertile, and female knockout mice were unable to suckle their litters. Overall, these mice exhibited abnormalities in the testis and salivary gland, thinning of compact bone, calcium deposits in the brain, chronic gastritis, and hyperplasia of the gastric epithelium. In addition, recent studies using siRNA to occludin demonstrate that occludin forms a barrier to organic acids up to 6.96 Å, such as arginine and choline (60). Thus, these studies have led to the hypothesis that occludin contributes to the regulation of barrier properties by creating a doorway or regulated pore through the tight junction.

Occludin associates with a number of structural and regulatory molecules supporting a model in which occludin contributes to regulation of barrier properties. The C-terminal cytoplasmic domain of occludin binds to ZO-1 in vitro (61), and this interaction may serve to link occludin to the actin cytoskeleton (62). Similarly, ZO-2 and ZO-3 bind to the C-terminus of occludin in vitro (63,64). In addition to this link to the cell cytoskeleton, occludin interacts with a number of regulatory proteins at tight junctions. Use of a 27 amino acid region of the C-terminus of occludin that encodes a putative coiled–coiled domain helped identify several occludin-binding proteins: protein kinase C-ζ, c-Yes, connexin-26, and p85, the regulatory subunit of phosphatidylinositol 3-kinase (65). Occludin may also interact with proteins via its N-terminal cytoplasmic region. The E3 ubiquitin–protein ligase, Itch, was found to associate with the N-terminus of occludin in vitro and in vivo, suggesting that occludin content or localization may be regulated by ubiquitination (66). These protein–protein interactions may regulate junction formation and barrier properties.

Occludin phosphorylation may provide a molecular mechanism to control barrier properties. Studies from our group have demonstrated that both VEGF and shear stress induce permeability across endothelial monolayers associated with a rapid phosphorylation of occludin (67,68). The occludin phosphorylation was attenuated by a non-hydrolyzable cAMP analog that also inhibits shear-induced permeability (68). This phosphorylation of occludin appears to be serine or threonine directed since immunoprecipitation of occludin and phosphotyrosine blotting did not reveal any evidence of occludin tyrosine phosphorylation in this cell system (unpublished observation). However, in epithelial cells, evidence of occludin tyrosine phosphorylation exists (69). In addition, others have identified occludin phosphorylation in response to histamine (70) and use of brain extracts has helped identify casein kinase II as an occludin kinase (71). Collectively, this work demonstrates a close association of occludin phosphorylation with permeability. Future studies identifying specific occludin phosphorylation sites, followed by mutational analysis, should reveal the functional significance of occludin phosphorylation.

In addition to occludin phosphorylation, redistribution of occludin may contribute to loss of the blood–retinal barrier. Both VEGF and diabetes induce a redistribution of occludin from the plasma membrane to the cell cytoplasm (43,54). A similar change in junction organization was observed in retinal-pigmented epithelial cells in response to hepatocyte growth factor (72). In an epithelial cell culture system, platelet-derived growth factor, a growth factor closely related to VEGF, stimulated the redistribution of occludin and other tight junction proteins from the plasma membrane to an early endosome compartment (73). Recent experiments support a model in which occludin recycles through an endosomal compartment (74) and that endocytosis occurs through a clathrin-mediated pathway in epithelial cells (75). One potential molecular mechanism for VEGF-regulated permeability includes occludin phosphorylation releasing occludin from a neighboring endothelial tight junction. Next, endocytosis of occludin leads to its translocation from the cell plasma membrane to an internal compartment. However, many other possible models exist to describe the data and future studies on both phosphorylation and recycling of occludin and are necessary to elucidate the pathological mechanisms for loss of endothelial barrier properties.

RESTORING BARRIER PROPERTIES

A number of therapies are currently under trial for diabetic retinopathy, therapies that have been developed to prevent loss of vascular barrier function. These methods include binding VEGF and preventing receptor activation through the use of a VEGF aptamer (76) or a modified, soluble VEGF receptor, the VEGF trap (77,78), or preventing VEGF signal transduction with the use of a protein kinase C inhibitor (79). However, little has been done to consider induction of barrier properties once lost. Our laboratory and others have demonstrated that VEGF and diabetes reduce occludin content (80,81), increase occludin phosphorylation, and stimulate occludin redistribution as described earlier. Glucocorticoids have been used to treat brain tumors for over 35 years (82,83). Brain tumors possess a number of similarities to diabetic retinopathy in relation to vascular changes. In both cases, a blood–neural barrier characterized by a high degree of well-developed tight junctions is altered leading to increased permeability. An increase in VEGF or inflammatory cytokines is believed to contribute to the loss of barrier function. Given the success of steroids to reverse vascular permeability, it is hypothesized that this steroid hormone acts on the endothelial cells to induce formation of the tight junctions. Indeed, our studies demonstrate that glucocorticoids directly act on endothelial cells to increase expression of occludin and its assembly at the cell border, reduce occludin phosphorylation, and increase barrier properties (84). The effect of glucocorticoids on endothelial cells was also observed by Hoheisel et al. (85), who demonstrated that hydrocortisone treatment increases TER nearly threefold and reduces sucrose permeability fivefold in pig brain capillary endothelial cells in a dose-responsive manner.

A positive effect of glucocorticoids on barrier properties has also been observed in epithelial cells. Dexamethasone treatment for four days increases the electrical resistance and reduces radiolabeled mannitol and insulin flux across 31EG4 nontransformed epithelial cells (86) and the Con8 mammary epithelial tumor cell line (87). Dexamethasone treatment increased ZO-1 content in the 31EG4 cells by slightly more than twofold after four days treatment while RNA content did not change (88). This is in contrast with the finding in bovine retinal endothelial cells in which ZO-1 content did not change but its redistribution to the cell border dramatically increased with hydrocortisone treatment (84). The redistribution of ZO-1 was also observed in epithelial cells and may be related to fascin expression, which is thought to bind to ZO-1 and retain the protein in the cytoplasm (89,90). Glucocorticoids downregulate fascin and allow

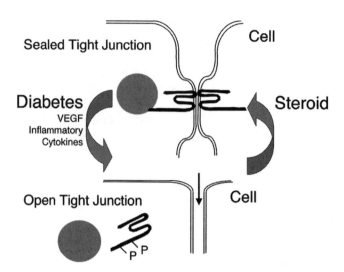

Figure 2 Diabetes leads to loss of the tight junctions while glucocorticoids induce assembly of tight junctions. In diabetes, VEGF produced from the neural retina as well as inflammatory cytokines cause phosphorylation of occludin, tight junction disassembly, and loss of tight junction proteins. Steroids induce the synthesis of tight junction proteins, assembly of tight junctions at the cell border, and dephosphorylation of occludin associated with increased barrier properties in vitro and may induce barrier formation in vivo. *Abbreviation*: VEGF, vascular endothelial growth factor.

redistribution of ZO-1 to the cell border and organization of tight junctions. Whether a similar mechanism contributes to endothelial barrier induction in response to glucocorticoids remains unknown at present. Furthermore, others have demonstrated an increase in occludin in response to glucocorticoids in epithelial cells (91). Thus, steroids induce tight junction protein expression and redistribution to the plasma membrane in epithelial and endothelial cell systems. Localized delivery of glucocorticoids may provide a means to restore barrier integrity and reduce inflammation in diabetic retinopathy (Fig. 2). However, given the risks associated with prolonged steroid use, it is imperative to determine the molecular mechanisms by which glucocorticoids control barrier properties so that novel, more specific therapies may be developed.

In conclusion, recent evidence indicates that permeability at the vascular blood–retinal barrier is regulated by a number of tight junction proteins that act together to protect the neural tissue. Diabetes leads to loss of the blood–retinal barrier by altering the content, phosphorylation state, and localization of tight junction proteins such as occludin. New treatment approaches are designed to target the regulation of the tight junction proteins in order to prevent macular edema and preserve vision in people with diabetes.

REFERENCES

1. Reese TS, Karnovsky MJ. Fine structural localization of a blood–brain barrier to exogenous peroxidase. J Cell Biol 1967; 34:207–217.
2. Shakib M, Cunha-Vaz JG. Studies on the permeability of the blood–retinal barrier. IV. Junctional complexes of the retinal vessels and their role in the permeability of the blood–retinal barrier. Exp Eye Res 1966; 5:229–234.

3. Cunha-Vaz JG, Shakib M, Ashton N. Studies on the permeability of the blood–retinal barrier. I. On the existence, development, and site of a blood–retinal barrier. Br J Ophthalmol 1966; 50:441–453.

4. Demeuse P, Kerkhofs A, Struys-Ponsar C, Knoops B, Remacle C, van den Bosch de Aguilar P. Compartmentalized coculture of rat brain endothelial cells and astrocytes: a syngenic model to study the blood–brain barrier. J Neurosci Methods 2002; 121: 21–31.

5. Duport S, Robert F, Muller D, Grau G, Parisi L, Stoppini L. An in vitro blood–brain barrier model: cocultures between endothelial cells and organotypic brain slice cultures. Proc Natl Acad Sci USA 1998; 95:1840–1845.

6. Stanness KA, Guatteo E, Janigro D. A dynamic model of the blood–brain barrier "in vitro." Neurotoxicology 1996; 17:481–496.

7. Rubin LL, Hall DE, Porter S, et al. A cell culture model of the blood–brain barrier. J Cell Biol 1991; 115:1725–1735.

8. Abbott NJ. Astrocyte–endothelial interactions and blood–brain barrier permeability. J Anat 2002; 200:629–638.

9. Tout S, Chan-Ling T, Hollander H, Stone J. The role of Müller cells in the formation of the blood–retinal barrier. Neuroscience 1993; 55:291–301.

10. Klein R, Klein BEK, Moss SE, Cruscishanks KJ. The wisconsin epidemiologic study of diabetid retinopathy. XV. The long-term incidence of macular edema. Ophthalmology 1995; 102:7–16.

11. Vitale S, Maguire MG, Murphy RP, et al. Clinically significant macular edema in type I diabetes. Incidence and risk factors. Ophthalmology 1995; 102:1170–1176.

12. Anonymous. Progression of retinopathy with intensive versus conventional treatment in the diabetes control and complications trial. Diabetes control and complications trial research group. Ophthalmology 1995; 102:647–661.

13. Moss SE, Klein R, Klein BE. The 14-year incidence of visual loss in a diabetic population. Ophthalmology 1998; 105:998–1003.

14. Do Carmo A, Ramos P, Reis A, Proenca R, Cunha-vaz JG. Breakdown of the inner and outer blood retinal barrier in streptozotocin-induced diabetes. Exp Eye Res 1998; 67: 569–575.

15. Aiello LP. Vascular endothelial growth factor and the eye: biochemical mechanisms of action and implications for novel therapies. Ophthal Res 1997; 29:354–362.

16. Miller JW. Vascular endothelial growth factor and ocular neovascularization. Am J Pathol 1997; 151:13–23.

17. Miller JW, Adamis AP, Aiello LP. Vascular endothelial growth factor in ocular neovascularization and proliferative diabetic retinopathy. Diabetes Metab Rev 1997; 13: 37–50.

18. Paques M, Massin P, Gaudric A. Growth factors and diabetic retinopathy. Diabetes Metab 1997; 23:125–130.

19. Gilbert RE, Vranes D, Berka JL, et al. Vascular endothelial growth factor and its receptors in control and diabetic rat eyes. Lab Invest 1998; 78:1017–1027.

20. Hammes HP, Lin J, Bretzel RG, Brownlee M, Breier G. Upregulation of the vascular endothelial growth factor/vascular endothelial growth factor receptor system in experimental background diabetic retinopathy of the rat. Diabetes 1998; 47:401–406.

21. Murata T, Nakagawa K, Khalil A, Ishibashi T, Inomata H, Sueishi K. The relation between expression of vascular endothelial growth factor and breakdown of the blood-retinal barrier in diabetic rat retinas. Lab Invest 1996; 74:819–825.

22. Adamis AP, Miller JW, Bernal MT, et al. Increased vascular endothelial growth factor levels in the vitreous of eyes with proliferative diabetic retinopathy. Am J Ophthalmol 1994; 118:445–450.

23. Aiello LP, Avery RL, Arrigg PG, et al. Vascular endothelial growth factor in ocular fluid of patients with diabetic retinopathy and other retinal disorders. N Engl J Med 1994; 331:1480–1487.

24. Chen YS, Hackett SF, Schoenfeld CL, Vinores MA, Vinores SA, Campochiaro PA. Localization of vascular endothelial growth factor and its receptors to cells of vascular and avascular epiretinal membranes. Br J Ophthalmol 1997; 81:919–926.

25. Amin RH, Frank RN, Kennedy A, Eliott D, Puklin JE, Abrams GW. Vascular endothelial growth factor is present in glial cells of the retina and optic nerve of human subjects with nonproliferative diabetic retinopathy. Invest Ophthalmal Vis Sci 1997; 38:36–47.

26. Lutty GA, McLeod DS, Merges C, Diggs A, Plouét J. Localization of vascular endothelial growth factor in human retina and choroid. Arch Ophthalmol 1996; 114: 971–977.

27. Miyamoto K, Khosrof S, Bursell SE, et al. Prevention of leukostasis and vascular leakage in streptozotocin-induced diabetic retinopathy via intercellular adhesion molecule-1 inhibition. Proc Natl Acad Sci USA 1999; 96:10,836–10,841.

28. Luna JD, Chan CC, Derevjanik NL, et al. Blood-retinal barrier (BRB) breakdown in experimental autoimmune uveoretinitis: comparison with VEGF, TNF, and IL1-B mediated breakdown. J Neurosci Res 1997; 49:268–280.

29. Bailey TA, Kanuga N, Romero IA, Greenwood J, Luthert PJ, Cheetham ME. Oxidative stress affects the junctional integrity of retinal pigment epithelial cells. Invest Ophthalmol Vis Sci 2004; 45:675–684.

30. Kevil CG, Oshima T, Alexander JS. The role of p38 MAP kinase in hydrogen peroxide mediated endothelial solute permeability. Endothelium 2001; 8:107–116.

31. Kevil CG, Oshima T, Alexander B, Coe LL, Alexander JS. H(2)O(2)-mediated permeability: role of MAPK and occludin. Am J Physiol Cell Physiol 2000; 279:C21–C30.

32. D'Atri F, Nadalutti F, Citi S. Evidence for a functional interaction between cingulin and ZO-1 in cultured cells. J Biol Chem 2002; 277:27,757–27,764.

33. Cordenonsi M, D'Atri F, Hammar E, et al. Cingulin contains globular and coiled-coil domains and interacts with ZO-1, ZO-2, ZO-3, and myosin. J Cell Biol 1999; 147: 1569–1582.

34. Nishimura M, Kakizaki M, Ono Y, et al. JEAP, a novel component of tight junctions in exocrine cells. J Biol Chem 2002; 277:5583–5587.

35. Mitic LL, van Itallie CM, Anderson JM. Molecular physiology and pathophysiology of tight junctions—I. Tight junction structure and function: lessons from mutant animals and proteins. Am J Physiol Gastrointest Liver Physiol 2000; 279:G250–G254.

36. Kniesel U, Wolburg H. Tight junctions of the blood–brain barrier. Cell Mol Neurobiol 2000; 20:57–76.

37. Fanning AS, Mitic LL, Anderson JM. Transmembrane proteins in the tight junction barrier. J Am Soc Nephrol 1999; 10:1337–1345.

38. Tsukita S, Furuse M. Occludin and claudins in tight-junction strands: leading or supporting players? Trends Cell Biol 1999; 9:268–273.

39. Morita K, Furuse M, Fujimoto K, Tsukita S. Claudin multigene family encoding four-transmembrane domain protein components of tight junction strands. Proc Natl Acad Sci USA 1999; 96:511–516.

40. Li WY, Huey CL, Yu AS. Expression of claudin-7 and -8 along the mouse nephron. Am J Physiol Renal Physiol 2004; 286:F1063–F1071.

41. Rahner C, Mitic LL, Anderson JM. Heterogeneity in expression and subcellular localization of claudins 2, 3, 4, and 5 in the rat liver, pancreas, and gut. Gastroenterology 2001; 120:411–422.

42. Morita K, Sasaki H, Furuse M, Tsukita S. Endothelial claudin: claudin-5/TMVCF constitutes tight junction strands in endothelial cells. J Cell Biol 1999; 147:185–194.

43. Barber AJ, Antonetti DA. Mapping the blood vessels with paracellular permeability in the retinas of diabetic rats. Invest Ophthalmol Vis Sci 2003; 44:5410–5416.

44. Lippoldt A, Kniesel U, Liebner S, et al. Structural alterations of tight junctions are associated with loss of polarity in stroke-prone spontaneously hypertensive rat blood–brain barrier endothelial cells. Brain Res 2000; 885:251–261.

45. Furuse M, Sasaki H, Fujimoto K, Tsukita S. A single gene product, claudin-1 or -2, reconstitutes tight junction strands and recruits occludin in fibroblasts. J Cell Biol 1998; 143:391–401.

46. Furuse M, Sasaki H, Tsukita S. Manner of interaction of heterogeneous claudin species within and between tight junction strands. J Cell Biol 1999; 147:891–903.

47. Furuse M, Hata M, Furuse K, et al. Claudin-based tight junctions are crucial for the mammalian epidermal barrier: a lesson from claudin-1-deficient mice. J Cell Biol 2002; 156:1099–1111.

48. Nitta T, Hata M, Gotoh S, et al. Size-selective loosening of the blood–brain barrier in claudin-5-deficient mice. J Cell Biol 2003; 161:653–660.

49. Van Itallie CM, Fanning AS, Anderson JM. Reversal of charge selectivity in cation or anion-selective epithelial lines by expression of different claudins. Am J Physiol Renal Physiol 2003; 285:F1078–F1084.

50. Muresan Z, Paul DL, Goodenough DA. Occludin 1B, a variant of the tight junction protein occludin. Mol Biol Cell 2000; 11:627–634.

51. Mankertz J, Waller JS, Hillenbrand B, et al. Gene expression of the tight junction protein occludin includes differential splicing and alternative promoter usage. Biochem Biophys Res Commun 2002; 298:657–666.

52. Kevil CG, Okayama N, Trocha SD, et al. Expression of zonula occludens and adherens junctional proteins in human venous and arterial endothelial cells: role of occludin in endothelial solute barriers. Microcirculation 1998; 5:197–210.

53. Hirase T, Staddon JM, Saitou M, et al. Occludin as a possible determinant of tight junction permeability in endothelial cells. J Cell Sci 1997; 110:1603–1613.

54. Barber AJ, Antonetti DA, Gardner TW. Altered expression of retinal occludin and glial fibrillary acidic protein in experimental diabetes. Invest Ophthalmol Vis Sci 2000; 41:3561–3568.

55. Balda MS, Whitney JA, Flores S, Gonzalez M, Cereijido M, Matter K. Functional dissociation of paracellular permeability and transepithelial electrical resistance and disruption of the apical-basolateral intramembrane diffusion barrier by expression of a mutant tight junction membrane protein. J Cell Biol 1996; 134:1031–1049.

56. Wong V, Gumbiner BM. A synthetic peptide corresponding to the extracellular domain of occluding perturbs the tight junction permeability barrier. J Cell Biol 1997; 136:399–409.

57. Lacaz-Vieira F, Jaeger MM, Farshori P, Kachar B. Small synthetic peptides homologous to segments of the first external loop of occludin impair tight junction resealing. J Membr Biol 1999; 168:289–297.

58. Saitou M, Fujimoto K, Doi Y, et al. Occludin-deficient embryonic stem cells can differentiate into polarized epithelial cells bearing tight junctions. J Cell Biol 1998; 141: 397–408.

59. Saitou M, Furuse M, Sasaki H, et al. Complex phenotype of mice lacking occludin, a component of tight junction strands. Mol Biol Cell 2000; 11:4131–4142.

60. Yu AS, McCarthy KM, Francis SA, et al. Knock down of occludin expression leads to diverse phenotypic alterations in epithelial cells. Am J Physiol Cell Physiol 2005; 288(6):C1231–C1241.

61. Furuse M, Itoh M, Hirase T, et al. Direct association of occludin with ZO-1 and its possible involvement in the localization of occludin at tight junctions. J Cell Biol 1994; 127:1617–1626.

62. Fanning AS, Jameson BJ, Jesaitis LA, Anderson JM. The tight junction protein ZO-1 establishes a link between the transmembrane protein occludin and the actin cytoskeleton. J Biol Chem 1998; 273:29,745–29,753.

63. Haskins J, Gu L, Wittchen ES, Hibbard J, Stevenson BR. ZO-3, a novel member of the MAGUK protein family found at the tight junction, interacts with ZO-1 and occludin. J Cell Biol 1998; 141:199–208.

64. Itoh M, Morita K, Tsukita S. Characterization of ZO-2 as a MAGUK family member associated with tight as well as adherens junctions with a binding affinity to occludin and alpha catenin. J Biol Chem 1999; 274:5981–5986.

65. Nusrat A, Chen JA, Foley CS, et al. The coiled-coil domain of occludin can act to organize structural and functional elements of the epithelial tight junction. J Biol Chem 2000; 275:29,816–29,822.

66. Traweger A, Fang D, Liu YC, et al. The tight junction-specific protein occludin is a functional target of the E3 ubiquitin-protein ligase itch. J Biol Chem 2002; 277:10,201–10,208.

67. Antonetti DA, Barber AJ, Hollinger LA, Wolpert EB, Gardner TW. Vascular endothelial growth factor induces rapid phosphorylation of tight junction proteins occludin and zonula occluden 1. A potential mechanism for vascular permeability in diabetic retinopathy and tumors. J Biol Chem 1999; 274:23,463–23,467.

68. DeMaio L, Chang YS, Gardner TW, Tarbell JM, Antonetti DA. Shear stress regulates occludin content and phosphorylation. Am J Physiol Heart Circ Physiol 2001; 281: H105–H113.

69. Chen YH, Lu Q, Goodenough DA, Jeansonne B. Nonreceptor tyrosine kinase c-Yes interacts with occludin during tight junction formation in canine kidney epithelial cells. Mol Biol Cell 2002; 13:1227–1237.

70. Hirase T, Kawashima S, Wong EY, et al. Regulation of tight junction permeability and occludin phosphorylation by Rhoa-p160ROCK-dependent and -independent mechanisms. J Biol Chem 2001; 276:10,423–10,431.

71. Smales C, Ellis M, Baumber R, Hussain N, Desmond H, Staddon JM. Occludin phosphorylation: identification of an occludin kinase in brain and cell extracts as CK2. FEBS Lett 2003; 545:161–166.

72. Jin M, Barron E, He S, Ryan SJ, Hinton DR. Regulation of RPE intercellular junction integrity and function by hepatocyte growth factor. Invest Ophthalmol Vis Sci 2002; 43:2782–2790.

73. Harhaj NS, Barber AJ, Antonetti DA. Platelet-derived growth factor mediates tight junction redistribution and increases permeability in MDCK cells. J Cell Physiol 2002; 193:349–364.

74. Morimoto S, Nishimura N, Terai T, et al. Rab13 mediates the continuous endocytic recycling of occludin to the cell surface. J Biol Chem 2005; 280:2220–2228.

75. Ivanov AI, Nusrat A, Parkos CA. Endocytosis of epithelial apical junctional proteins by a clathrin-mediated pathway into a unique storage compartment. Mol Biol Cell 2004; 15: 176–188.

76. Carrasquillo KG, Ricker JA, Rigas IK, Miller JW, Gragoudas ES, Adamis AP. Controlled delivery of the anti-VEGF aptamer EYE001 with poly(lactic-co-glycolic)acid microspheres. Invest Ophthalmol Vis Sci 2003; 44:290–299.

77. Qaum T, Xu Q, Joussen AM, et al. VEGF-initiated blood–retinal barrier breakdown in early diabetes. Invest Ophthalmol Vis Sci 2001; 42:2408–2413.

78. Saishin Y, Saishin Y, Takahashi K, et al. VEGF-TRAP(R1R2) suppresses choroidal neovascularization and VEGF-induced breakdown of the blood–retinal barrier. J Cell Physiol 2003; 195:241–248.

79. Aiello LP, Bursell SE, Clermont A, et al. Vascular endothelial growth factor-induced retinal permeability is mediated by protein kinase C in vivo and suppressed by an orally effective ß-isoform-selective inhibitor. Diabetes 1997; 46:1473–1480.

80. Antonetti D, Barber A, Khin S, et al. Vascular permeability in experimental diabetes is associated with reduced endothelial occludin content. Diabetes 1998; 47:1953–1959.

81. Wang W, Dentler WL, Borchardt RT. VEGF increases BMEC monolayer permeability by affecting occludin expression and tight junction assembly. Am J Physiol Heart Circ Physiol 2001; 280:H434–H440.

82. French L, Galicich J. The use of steriods for control of cerebral edema. Clin Neurosurg 1962; 10:212–223.

83. Ruderman NB, Hall TC. Use of glucocorticoids in the palliative treatment of metastatic brain tumors. Cancer 1965; 18:298–306.

84. Antonetti DA, Wolpert EB, DeMaio L, Harhaj NS, Scaduto RC. Hydrocortisone decreases retinal endothelial cell water and solute flux coincident with increased content and decreased phosphorylation of occludin. J Neurochem 2002; 80:667–677.

85. Hoheisel D, Nitz T, Franke H, et al. Hydrocortisone reinforces the blood–brain properties in a serum free cell culture system [corrected and republished article originally printed in Biochem Biophys Res Commun 1998; 244(1):312–316]. Biochem Biophys Res Commun 1998; 247:312–315.

86. Zettl KS, Sjaastad MD, Riskin PM, Parry G, Machen TE, Firestone GL. Glucocorticoid-induced formation of tight junctions in mouse mammary epithelial cells in vitro. Proc Natl Acad Sci USA 1992; 89:9069–9073.

87. Buse P, Woo PL, Alexander DB, et al. Transforming growth factor-alpha abrogates glucocorticoid-stimulated tight junction formation and growth suppression in rat mammary epithelial tumor cells. J Biol Chem 1995; 270:6505–6514.

88. Singer KL, Stevenson BR, Woo PL, Firestone GL. Relationship of serine/threonine phosphorylation/dephosphorylation signaling to glucocorticoid regulation of tight junction permeability and ZO-1 distribution in nontransformed mammary epithelial cells. J Biol Chem 1994; 269:16,108–16,115.

89. Guan Y, Woo PL, Rubenstein NM, Firestone GL. Transforming growth factor-alpha abrogates the glucocorticoid stimulation of tight junction formation and reverses the steroid-induced down-regulation of fascin in rat mammary epithelial tumor cells by a Ras-dependent pathway. Exp Cell Res 2002; 273:1–11.

90. Wong V, Ching D, McCrea PD, Firestone GL. Glucocorticoid down-regulation of fascin protein expression is required for the steroid-induced formation of tight junctions and cell–cell interactions in rat mammary epithelial tumor cells. J Biol Chem 1999; 274: 5443–553.

91. Stelwagen K, McFadden HA, Demmer J. Prolactin, alone or in combination with glucocorticoids, enhances tight junction formation and expression of the tight junction protein occludin in mammary cells. Mol Cell Endocrinol 1999; 156:55–61.

3
Neuroprotection

Dennis W. Rickman and Melissa J. Mahoney
Departments of Ophthalmology and Neurobiology, Duke University Medical Center, Durham, North Carolina, U.S.A.

INTRODUCTION

The mammalian retina comprises a rich, heterogeneous mosaic of neuronal morphological phenotypes intermeshed in an intricate pattern of synaptic connectivity. This cellular diversity is further amplified by a wide variety of neurochemical phenotypes, defined by specific expression patterns of numerous neurotransmitters, receptors, and transporters, as well as intracellular regulators such as calcium binding proteins. This complexity confers, in part, regional specializations in the retinal wiring and reflects distinct regional metabolic requirements of retinal neurons. An implication of this cellular diversity is that populations of retinal neurons exhibit differential vulnerability to a variety of diseases or injuries, including genetic, environmental, and metabolic insults. The endpoint for all of these insults is neuronal cell death (either necrotic or apoptotic), and a major challenge in ophthalmology is to prevent or delay retinal neuron loss, even in the face of a continued disease process—hence, neuroprotection. Differential, or selective, vulnerability of retinal neurons also suggests multiple potential targets for cell- or pharmacological-based neuroprotective interventions. The goal of this chapter is to review the expression patterns of a number of cellular and molecular targets (i.e., cell surface receptors, intracellular regulators of cell death, and differential sensitivity to trophic factors) that may underlie a particular nerve cell's predilection for survival or death in the face of disease.

EXCITOTOXICITY AS A STIMULUS FOR NEURONAL CELL DEATH

Excitatory neurotransmission in the central nervous system (CNS), including the retina, is accomplished primarily by the amino acid glutamate. In the retina, glutamate is released by photoreceptors, bipolar cells, and ganglion cells, presynaptically, in normal neurotransmission (1–6). Normally, glutamate is rapidly removed from the extracellular space following its release at the synapse. Glutamate removal is accomplished both by binding to specific postsynaptic receptors that mediate activation of the postsynaptic cell and removal by glutamate transporters located in the plasma membranes of both neurons and Müller glial cells. Thus, glutamate in the synaptic cleft is typically maintained at a low level (7,8).

Multiple glutamate receptor subtypes have been identified, and these are characterized based on their sensitivities to different glutamate receptor analogs (Table 1) (1). Glutamate binds to both ionotropic receptors, which, in heteromeric association, form ion channels in the neuronal plasma membrane, and metabotropic receptors, which are coupled to G-protein-mediated pathways. Ionotropic glutamate receptors (iGluR) can be further subdivided into those that bind glutamate and its analog N-methyl-D-aspartate (NMDA, i.e., NMDA receptors) and those that are sensitive to kainate, alpha-amino-3-hydroxy-5-methyl-4-isoxazoleproprionic acid (AMPA), and quisqualate (i.e., non-NMDA receptors). Binding to the NMDA receptor is further characterized by a preferential increase in Ca^{2+} permeability. In the retina, NMDA-type glutamate receptors are localized to ganglion cells and to some amacrine cells (9–13), and their responses can be blocked by the selective antagonists MK-801, AP-5 (2-amino-5-phosphonopentanoic acid), and AP-7 (2-amino-7-phosphoheptanoic acid) (14,15).

There is considerable evidence that overstimulation of glutamate receptors promotes cell death in a number of retinal disease processes. Glutamate overstimulation may be particularly important in acute ischemic injuries, but it also may play a role in diabetic retinopathy (16) and chronic neurodegenerative processes such as glaucoma (17). Evidence for this includes the observation that glutamate levels are elevated in the vitreous of patients with these conditions (16,17).

Retinal ischemic injury, for example, has been shown to result in the overstimulation of ionotropic glutamate receptors following the extracellular accumulation of glutamate (18,19). This effect appears to involve, in particular, the NMDA class of glutamate receptors. The subsequent excessive influx of calcium results in

Table 1 Glutamate Receptor Functional Subtypes and Gene Subunits

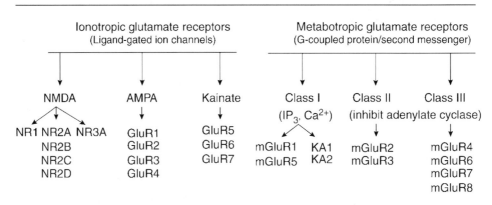

Ionotropic glutamate receptors (Ligand-gated ion channels)			Metabotropic glutamate receptors (G-coupled protein/second messenger)		
NMDA	AMPA	Kainate	Class I (IP_3, Ca^{2+})	Class II (inhibit adenylate cyclase)	Class III
NR1 NR2A NR3A	GluR1	GluR5			mGluR4
NR2B	GluR2	GluR6	mGluR1 KA1	mGluR2	mGluR6
NR2C	GluR3	GluR7	mGluR5 KA2	mGluR3	mGluR7
NR2D	GluR4				mGluR8

Note: Glutamate receptors are subdivided into two main functional classes, those that regulate ion channels (ionotropic receptors) and those that activate second messenger systems through activation of G-coupled proteins (metabotropic receptors). The ionotropic receptors are further subdivided into subtypes, based on the specificity of their activation by specific ligands (NMDA, AMPA, or kainate). The molecular structure of these subtypes is conferred by the patterns of expression of specific genes. Similarly, the metabotropic receptors either stimulate the generation of IP_3 (and diacylglycerol) resulting in an increase in intracellular Ca^{2+} (Class I) or, alternatively, inhibit adenylate cyclase and stimulate the generation of cAMP (Classes II and III). Likewise, these classes are conferred by the specific gene expression.
Abbreviations: NMDA, N-methyl-D-aspartate; AMPA, alpha-amino-3-hydroxy-5-methyl-4-isozazole proprionic acid.

the activation of intracellular pathways that trigger cell death, including the apoptotic cascade and generation of free radicals (described later).

Based on evidence that glutamate overstimulation is important in ocular neuronal cell death, an initial extracellular target for neuroprotective intervention is the glutamate receptor. In the retina, a component of neurochemical diversity is conferred by the differential distribution of glutamate receptor subunits to a variety of retinal neurons. Thus, retinal neurons may have differential vulnerabilities to injury. This suggests specific cellular targets for neuroprotection. Indeed, glutamate receptor antagonists, such as memantine and MK-801, prolong survival of neurons in glaucoma and ischemic injury, respectively (20–22). Memantine may have more therapeutic benefit, in as much as it is a more specific uncompetitive antagonist and has a voltage-dependent fast off rate, making it therapeutically safer.

Experimentally and therapeutically, a delicate balance must be reached between blocking excitotoxicity while maintaining normal neurotransmitter receptor–ligand interaction. Ultimately, this balance may be achieved by specifically targeting glutamate receptor subunits at the molecular level as opposed to a "shotgun" approach with more generalized pharmacological antagonists. Therefore, delivery of agents that target gene expression (e.g., viral constructs) or mRNA translation (e.g., antisense oligonucleotides) may prove to offer more selective neuroprotection while preserving overall neurotransmitter function. This approach assumes, however, a more complete characterization of the populations of retinal neurons that display particular vulnerabilities to excitotoxic damage.

INTRACELLULAR EFFECTORS OF CELL DEATH

Glutamate receptor overactivation that results from ischemic injury, and perhaps from chronic neurodegenerative disease as well, is enhanced during reperfusion. Furthermore, there may be activation of effector pathways that result in the stimulation of the apoptotic cascade (23,24,30). In particular, ionotrophic receptor overstimulation appears to activate pathways that lead to cell death modulated at the level of the mitochondria. The process of programmed cell death, or apoptosis, is in part regulated by a family of molecules related to the B cell leukemia-2 (*bcl*-2) gene product (Bcl-2) (21). This proposed regulatory scheme is summarized in Figure 1. These molecules share limited sequences within three Bcl-2 homology domains (BH1, BH2, and BH3). Regulators of cell death that enhance survival are Bcl-2, Bcl-X, and its splice variant Bcl-X_L. Those that induce apoptosis include Bax, Bak, and Bad (25–29). Molecules that interact at the level of the inner mitochondrial membrane include the cell death promoter, Bax, and the cell survival promoter, Bcl-2 (and closely related family members). In general, homodimerization of Bax at the inner mitochondrial membrane creates ion channels that allow the influx of ions into the mitochondrial matrix, swelling of the mitochondria, and the subsequent release of cytochrome c into the cytoplasm. This further activates cascades of cysteine proteases (caspases) that causes DNA fragmentation and disrupts cytoskeletal integrity—hallmarks of apoptosis (23,24).

Additionally, Bax can form heterodimers with Bcl-2 or Bcl-X_L. This leads to the subsequent inability of Bcl-2 or Bcl-X_L to homodimerize, resulting in a suppression of their protective effects against cell death. Recently, further characterization of the Bcl-2 family has revealed that unbound (unphosphorylated) cytoplasmic Bad also selectively heterodimerizes with Bcl-2 or Bcl-X_L, displacing Bax and promoting cell death by creating a cytoplasmic pool of free Bax (25–29).

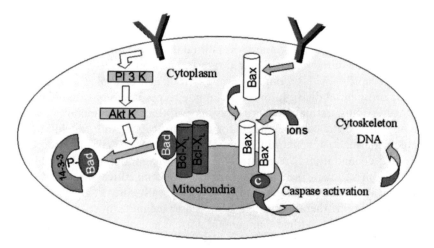

Figure 1 Mechanisms of neuronal cell death. Homodimerization of a Bcl-2 family member, Bax, in the mitochondrial membrane allows the influx of ions and the subsequent release of cytochrome c from mitochondria. This activates cytoplasmic caspases, resulting in DNA fragmentation and disruption of the cytoskeleton. The deleterious action of Bax can be interrupted by the heterodimerization of Bax with Bcl-2 (or as shown here a related family member Bcl-X_L), resulting in cell survival. The availability of Bcl-2 is regulated, in part, by its heterodimerization with Bad. This association promotes cell death by creating a free cytoplasmic pool of Bax. See text for a complete description.

On the other hand, Bcl-2 serves as a checkpoint for Bax activation by conferring a protective effect through its heterodimerization with Bax, thus preventing formation of ion channels in the mitochondrial membrane. Indeed, overexpression of Bcl-2 results in abnormally high numbers of neurons surviving beyond the perinatal period, while Bcl-2-targeted deletion results in increased neuronal cell death (described later).

Several members of the Bcl-2 family have been characterized in the mammalian retina (32–42), and analyses of mice with targeted deletion or overexpression of *bcl*-2-related genes are consistent with the model described previously. For instance, *bcl*-2-deficient mice display a protracted loss of retinal ganglion cell axons well after the period of developmental programmed cell death (35), whereas deletion of *bax* results in a substantial increase in the number of ganglion cell axons in adult mice (36,37). In contrast, *bcl*-2 overexpression increases the number of ganglion cells that survive to adulthood and prevents ganglion cell death following optic nerve injury (38–41). Furthermore, Isenmann et al. (33) reported that, following optic nerve injury in the rat, there is an upregulation of Bax protein by retinal ganglion cells. This increase preceded DNA fragmentation, supporting the notion that Bax is a regulator of retinal ganglion cell death. In this system, Bcl-2 expression by ganglion cells appeared unchanged, further suggesting that the ratio of Bax to Bcl-2 favors cell death in an optic nerve crush model.

Several lines of evidence suggest that an additional, finer level of control over Bcl-2–Bax interaction may be achieved by the participation of related family members such as Bad. In the retina, Bad is expressed predominantly by ganglion cells (42). Normally, Bad is sequestered in the cytoplasm by the 14-3-3 class of proteins. Extracellular growth or survival factors appear to affect survival, in part, through activation of a pathway involving phosphoinositide 3 kinase (PI3-K) (43,44). PI3-K, in turn, activates a serine–threonine protein kinase, Akt, that phosphorylates Bad (45–47) and promotes its association with the 14-3-3 protein family (29). Thus sequestered, Bad is unable to prevent the heterodimerization of Bcl-X_L with Bax, favoring cell survival.

In the adult rat brain, Bad is expressed exclusively by epithelial cells of the choroid plexus (48), suggesting that Bad may play a critical role in regulating the brain's sensitivity to vascular-mediated environmental changes, including alterations in the oxygen concentration of the blood. In the developing brain, however, Bad is expressed in neurons throughout the hippocampus and cerebral cortex, neuronal populations that are particularly sensitive to ischemic insult, even in the adult. Likewise, in the developing retina, Bad is highly expressed in ganglion cells and numerous neurons in the inner nuclear layer—cells that may be particularly vulnerable in ischemic retinopathies.

Interestingly, the pro-apoptotic effects of Bad are blocked by the immunosuppressants cyclosporin (CsA) and FK506. In a model of transient ischemia/reperfusion following middle cerebral artery occlusion, both compounds reduced cerebral infarct volume to 30% of control (49). Blockade of calcineurin-mediated dephosphorylation of Bad is a potential mechanism for this effect. Thus phosphorylated, Bad remains sequestered in the cytoplasm and is unable to bind to Bcl-2 (or Bcl-X_L), thereby allowing Bcl-2 to exert its protective effect. Regulating expression of specific members of the Bcl-2 family by targeted gene expression is another potential therapeutic tool. This approach is covered in Chapter 11.

OXIDATIVE STRESS AND THE GENERATION OF FREE RADICALS

One downstream consequence of neuronal injury, either from an acute ischemic event, as the result of a chronic neurodegenerative process (such as a long-term elevation of intraocular pressure) (50), or even from the normal aging process, is the generation of reactive oxygen species (ROS) and free oxygen radicals (i.e., oxygen-containing species with an unpaired electron, including $^\bullet O_2^-$, OH, $^\bullet NO$, and $ONOO^-$). Normally, natural antioxidant mechanisms prevent interaction of free radicals with cellular constituents, such as fatty acid side chains of membrane lipids (that could be subjected to lipid peroxidation). They also protect the cell from nucleic acid breakdown and damage to cellular proteins (51). These natural defenses include the antioxidant enzyme superoxide dismutase (SOD), glutathione reductase and catalase, and vitamins E and C. Indeed, treatments with various antioxidant compounds have proved effective in maintaining retinal function following ischemia/reperfusion injury (52–54). In fact, natural nutritional and antioxidant supplements have been suggested to protect against photoreceptor loss in age-related macular degeneration and other degenerative processes of aging (55–57).

Several murine models support a role for oxidative stress in neuronal degeneration. For example, overexpression of SOD isoenzymes reduces both global and focal ischemic injury in models of traumatic brain injury (58–60). Conversely, targeted deletion of Cu, Zn-SOD and extracellular (EC)-SOD worsens the outcome of focal ischemia (61,62). Recently, an especially intriguing protective effect has been observed in a model of cerebral ischemia using middle cerebral artery occlusion (63–65). Application of the EC-SOD mimic, AEOL10113 (a metalloporphyrin catalytic antioxidant) [manganese(III) *meso*-tetrakis (*N*-ethylpyridinium-2-yl) porphyrin], even six hours postischemia resulted in marked reduction in cerebral infarct volume versus control (63). Similar results have been observed in a model of transient retinal ischemia/reperfusion (Rickman et al., unpublished results).

NEUROTROPHINS AND NEUROTROPHIN DEPRIVATION AS A STIMULUS FOR RETINAL CELL DEATH

The continued availability of adequate trophic support appears to be crucial not only for the development of nerve cells and their interconnecting circuitry, but also for the maintenance of neurons and their synapses in the adult (31). There is considerable evidence that diffusible, target-derived trophic factors play important roles in the development of specific retinal cell types. In particular, the neurotrophins [nerve growth factor (NGF), brain-derived neurotrophic factor (BDNF), neurotrophin (NT)-3 and NT-4/5] have received considerable attention for their potential roles in both developing and adult nervous systems (66,67).

The neurotrophins bind to both low-affinity receptors and to distinct high-affinity receptors. The low-affinity receptor (p75) is a transmembrane glycoprotein that binds all of the neurotrophins with similar kinetics (68–70). The high-affinity receptors are isoforms of the protooncogene, Trk, a tyrosine kinase that shows some degree of neurotrophin binding specificity (71). For instance, the TrkA isoform preferentially recognizes NGF, while the TrkB isoform binds both BDNF and NT-4. Thus, TrkC prefers NT-3. Generally, the patterns of neurotrophin expression in neuronal targets coincide spatially and temporally with the expression of their cognate high-affinity receptors in the responsive neurons (72–76).

NEUROTROPHINS SUPPORT THE DEVELOPMENT AND MAINTENANCE OF RETINAL GANGLION CELLS

The initial stage in the development of functional retinal circuitry is the differentiation of retinal ganglion cells. Arguably, the survival and differentiation of ganglion cells is dependent upon adequate trophic support from central target sources (77). This hypothesis is supported by the findings that (i) BDNF supports the survival of dissociated ganglion cells from the perinatal retina (78–80), (ii) neurotrophins and their receptors are expressed concordantly in the developing visual system (67,72,76), (iii) following optic nerve transection, intraocular injection of BDNF (81) or NGF (82) prolongs, though only modestly, the survival of a subpopulation of ganglion cells—even with long-term delivery by viral transfection (83), and (iv) application of exogenous BDNF to the superior colliculus results in reduced developmental ganglion cell death (84). Arguably, neurotrophins contribute not only to the survival of retinal ganglion cells but also to their morphological maturation (85,86). In the adult retina it is likely that maintenance of ganglion cell morphological integrity is crucial for maintaining inner retinal circuitry and function. Indeed, retinal ganglion cells, themselves, express low levels of BDNF (75,87), and this can be upregulated following injury to the optic nerve or following administration of brimonidine, an agent commonly used to lower intraocular pressure (88). However, compromising retrograde axoplasmic transport along the optic nerve may lead to an interruption in sufficient trophic support to ganglion cells, resulting in remodeling of their dendritic arborizations and a subsequent breakdown of inner retinal circuitry.

NEUROTROPHINS SUPPORT THE DEVELOPMENT OF INNER RETINAL CIRCUITRY

Under scotopic conditions, mammalian visual processing is dominated by a circuit classically thought to involve only rod photoreceptors, a unique class of rod bipolar cells and ganglion cells. However, it is now clear from the observations of a number of investigators (89–91) that a distinctive, inhibitory interneuron, the *AII* amacrine cell, is interposed between the rod bipolar cell and ganglion cell. A role has been established for BDNF in the phenotypic differentiation of *AII* amacrine cells and, thus, the development of the neural pathway underlying scotopic visual processing (67,92,93). Furthermore, the network of *AII* amacrine cells is modulated by dopaminergic innervation from a population of sparsely distributed, wide-field amacrine cells (94–96). This cell, in the proximal inner nuclear layer (INL), is labeled with antibodies to tyrosine hydroxylase (TH), the rate-limiting enzyme in dopamine biosynthesis. Dendrites of the dopaminergic amacrine cell contribute to a moderately dense plexus in the inner plexiform layer (IPL) where they form "ring-like" structures surrounding the somata and initial dendrites of the *AII* amacrine cells are sites of synaptic contact (97). Generally, at scotopic light levels, the *AII*s are interconnected via gap junctions in sublamina *a* of the IPL, enhancing the overall sensitivity of the rod signaling pathway. In response to increased light levels, dopamine is released, uncoupling gap junctions and reducing the overall sensitivity of the rod pathway. Development of the dopaminergic amacrine cell also has been shown to be dependent on BDNF (98). In retinas from *BDNF* knockout mice there is a reduced number of TH-containing somata, and the density of the dopaminergic plexus in the IPL is greatly reduced, as compared to the wild type. Conversely, intraocular injection of BDNF in the normal retina results in precocious sprouting of dopaminergic processes throughout the IPL (87). These demonstrated roles for neurotrophins in the development and maintenance of the inner retinal circuitry are consistent with the well-documented role of neurotrophin-mediated survival following transient ischemia (99–102).

MODELS OF PHOTORECEPTOR DEGENERATION AND STRATEGIES FOR THEIR TREATMENT

Numerous genetic models of photoreceptor degeneration have been characterized. These include models where the primary defect is in the metabolic machinery of the photoreceptor cell (e.g., *rd* mouse) (103), mutations in genes encoding photopigments (e.g., *Pro23His* rat) (104), or in the adjacent retinal pigment epithelium [e.g., Royal College of Surgeons (RCS) rat] (105). These models all share a general feature: a relatively rapid loss of photoreceptors during the early postnatal period. The rate of photoreceptor loss ranges from a few weeks (*rd* mouse) to several months (rat models). Alternatively, the light damage model is of interest because it offers a degree of experimental control (106,107). Briefly, constant exposure of albino rats to ambient light for one week results in relatively rapid photoreceptor degeneration and accompanying outer nuclear layer thinning over a period of weeks. The most successful therapeutic approaches for all of these models have been based largely on retarding photoreceptor demise by either (i) intraocular injection of growth factors or cytokines (102,106,108–110), or (ii) transplantation of fetal retinal cells or RPE (111–118) cells. There is also evidence for upregulation of endogenous basis fibroblast growth factor (bFGF) and ciliary neurotrophic factor (CNTF) mRNAs following mechanical lesion to the retina and expansion of the subretinal space (119). In the light damage model of photoreceptor degeneration, there is evidence of invading, activated microglia that release BNDF, CNTF, and glial derived neuro-trophic factor (GDNF) that enhance photoreceptor survival. This observation is intri-guing since photoreceptors, themselves, do not express receptors for neurotrophic factors, suggesting that their protective effects are mediated through interactions with Müller cells (120). This hypothesis is further supported by Wahlin et al. (121), who demonstrated that treatment with BDNF, CNTF, or FGF2 resulted in the upregula-tion of downstream effectors only in cells of the inner retina, but not in photorecep-tors. It should be noted, however, that there is a recent report demonstrating the presence of BDNF and its receptor, TrkB, in green-red cones of the rat retina (122).

Recent gene therapy strategies have modified the growth factor approach by targeting neurotrophin genes to retinal neurons or Müller glial cells in an attempt to provide continuous trophic support (72,123). Unfortunately, the long-term result of these efforts only slows the progression of photoreceptor degeneration and delays the onset of blindness.

NEUROTROPHIN DELIVERY TO CNS TISSUE

Despite the promise of neurotrophin-based therapies, targeted delivery of proteins to specific neurons is difficult to achieve. For example, most proteins do not efficiently cross the blood–brain and blood–retinal barriers and are therefore not effectively delivered to the brain or retinal tissue via systemic administration (124,125). Direct intraocular injection is an alternative method to deliver proteins to the retina. Gener-ally, following intraocular protein injection, molecules are rapidly cleared from the eye. Elimination half-lives for proteins range from hours (126) to days depending on several factors, including the molecular weight of the injected agent (127). Because the half-life of most proteins in the vitreous is short, repeated injections may be neces-sary to maintain survival and differentiation effects on retinal cells (70). However, multiple intraocular injections increase the risk of cataract formation, retinal detach-ment, and endophthalmitis (128). Alternatively, implantation of pumps into the vitr-eous may extend the period over which neurotrophin is delivered. However, delivery is nonlocalized and high doses (microgram levels) of neurotrophin may need to be deliv-ered to achieve bioactive effects within the retinal tissue (128). Unfortunately, undesir-able side effects have been observed in human patients who received daily microgram levels of nerve growth factor (NGF) by chronic infusion to treat neurodegenerative disease (129). Therefore, localized methods of delivery are preferred to safely supply therapeutic levels of neurotrophin to targeted cell populations in the brain and retina.

Controlled delivery systems may offer safer, more localized, long-term delivery of proteins via a single administration. In addition, they can protect unreleased protein from degradation and they reduce the number of necessary surgical procedures to a single intervention. Several methods of controlled neurotrophin delivery to the brain have been developed. Controlled protein release from biodegradable spherical microparticles that encapsulate protein is one such example (128,130–134). When dispersed in an aqueous environment the microparticles, which are usually formed from biodegradable polymers, begin to degrade. The rate of protein release is

controlled by the rate of polymer degradation and the rate of diffusion through a
porous polymer microsphere network. The kinetics of protein release from polymeric
microspheres, particularly those composed of poly-(D,L-lactic-co-glycolide), have been
characterized (134). Release profiles usually reveal an initial burst of neurotrophin at
short times followed by a longer period of continuous release (Fig. 2).

When NGF is delivered to brain tissue from polymeric microspheres
compressed into a small pellet, NGF concentration is highest at the polymer device
surface; concentration drops 10-fold within 2 mm of the implant. Ninety percent of
exogenously supplied NGF is localized to a region 1–2 mm from the polymeric device
(135–138). As a result, in tissue located near the polymer matrix surface, cells separated by tens of micrometers consistently experience different neurotrophin levels.
These concentration differences result in differences in spatial variations of NGFs
biological effects. The transport of NGF through the brain in the region near the
delivery device can be described by a mathematical model that encompasses diffusion
and first-order elimination. Mathematical models predict that NGF can be more
uniformly distributed throughout a tissue volume when it is delivered from multiple
dispersed sources. The injection of microspheres loaded with neurotrophin may offer
an alternative mode of treatment where the effective area of therapeutic NGF delivery
is dependent on the spacing between microspheres. Microspheres encapsulating other
neurotrophic factors, such as BDNF, CNTF, and GDNF, are under study and can
efficiently be delivered to either the vitreous or the subretinal space by intraocular
injection (Rickman, unpublished studies).

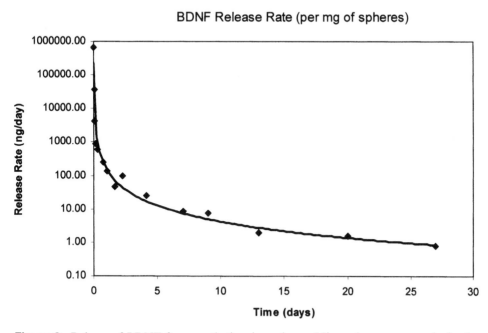

Figure 2   Release of BDNF from synthetic microspheres. Microspheres composed of poly-
(D,L-lactic-co-glycolide) were engineered to release BDNF over a sustained period. The release
of total protein was measured by protein assay and ELISA. There is an initial burst of release
over the first 48 hours, followed by sustained release of nanogram quantities for up to 28 days.
Abbreviations: BDNF, brain derived neurotrophic factor; ELISA, enzyme linked immunosorbent assay.

Hydrogels, insoluble yet water-swellable cross-linked polymer networks have also received considerable interest as three-dimensional matrices supporting cell growth and differentiation. In most cases, gelation can be induced directly, in the presence of cells, resulting in uniform cell density throughout the implant. The combined high water content and elasticity of polymer hydrogels lead to many tissue-like properties of these materials, making them ideal candidates for tissue engineering. For example, hydrogels of poly[N-2-(hydroxypropyl)methacrylamide] (PHPMA) loaded with BDNF producing fibroblasts inserted into cavities made in the optic tract resulted in increased in-growth of axons into implants. Retinal axons exhibited a complex branching pattern and they regrew the greatest distances within implants containing BDNF after four to eight weeks (139). Similar effects have been observed for gels implanted into lesioned cavities in the cerebral hemispheres and spinal cord (140).

A method to deliver CNTF from a genetically engineered encapsulated cell-based delivery system has also been developed. This system is described in detail in Chapter 8.

SUMMARY

Three of the leading causes of blindness in the world (glaucoma, diabetic retinopathy, and age-related macular degeneration) are chronic, degenerative processes whose precise etiologies may be unclear due to a multiplicity of factors. For instance, although glaucoma may be associated with elevated intraocular pressure due to impeded outflow of aqueous humor in the anterior chamber, the ultimate mortality of retinal ganglion cells may be more directly attributed to a constellation of factors in the posterior pole, including ischemia at the optic nerve head, excitotoxin exposure, oxidative stress, and neurotrophin deprivation. It is likely that cascades of these events confound the targeting and, perhaps more importantly, the timing of therapeutic neuroprotective intervention. Another important consideration is the vulnerability of a particular population of retinal neurons. Certainly, all retinal ganglion cells do not undergo cell death at, or even near, the same time in glaucoma. On the contrary, the demise is usually prolonged over many months to years. Thus, the optimal time for neuroprotective intervention is problematic, and sustained, targeted delivery without systemic side effects is preferable, albeit difficult to achieve. It is also important to consider the presumed underlying condition (in this case, an often-associated elevated intraocular pressure) and to eliminate or control the initiating insult. A similar argument can be made for other neurodegenerative diseases of the retina. If, for example, the degeneration of photoreceptors is due to the dysfunction of the adjacent retinal pigment epithelium (RPE; as in Best's disease or Leber's congenital amaurosis), neuroprotective strategies alone may prove futile and, ultimately, repair or replacement strategies for RPE may be necessary. Likewise, retinal degenerations that result from vascular insufficiency, abnormal vascular permeability, or neovascularization will certainly require adjunctive therapies (surgical, pharmacological, or both) to, at best, equilibrate the retinal blood supply.

In conclusion, the future of neuroprotectant drug delivery is exciting. Multiple targets have been identified, and novel, sustained delivery systems are under development. Difficulties remain, many at the cellular level, in better defining the selective vulnerabilities and requirements of specific populations of neurons. Nevertheless, it is likely that more selective neuroprotectants will someday be added to the therapeutic arsenal.

REFERENCES

1. Thoreson WB, Witkovsky P. Glutamate receptors and circuits in the vertebrate retina. Prog Retin Eye Res 1999; 18:765–810.
2. Ehinger B, Ottersen OP, Storm-Mathisen J, Dowling JE. Bipolar cells in the turtle retina are strongly immunoreactive for glutamate. Proc Natl Acad Sci USA 1988; 85:8321–8325.
3. Marc RE, Liu W-LS, Kalloniatis M, Raiguel SF, Van Hasendonck E. Patterns of glutamate immunoreactivity in the goldfish retina. J Neurosci 1990; 10:4006–4034.
4. Kalloniatis M, Fletcher EL. Immunocytochemical localization of the amino acid neurotransmitters in the chicken retina. J Comp Neurol 1993; 336:174–193.
5. Yang C-Y, Yazulla S. Glutamate-, GABA-, and GAD-immunoreactivities co-localize in bipolar cells of tiger salamander retina. Vis Neurosci 1994; 11:1193–1203.
6. Jojich L, Porcho RG. Glutamate immunoreactivity in the cat retina: a quantitative study. Vis Neurosci 1996; 13:117–133.
7. Kanai, Y, Hediger MA. Primary structure and functional characterization of a high-affinity glutamate transporter. Nature 1992; 360:467–471.
8. Kanai Y, Trotti D, Nussberger S, Hediger MA. A new family of neurotransmitter transporters: the high-affinity glutamate transporters. FASEB J 1994; 8:1450–1459.
9. Massey SC, Miller RF. Glutamate receptors of ganglion cells in the rabbit retina: evidence for glutamate as a bipolar cell transmitter. J Physiol 1988; 405:635–655.
10. Massey SC, Miller RF. N-Methyl-D-aspartate receptors of ganglion cells in rabbit retina. J Neurophysiol 1990; 63:16–30.
11. Dixon DB, Copenhagen DR. Two types of glutamate receptors differentially excite amacrine cells in the tiger salamander retina. J Physiol 1992; 449:589–606.
12. Diamond JA, Copenhagen DR. The contribution of NMDA and non-NMDA receptors to the light-evoked input-output characteristics of retinal ganglion cells. Neuron 1993; 11:725–738.
13. Cohen ED, Miller RF. The role of NMDA and non-NMDA excitatory amino acid receptors in the functional organization of primate retinal ganglion cells. Vis Neurosci 1994; 11:317–332.
14. Laabich A, Cooper NG. Regulation of calcium/calmodulin-dependent protein kinase II in the adult rat retina is mediated by ionotropic glutamate receptors. Exp Eye Res 1999; 68:703–713.
15. Boos R, Muller F, Wassle H. Actions of excitatory amino acids on brisk ganglion cells in the cat retina. J Neurophysiol 1990; 64:1368–1379.
16. Ambati J, Chalam KV, Chawla DK, et al. Elevated gamma-aminobutyric acid, glutamate, and vascular endothelial growth factor levels in the vitreous of patients with proliferative diabetic retinopathy. Arch Ophthalmol 1997; 115:1161–1166.
17. Dreyer EB, Zurakowski D, Schumer RA, Podos SM, Lipton SA. Elevated glutamate levels in the vitreous body of humans and monkeys with glaucoma. Arch Ophthalmol 1996; 114:299–305.
18. Osborne NN, Ugarte M, Chao M, et al. Neuroprotection in relation to retinal ischemia and relevance to glaucoma. Surv Ophthalmol 1999; 43(suppl 1):S102–S128.
19. Sucher NJ, Lipton SA, Dryer EB. Molecular basis of glutamate toxicity in retina ganglion cells. Vision Res 1997; 37:3483–3493.
20. WoldeMussie E, Yoles E, Schwartz M, Ruiz G, Wheeler LA. Neuroprotective effect of memantine in different retinal injury models in rats. J Glaucoma 2002; 11:474–480.
21. Osborne NN. Memantine reduces alterations to the mammalian retina, in situ, induced by ischemia. Vis Neurosci 1999; 16:45–52.
22. Lagreze WA, Knorle R, Bach M, Feuerstein TJ. Memantine is neuroprotective in a rat model of pressure-induced retinal ischemia. Invest Ophthalmol Vis Sci 1998; 39:1063–1066.
23. Cellerino A, Bahr M, Isenmann S. Apoptosis in the developing visual system. Cell Tissue Res 2000; 301:53–69.
24. Pettmann B, Henderson CE. Neuronal cell death. Neuron 1998; 20:633–647.

25. Kelekar A, Chang BS, Harlan JE, Fesik SW, Thompson CB. Bad is a BH3 domain-containing protein that forms an inactivating dimer with Bcl-X$_L$. Mol Cell Biol 1977; 17:7040–7046.

26. Ottilie S, Diaz J-L, Horne W, et al. Dimerization properties of human BAD. J Biol Chem 1977; 272:30,866–30,872.

27. Yang E, Zha J, Jockel J, Boise LH, Thompson CB, Korsmeyer SJ. Bad, a heterodimeric partner for Bcl-X$_L$ and Bcl-2, displaces Bax and promotes cell death. Cell 1995; 80: 285–291.

28. Zha J, Harada H, Osipov K, Jockel J, Waksman G, Korsmeyer SJ. GH3 domain of BAD is required for heterodimerization with BCL-X$_L$ and pro-apoptotic activity. J Biol Chem 1977; 272:24101–24104.

29. Zha J, Harada H, Yang E, Jockel J, Korsmeyer SJ. Serine phosphorylation of death agonist BAD in response to survival factor results in binding to 14-3-3 not BCL-X. Cell 1996; 87:619–628.

30. Friedlander RM. Apoptosis and caspases in neurodegenerative diseases. N Eng J Med 2003; 348:1365–1375.

31. Oppenheim RW. Cell death during development of the nervous system. Ann Rev Neurosci 1991; 14:453–501.

32. Chen ST, Garey LJ, Jen LS. Bcl-2 proto-oncogene protein immunoreactivity in normally developing and axotomised rat retinas. Neurosci Lett 1994; 172:11–14.

33. Isenmann S, Wahl S, Krajewski S, Reed JC, Bahr M. Up-regulation of Bax protein in degenerating retinal ganglion cells precedes apoptotic cell death after optic nerve lesion in the rat. Eur J Neurosci 1997; 9:1763–1772.

34. Levin LA, Schlamp CL, Spieldoch RL, Geszvain KM, Nickells RW. Identification of the *bcl*-2 family of genes in the rat retina. Invest Ophthal Vis Sci 1997; 38:2545–2553.

35. Cellerino A, Michaelidis T, Meyer M, Bahr M, Thoenen H. Protracted loss of retinal ganglion cells following the period of naturally occurring cell death in mice lacking the *bcl*-2 gene [abstr]. Soc Neurosci Abstr 1996; 22:1977.

36. Olgilvie JM, Deckwerth TL, Kundson CM, Korsmeyer SJ. Suppression of developmental retinal cell death but not of photoreceptor degeneration in *Bax*-deficient mice. Invest Ophthal Vis Sci 1998; 39:1713–1720.

37. White FA, Keller-Peck CR, Knudson CM, Korsmeyer SJ, Snider WD. Widespread elimination of naturally occurring neuronal death in *Bax*-deficient mice. J Neurosci 1998; 18:1428–1439.

38. Bonfanti L, Strettoi E, Chierzi S, et al. Protection of retinal ganglion cells from natural and axotomy-induced cell death in neonatal transgenic mice overexpressing bcl-2. J Neurosci 1996; 16:4186–4194.

39. Cenni MC, Bonfanti L, Martinou JC, Ratto GM, Strettoi E, Maffei L. Long-term survival of retinal ganglion cells following optic nerve section in adult *bcl*-2 transgenic mice. Eur J Neurosci 1996; 8:1735–1745.

40. Martinou JC, Dubois-Dauphin M, Staple JK, et al. Overexpresson of BCL-2 in transgenic mice protects neurons from naturally occurring cell death and experimental ischemia. Neuron 1994; 13:1017–1030.

41. Prociatti V, Pizzorusso T, Cenni MC, Maffei L. The visual response of retinal ganglion cells is not altered by optic nerve section in adult *bcl*-2 transgenic mice. Proc Natl Acad Sci USA 1996; 93:14,955–14,959.

42. Rickman DW, Nacke RE, Bowes Rickman C. Characterization of the cell death promoter, Bad, in the developing rat retina and forebrain. Dev Brain Res 1999; 115:41–47.

43. D'Mello SR, Borodezt K, Soltoff SP. Insulin-like growth factor and potassium depolarization maintain neuronal survival by distinct pathways: possible involvement of PI-3-kinase in IGF-1 signaling. J Neurosci 1997; 275:661–665.

44. Yao R, Cooper GM. Requirement for phosphathdylinositol-3-kinase in the prevention of apoptosis by nerve growth factor. Science 1995; 267:2003–2006.

45. del Peso L, Gonzalez-Garcia M, Page C, Herrera R, Nunez G. Interleukin-3-induced phosphorylation of BAD through the protein kinase Akt. Science 1997; 278:687–689.
46. Dudek H, Datta SR, Franke TF, et al. Regulation of neuronal survival by the serine–threonine protein kinase Akt. Science 1997; 275:661–665.
47. Franke TF, Kaplan DR, Cantley LC, Toker A. Direct regulation of the *Akt* proto-oncogene product by phosphatidylinositol-3,4-bis-phosphate. Science 1997; 275:665–668.
48. D'Agata V, Magro G, Travali S, Musco S, Cavallaro S. Cloning and expression of the programmed cell death regulator Bad in the rat brain. Neurosci Lett 1998; 243:137–140.
49. Yoshimoto T, Uchino H, He QP, Li PA, Siesjo BK. Cyclosporin A, but not FK506, prevents the downregulation of phosphorylated Akt after transient focal ischemia in the rat. Brain Res 2001; 899:148–158.
50. Muller A, Pietri S, Villain M, Frejaville C, Bonne C, Culcas M. Free radicals in rabbit retina under ocular hyperpressure and functional consequences. Exp Eye Res 1997; 64:637–643.
51. Garcia-Valenzuela E, Shareef S, Walsh J, Sharma SC. Programmed cell death of retinal ganglion cells during experimental glaucoma. Exp Eye Res 1995; 61:33–44.
52. Block F, Schwarz M. Effects of antioxidants on ischemic retinal function. Exp Eye Res 1997; 64:559–564.
53. Nayak MS, Kita M, Marmor MF. Protection of rabbit retina from ischemic injury by superoxide dismutase and catalase. Invest Ophthalmol Vis Sci 1993; 34:2018–2022.
54. Kuriyama H, Waki M, Nakagawa M, Tsuda M. Involvement of oxygen free radicals in experimental retinal ischemia and the selective vulnerability of retinal damage. Ophthalmic Res 2001; 33:196–202.
55. Mares-Perlman JA, Millen AE, Ficek TL, Hankinson SE. The body of evidence to support a protective role for lutein and zeaxanthin in delaying chronic disease. Overview. J Nutr 2002; 132:518S–524S.
56. Beatty S, Koh H, Phil M, Henson D, Boulton M. The role of oxidative stress in the pathogenesis of age-related macular degeneration. Surv Ophthalmol 2000; 45:115–134.
57. Belda JI, Roma J, Vilela C, et al. Serum vitamin E levels negatively correlate with severity of age-related macular degeneration. Mech Ageing Dev 1999; 107:159–164.
58. Murakami K, Kondo T, Epstein CJ, Chan PH. Overexpression of CuZn-superoxide dismutase reduces hippocampal injury after global ischemia in transgenic mice. Stroke 1997; 28:1797–1804.
59. Sheng H, Kudo M, Mackensen GB, Pearlstein RD, Crapo JD, Warner DS. Mice over-expressing extracellular superoxide dismutase have increased resistance to global cerebral ischemia. Exp Neurol 2000; 163:392–398.
60. Sheng H, Bart RD, Oury TD, Pearlstein RD, Crapo JD, Warner DS. Mice overexpressing extracellular superoxide dismutase have increased resistance to focal cerebral ischemia. Neuroscience 1999; 88:185–191.
61. Kondo T, Reaume AG, Huang TT, et al. Reduction of CuZn-superoxide dismutase activity exacerbates neuronal cell injury and edema formation after transient focal cerebral ischemia. J Neurosci 1997; 17:4180–4189.
62. Sheng H, Brady TC, Pearlstein RD, Crapo JD, Warner DS. Extracellular superoxide dismutase deficiency worsens outcome from focal ecrebral ischemia in the mouse. Neurosci Lett 1999; 267:13–16.
63. Mackensen GB, Patel M, Sheng H, et al. Neuroprotection from delayed postischemic administration of a metalloporphyrin catalytic antioxidant. J Neurosci 2001; 21:4582–4592.
64. Batinic-Haberle I. Manganese porphyrins and related compounds as mimics of superoxide dismutase. Methods Enzymol 2002; 349:223–233.
65. Batinic-Haberle I, Liochev SI, Spasojevic I, Fridovich I. A potent superoxide dismutase mimic: manganese beta-octabromo-meso-tetrakis-(*N*-methylpyridinium-4-yl) porphyrin. Arch Biochem Biophys 1997; 343:225–233.
66. Davies AM. The role of neurotrohins in the developing nervous system. J Neurobiol 1994; 25:1334–1348.

67. Rickman DW. Neurotrophins and development of the rod pathway: an elementary deduction. Micro Res Tech 2000; 50:124–129.

68. Hempstead BL, Dionisio M-Z, Kaplan DR, Parada LF, Chao MV. High-affinity NGF binding requires coexpression of the *trk* proto-oncogene and the low-afinitiy NGF receptor. Nature 1991; 350:678–683.

69. Rodriguez-Tebar A, Dechant G, Barde Y-A. Binding of brain-derived neurotrophic factor to the nerve growth factor receptor. Neuron 1990; 4:487–492.

70. Rodriguez-Tebar A, Dechant G, Gotz R, Barde Y-A. Binding of neurotrophin-3 receptors in the developing chicken retina. EMBO J 1992; 11:917–922.

71. Barbacid M. The trk family of neurotrophin receptors. J Neurobiol 1994; 25:1386–1403.

72. Cohen-Cory S, Fraser SE. BDNF in the development of the visual system of *Xenopus*. Neuron 1994; 12:747–761.

73. Jelsma TN, Friedman HH, Berkelaar H, Bray GM, Aguayo AJ. Different forms of the neurotrophin receptor *trk*B mRNA predominate in rat retina and optic nerve. J Neurobiol 1993; 24:1207–1214.

74. Maisonpierre PC, Bellusicio L, Squinto S, et al. NT-3, BDNF and NGF in the developing rat nervous system: parallel as well as reciprocal patterns of expression. Neuron 1990; 5:501–509.

75. Perez M-TR, Caminos E. Localization of trkB and BDNF mRNAs in neonatal and adult rat retina. Neurosci Lett 1995; 183:96–99.

76. Rickman DW, Brecha NC. Expression of the proto-oncogene, trk, receptors in the developing rat retina. Vis Neurosci 1995; 12:215–222.

77. Yip HK, So K-F. Axonal regeneration of retinal ganglion cells: effect of trophic factors. Prog Retin Eye Res 2000; 19:559–575.

78. Johnson JE, Barde Y-A, Schwab M, Thoenen H. Brain-derived neurotrophic factor supports the survival of cultured rat retina ganglion cells. J Neurosci 1986; 6:3031–3038.

79. Rodriguez-Tebar A, Jeffery PL, Thoenen H, Barde Y-A. The survival of chick retinal ganglion cells in response to brain-derived neurotrophic factor depends on their embryonic age. Dev Biol 1989; 136:296–303.

80. Castillo JB, del Cerro M, Breakefield XO, et al. Retinal ganglion cell survival is promoted by genetically modified astrocytes designed to secrete brain-derived neurotrophic factor (BDNF). Brain Res 1994; 647:30–36.

81. Mansour-Robaey S, Clarke DB, Wang Y-C, Bray GM, Aguayo AJ. Effects of ocular injury and administration of brain-derived neurotropohic factor on survival and regrowth of axotomized retinal ganglion cells. Proc Natl Acad Sci USA 1994; 91:1632–1636.

82. Carmignoto G, Maffei L, Candeo P, Canella R, Comelli. Effect of NGF on the survival of rat retinal ganglion cells following optic nerve section. J Neurosci 1989; 9:1263–1272.

83. DiPolo A, Aigner J, Dunn RJ, Bray GM, Aguayo AJ. Prolonged delivery of brain-derived neurotropohic factor by adenovirus-infected Muller cells temporarily rescues injured retinal ganglion cells. Proc Natl Acad Sci USA 1998; 95:3978–3983.

84. Ma Y-T, Hsieh T, Forber ME, Johnson JE, Frost DO. BDNF injected into the superior colliculus reduces developmental retinal ganglion cell death. J Neurosci 1998; 18: 2097–2107.

85. Cohen-Cory S. BDNF modulates, but does not mediate, activity-dependent branching and remodeling of optic axon arbors in vivo. J Neurosci 1999; 19:9996–10003.

86. Lom B, Cohen-Cory S. Brain-derived neurotrophic factor differentially regulates retinal ganglion cell dendritic and axonal arborizations in vivo. J Neurosci 1999; 19:9928–9938.

87. Gao H, Qiao X, Hafti F, Hollyfield JG, Knusel B. Elevated mRNA expression of brain-derived neurotrophic factor in retinal ganglion cell layer after optic nerve injury. Invest Ophthalmol Vis Sci 1997; 38:1840–1847.

88. Gao H, Qiao X, Cantor LB, WuDunn D. Up-regulation of brain-derived neurotrophic factor expression by brimonidine in rat retinal ganglion cells. Arch Ophthalmol 2002; 120:797–803.

89. Famiglietti EV, Kolb H. A bistratified amacrine cell and synaptic circuitry in the inner plexiform layer of the retina. Brain Res 1975; 84:293–300.

90. Sterling P. Microcircuitry of the cat retina. Ann Rev Neurosci 1983; 6:149–185.

91. Wassle H, Boycott BB. Functional architecture of the mammalian retina. Physiol Rev 1991; 71:4115–4128.

92. Rickman DW, Bowes Rickman C. Suppression of *trk*B expression by antisense oligonucleotides alters a neuronal phenotype in the rod pathway of the developing rat retina. Proc Natl Acad Sci USA 1996; 93:12,564–12,569.

93. Rickman DW. Parvalbumin immunoreactivity is enhanced by brain-derived neurotrophic factor in organotypic cultures of rat retina. J Neurobiol 1999; 41:376–384.

94. Dacey DM. The dopaminergic amacrine cell. J Comp Neurol 1990; 301:461–489.

95. Brecha NC, Oyster CW, Takahashi ES. Identification and characterization of tyrosine hydroxylase immunoreactive amacrine cells. Invest Ophthalmol Vis Sci 1984; 25:66–70.

96. Oyster CW, Takahashi ES, Cilluffo M, Brecha NC. Morphology and distribution of tyrosine hydroxylase-like immunoreactive neurons in the cat retina. Proc Natl Acad Sci USA 1985; 82:6335–6339.

97. Voigt T, Wassle H. Dopaminergic innervation of AII amacrine cells in mammalian retina. J Neurosci 1987; 7:4115–4128.

98. Cellerino A, Pinzon-Duarte G, Carroll P, Kohler K. Brain-derived neurotrophic factor modulates the development of the domainergic network in the rodent retina. J Neurosci 1998; 18:3351–3362.

99. Han BH, Holtzman DM. BDNF protects the neonatal brain from hypoxic-ischemic injury *in vivo*, via the ERK pathway. J Neurosci 2000; 20:5775–5781.

100. Klocker N, Kermer P, Weishaupt JH, Labes M, Ankerhold R, Bahr M. Brain-derived neurotrophic factor-mediated neuroprotection of adult rat retinal ganglion cells in vivo does not exclusively depend on phosphatidyl-inositol-3′-kinase/protein kinase B signaling. J Neurosci 2000; 20:6962–6967.

101. Kurokawa T, Katai N, Shibuki H, et al. BDNF diminishes caspase-2 but not c-Jun immunoreactivity of neurons in retinal ganglion cell layer after transient ischemia. Invest Ophthalmol Vis Sci 1999; 40:3006–3011.

102. Unoki K, LaVail MM. Protection of the rat retina from ischemic injury by brain-derived neurotrophic factor, ciliary neurotrophic factor, and basic fibroblast growth factor. Invest Ophthalmol Vis Sci 1994; 35:907–915.

103. Bowes C, Li T, Danciger M, Baxter LC, Applebury ML, Farber DB. Retinal degeneration in the rd mouse is caused by a defect in the beta subunit of rod cGMP-phosphodiesterase. Nature 1990; 347:677–680.

104. Olsson JE, Gordon JW, Pawlyk BS, et al. Transgenic mice with a rhodopsin mutation (Pro23His): a mouse model of autosomal dominant retinitis pigmentosa. Neuron 1992; 9:815–830.

105. LaVail MM, Sidman RL, Gerhardt CO. Congenic strains of RCS rats with inherited retinal dystrophy. J Hered 1975; 66:242–244.

106. LaVail MM, Unoki K, Yasumura D, Matthes MT, Yancopoulos GD, Steinberg RH. Multiple growth factors, cytokines, and neurotrophins rescue photoreceptors from the damaging effects of constant light. Proc Natl Acad Sci USA 1992; 89:11,249–11,253.

107. Organisciak DT, Jiang YL, Wang HM, Pickford M, Blanks JC. Retinal light damage in rats exposed to intermittent light. Comparison with continuous light exposure. Invest Ophthalmol Vis Sci 1989; 30:795–805.

108. LaVail MM, Yasumura D, Matthes MT, et al. Protection of mouse photoreceptors by survival factors in retinal degenerations. Invest Ophthalmol Vis Sci 1998; 39: 592–602.

109. Masuda K, Watanabe I, Unoki K, Ohba N, Muramatsu T. Functional rescue of photoreceptors from the damaging effects of constant light by survival-promoting factors in the rat. Invest Ophthalmol Vis Sci 1995; 36:2142–2146.

110. Faktorovich EF, Steinberg RH, Yasumura D, Matthes MT, LaVail MM. Basic fibro-blast growth factor and local injury protect photoreceptors from light damage in the rat. J Neurosci 1992; 12:3554–3567.
111. Ghosh F, Johansson K, Ehinger B. Long-term full-thickness embryonic rabbit retinal transplants. Invest Ophthalmol Vis Sci 1999; 40:133–142.
112. Ghosh F, Bruun A, Ehinger B. Graft–host connections in long-term full-thickness embryonic rabbit retinal transplants. Invest Ophthalmol Vis Sci 1999; 40:126–132.
113. Seiler MJ, Aramant RB. Intact sheets of fetal retina transplanted to restore damaged rat retinas. Invest Ophthalmol Vis Sci 1998; 39:2121–2131.
114. Seiler MJ, Aramant RB, Bergstrom A. Co-transplantation of embryonic retina and ret-inal pigment epithelial cells to rabbit retina. Curr Eye Res 1995; 14:199–207.
115. Aramant RB, Seiler MJ. Fiber and synaptic connections between embryonic retinal transplants and host retina. Exp Neurol 1995; 133:244–255.
116. Little CW, Castillo B, DiLoreto DA, et al. Transplantation of human fetal retinal pigment epithelium resuces photoreceptor cells from degeneration in the Royal College of Surgeons rat retina. Invest Ophthalmol Vis Sci 1996; 37:204–211.
117. Castillo BV Jr, del Cerro M, White RM, et al. Efficacy of nonfetal human RPE for photoreceptor rescue: a study in dystrophic RCS rats. Exp Neurol 1997; 146:1–9.
118. Lin N, Fan W, Sheedlo HJ, Aschenbrenner JE, Turner JE. Photoreceptor repair in response to RPE transplants in RCS rats: outer segment regeneration. Curr Eye Res 1996; 15:1069–1077.
119. Wen R, Song Y, Cheng T, et al. Injury-induced upregulation of bFGF and CNTF mRNAs in the rat retina. J Neurosci 1995; 15:7377–7385.
120. Harada T, Harada C, Kohsaka S, et al. Microglia-Muller glia cell interactions control neurotrophic factor production during light-induced retinal degeneration. J Neurosci 2002; 22:9228–9236.
121. Wahlin KJ, Campochiaro PA, Zack DJ, Adler R. Neurotrophic factors cause activation of intracellular signaling pathways in Muller cells and other cells of the inner retina, but not photoreceptors. Invest Ophthalmol Vis Sci 2000; 41:927–936.
122. DiPolo A, Cheng L, Bray GM, Aguayo AJ. Colocalization of TrkB and brain-derived neu-rotrophic factor proteins in green-red-sensitive cone outer segments. Invest Ophthalmol Vis Sci 2000; 41:401–421.
123. Lau D, McGee LH, Zhou S, et al. Retinal degeneration is slowed in transgenic rats by AAV-mediated delivery of FGF-2. Invest Ophthalmol Vis Sci 2000; 41:3622–3633.
124. Loy R, Taglialatela G, Angelucci L, Heyer D, Perez-Polo R. Regional CNS uptake of blood-borne nerve growth factor. J Neurosci Res 1994; 39:339–346.
125. Poduslo JF, Curran GL, Berg CT. Macromolecular permeability across the blood–nerve and blood–brain barriers. Proc Natl Acad Sci USA 1994; 91:5705–5709.
126. Jaffe GJ, Green GD, McKay BS, Hartz A, Williams GA. Intravitreal clearance of tissue plasminogen activator in the rabbit. Arch Ophthalmol 1988; 106:969–972.
127. Maurice DM. Injection of drugs into the vitreous body. In: Leopold J, Burns R, eds. Symposium of ocular therapy. New York: John Wiley & Sons, 1976:59–71.
128. Herrero-Vanrell R, Refojo MF. Biodegradable microspheres for vitreoretinal drug delivery. Adv Drug Deliv Rev 2001; 52:5–16.
129. Jonhagen ME. Nerve growth factor treatment in dementia. Alzheimer Dis Assoc Disord 2000; 14(suppl 1):S31–S38.
130. Alonso MJ, Gupta RK, Min C, Siber GR, Langer R. Biodegradable microspheres as controlled-release tetanus toxoid delivery systems. Vaccine 1994; 12:299–306.
131. Camarata PJ, Suryanarayanan R, Turner DA, Parker RG, Ebner TJ. Sustained release of nerve growth factor from biodegradable polymer microspheres. Neurosurgery 1992; 30:313–319.
132. Cohen S, Yoshioka T, Lucarelli M, Hwang LH, Langer R. Controlled delivery systems for proteins based on poly(lactic/glycolic acid) microspheres. Pharm Res 1991; 8: 713–720.

133. Eldridge JH, Staas JK, Meulbroek JA, McGhee JR, Tice TR, Gilley RM. Biodegradable microspheres as a vaccine delivery system. Mol Immunol 1991; 28:287–294.

134. Hora MS, Rana RK, Nunberg JH, Tice TR, Gilley RM, Hudson ME. Release of human serum albumin from poly(lactide-co-glycolide) microspheres. Pharm Res 1990; 7:1190–1194.

135. Mathiowitz E, Kline D, Langer R. Morphology of polyanhydride microsphere delivery systems. Scanning Microsc 1990; 4:329–340.

136. Saltzman WM, Mak MW, Mahoney MJ, Duenas ET, Cleland JL. Intracranial delivery of recombinant nerve growth factor: release kinetics and protein distribution for three delivery systems. Pharm Res 1999; 16:232–240.

137. Krewson CE, Saltzman WM. Transport and elimination of recombinant human NGF during long-term delivery to the brain. Brain Res 1996; 727:169–181.

138. Krewson CE, Klarman ML, Saltzman WM. Distribution of nerve growth factor following direct delivery to brain interstitium. Brain Res 1995; 680:196–206.

139. Loh NK, Woerly S, Bunt SM, Wilton SD, Harvey AR. The regrowth of axons within tissue defects in the CNS is promoted by implanted hydrogel matrices that contain BDNF and CNTF producing fibroblasts. Exp Neurol 2001; 170:72–84.

140. Woerly S, Petrov P, Sykova E, Roitbak T, Simonova Z, Harvey AR. Neural tissue formation within porous hydrogels implanted in brain and spinal cord lesions: ultrastructural, immunohistochemical, and diffusion studies. Tissue Eng 1999; 5:467–488.

4

Regulatory Issues in Drug Delivery to the Eye

Lewis J. Gryziewicz
Regulatory Affairs, Allergan, Irvine, California, U.S.A.

Scott M. Whitcup
Research and Development, Allergan, Irvine and Department of Ophthalmology, Jules Stein Eye Institute, David Geffen School of Medicine at UCLA, Los Angeles, California, U.S.A.

INTRODUCTION

In order for a drug product to be marketed in the United States, it must be approved by the U.S. Food and Drug Administration (FDA). The authority for the FDA was established by the Federal Food Drug and Cosmetic Act (FD&C Act). The Act requires FDA to approve new drug products that are the subject of a New Drug Application (NDA) containing adequate data and information on the drug's safety and substantial evidence of the product's effectiveness.

FD&C Act leaves it to the interpretive and discretionary power of the FDA to determine the legal requirement that a sponsor present substantial evidence of effectiveness prior to a drug's approval. Pharmaceutical companies should work closely with the FDA to assure that the development program they are pursuing will meet FDA's expectations and criteria (1,2).

FDA has promulgated regulations based on the FD&C Act and its amendments. These are found in Title 21 of the Code of Federal Regulations. The regulations establish the basic requirements for receiving approval of an NDA. Greater detail is provided in guidelines and guidance that represent the FDA's current thinking on a given topic. Information specific to the development of an individual new drug product can be obtained from meetings and correspondence with the FDA.

Most drug products are not developed for a single market such as the United States, but with the intent of marketing the product worldwide. A difficulty for pharmaceutical companies has been the differing requirements from Health Authorities around the world. In an effort to harmonize worldwide requirements for the approval of drug products, the International Conference on Harmonization (ICH) was established.

ICH is a project involving regulatory and industry representatives of the major pharmaceutical marketplaces in the world; the European Union, Japan, and the United States. The purpose of ICH is to make recommendations on ways to achieve greater harmonization in the interpretation and application of technical guidelines and requirements for product registration in order to reduce or obviate the need to duplicate the testing carried out during the research and development of new medicines. The objective of such harmonization is a more economical use of human, animal, and material resources, and the elimination of unnecessary delay in the global development and availability of new medicines while maintaining safeguards on quality, safety and efficacy, and regulatory obligations to protect public health (4).

The ICH has published a collection of guidelines attempting to standardize the requirements for establishing the safety, efficacy, and quality of pharmaceutical products. These guidelines currently have been adopted not only by the ICH participating countries (the European Union, Japan, and the United States) but also by countries that are monitoring the ICH process including Canada and Australia.

DRUG DEVELOPMENT IN THE UNITED STATES

As a result of increasing standardization of regulatory requirements for new drug approval, global development is becoming more feasible. This chapter will review drug development in the United States as an example of the regulatory requirements for bringing a new drug to market.

Prior to initiation of human studies with an investigational drug in the United States, an Investigational New Drug (IND) application must be in effect with the FDA. An initial IND submission contains the study protocol, the investigator's brochure, the nonclinical (animal, cell culture, etc.) data that support the conduct of the clinical study, and information on the manufacturing and control of the drug substance and the drug product (3). The FDA has 30 days to review the information and make a determination if the investigation can begin.

The study protocol defines the conduct of the study. It is the responsibility of the investigator not to deviate from the protocol except in circumstances where the study subject's safety is at issue. The investigator's brochure contains all of the information on the IND that the investigator needs to safely conduct the study. This document is much longer than the physician insert for a marketed product. It gives a summary of all nonclinical and clinical studies of the drug along with information on the chemistry and manufacturing of the drug.

After the initial IND submission, it is continually amended with additional information throughout the development cycle of the product. Subsequent clinical study protocols are submitted to the IND prior to initiation of the study. Newly generated nonclinical data supporting the proposed clinical studies are submitted to the IND for FDA review. Changes in formulation or method of manufacture for the drug substance or drug products are submitted to the IND. The IND is also continually updated to inform the FDA of new safety information from the clinical studies. The investigator's brochure is updated frequently to include newly generated information.

An adverse event in a clinical study that is unexpected, unlabeled, and associated with the investigational drug must be reported to the FDA within 15 days. If the adverse event is life-threatening, the FDA must be notified within seven days. On a yearly basis adverse event data on the most frequent and serious adverse events are submitted to the IND along with updates on all investigations with the drug.

Nonclinical Testing

Pharmacology Models

The first step in developing a drug is determining its pharmacologic action in in vitro and in vivo models and in a nonclinical or animal model. Screening new compounds in animals is one approach to new drug discovery. Compounds are screened using a wide range of relatively simple and inexpensive procedures primarily in mice or rats.

Another approach is the use of a disease model in animals that resembles the disease process in humans. Compounds are then screened using the model and candidates are selected based on their activity.

A more recent approach is the idea of high throughput screening. A receptor model is developed and a wide range of compounds is screened. Compounds are selected for further study based on their affinity for the receptor.

Although new drug candidates are selected based on their in vivo and in vitro pharmacologic activity, the true potential of a compound is only evident once human clinical trials are initiated. Compounds that respond well in an in vitro receptor pharmacology model must be absorbed in vivo through an acceptable route of administration and achieve the necessary concentrations at their intended site of action. Because of species to species variability, an agent that shows efficacy in a nonclinical model may not be efficacious in humans.

Toxicology Requirements

The next step in drug development is the toxicological characterization of the compound. Prior to human exposure to a new drug, it is imperative to characterize the potential adverse effects and safety profile of the investigational new drug. This is accomplished through nonclinical safety testing.

For drugs intended for local delivery, as in the eye, ICH requirements call for a complete nonclinical assessment of the toxicologic, pharmacokinetic, and toxicokinetic profile of the drug systemically, but also after ocular delivery. Studies must generally be performed in two species, one of which should be a nonrodent. These requirements give added complexity to the ocular development of an active pharmaceutical ingredient.

Acute toxicity studies are single dose or exposure studies followed by an observation period for an appropriate period of time; typically 14 days. Single exposure studies allow for the use of higher doses and give a good indication of the potential adverse events that can arise in chronic studies.

Repeat dose nonclinical testing is also necessary and should at least cover the period of time for the proposed clinical trial. For early Phase I safety testing, toxicology studies can run as little as two weeks and typically for one month. Prior to initiating Phase II studies, which can last for three months or longer for drugs intended for chronic use, toxicology studies of at least three months are required. Phase III studies for chronic drugs require chronic toxicology studies of six months in a rodent and nine months in a nonrodent (5).

Effects of the compound on specific organ systems, i.e., cardiovascular, respiratory, and nervous, are evaluated. These are referred to as safety pharmacology studies and are intended to investigate the potential undesirable pharmacodynamic effects of a substance on physiological functions in relation to exposure. Parameters that are evaluated include blood pressure, heart rate, electrocardiogram, motor activity, behavioral changes, coordination, sensory/motor reflex responses, respiratory rate and depth (6,7).

An understanding of the absorption, distribution, metabolism, and excretion of the drug is established in nonclinical studies prior to administering the drug to humans. Ocular drug delivery, whether topical, periocular, or intraocular inevitably results in systemic absorption and with it the risk of systemic adverse events. For example, topical beta-blockers are potent enough to cause systemic side effects that can be significant in vulnerable patients. Systemic pharmacokinetic and toxicokinetic studies must be included in the drug development plan. The potential for a new chemical entity with potential systemic activity to accumulate in the body should be known (9).

Toxicokinetic studies evaluate the pharmacokinetic profile of the drug during the nonclinical toxicologic testing. It is important to relate the findings in nonclinical studies not only to the dose administered, but also to the bioavailability of the drug in the test animal. The human dose should be determined based on tissue exposure levels, not only to the dose administered in nonclinical studies. Thus it is important to consider differences in bioavailability and biodistribution when preparing to initiate human trials. For example, if the drug is better absorbed in humans, equivalent dosing on a milligram per kilogram (mg/kg) basis may result in higher blood levels in human subjects with a corresponding greater potential for adverse events (8).

Genotoxicity tests are in vitro and in vivo tests designed to detect compounds that induce genetic damage directly or indirectly by various mechanisms. Compounds that are positive in tests that detect such kinds of damage have the potential to be human carcinogens and mutagens. In vitro genotoxicity studies for the evaluation of mutations and chromosomal damage are required prior to first human exposure of a drug. This is accomplished through a test for gene mutation in bacteria, or Ames test, and a cytogenetic evaluation of chromosomal damage with mammalian cells, typically Chinese hamster ovary cells. An in vivo test for chromosomal damage using rodent hematopoeitic cells is required prior to beginning Phase II clinical studies (10,11).

As a new drug moves through development, longer-term toxicology studies are required. The carcinogenic potential of drugs intended for chronic use is typically evaluated in parallel with Phase III clinical testing. Carcinogenicity studies are designed to identify tumorigenic potential in animals and assess the relevant risk in humans. These studies involve lifetime exposure of the test rodents to the test article. Due to the low systemic exposure, drugs intended for ocular delivery may not require carcinogenicity studies unless there is a cause for concern or unless there is significant systemic exposure to the drug. However, some compounds are so potent that even small levels in the blood may lead to systemic side effects (12).

Male subjects may be enrolled in Phase I and II studies based on the histologic evaluation of the male reproductive organs in toxicology studies. The conduct of a male fertility study is required prior to the initiation of Phase III studies.

The inclusion of women of childbearing potential in clinical trials creates great concern for the unintentional exposure of an embryo or fetus to a new drug before information is available on the potential risks. In the European Union and Japan, reproductive toxicology studies are required prior to the enrollment of women in any clinical trial. The United States allows the inclusion of women of childbearing potential in clinical trials prior to the conduct of reproductive toxicology studies, provided appropriate precautions are taken to warn and to minimize risk. The United States requires completion of reproductive toxicology prior to inclusion of women of childbearing potential in Phase III studies.

Three sets of reproductive toxicology studies are typically conducted in drug development; assessment of fertility and embryonic development, pre- and postnatal development, and embryo–fetal development. The study of fertility and embryonic development evaluates treatment of males and females from before mating to mating and implantation. The study of pre- and postnatal development assesses the effects of the drug on the pregnant/lactating female and on development of the conceptus and the offspring from the female from implantation through weaning. The embryo–fetal development study evaluates the pregnant female and the development of the embryo and fetus. These studies give a complete picture of the effects on ability to mate, effects on the fetus, and effects on the offspring after birth (13,14).

Formulation Development

Although early stage clinical trials are generally performed with very simple formulations of a test drug, before the drug can be approved it must be formulated into a product that a patient can use. The formulation of a new drug into an ophthalmic solution is a complicated endeavor. While drug manufacturers want to produce products that have a shelf life of at least two years, a drug in solution is in its most unstable state. Therefore, many topical ophthalmic drug candidates fail because of their instability in solution. Similarly, many drug candidates are rejected because of poor bioavailability after topical application, often due to low aqueous solubility. Topically applied low-solubility drug substances can be brought to market but they must be formulated either as suspensions or emulsions.

Inactive ingredients are incorporated into the formulation, which prevent oxidation or reduction of the drug substance in solution. Salts are added to make the solution isotonic and the pH is adjusted to most closely assimilate physiologic pH. These concerns are particularly important in the development of ophthalmic solutions.

Ophthalmic solutions are manufactured to be sterile and preservatives are incorporated to assure that the solution is not contaminated during its shelf life. It is desirable to formulate using the lowest level of preservative that will assure the product is able to prevent contamination. High levels of preservatives and surfactants may cause patient discomfort such as burning and stinging sensation and may even induce punctate keratitis. However, too low a level leaves the product vulnerable to microbial contamination, both during storage and during the consumer use period of a multidose bottle.

Prior to moving into clinical development, the sponsor must be certain that the drug product will meet its potency requirement throughout the duration of the study. During development the formulation may change as more data on the stability of the product is gathered to assure that the product that is brought to market has an acceptable shelf life. Prior to submitting an NDA in the United States, a manufacturer will generally have at least one year worth of stability data on the final formulation in the intended market package to submit to the FDA. This is supplemented by further stability data justifying the ultimate expiration date that is placed on every product (15).

Products can be manufactured as sterile by different methods known as aseptic processing and terminal sterilization. Aseptic processing involves passing the ophthalmic solution through a 0.2 μm filter in order to rid the solution of all bacteria. The solution is then filled into sterile ophthalmic containers under sterile conditions. This assures a sterile product.

Filling the product into its container and sterilizing it through autoclaving is known as terminal sterilization; or sterilization of the final product. While this may appear to be a better alternative because all organisms present in the final product are destroyed, the difficulty with terminal sterilization is that many drug substances cannot stand up to the heat required for terminal sterilization. Even when the product can withstand the autoclave environment, the materials that are used to produce the bottles, such as low-density polyethylene, cannot withstand autoclaving. Materials that can withstand autoclaving produce a bottle that is so rugged as to require a greater force than many older patients can apply to deliver the product through the tip. The pharmaceutical industry is in search of a material that is rugged enough to withstand autoclaving while being soft enough that a consumer is able to squeeze the final bottle and dispense one drop into their eye. Other forms of terminal sterilization include e-beam and gamma radiation. Although these too impose their own constraints on the drug and packaging system, over the last 20 years they have become increasingly popular.

Clinical Development

The objective of a clinical research program is to demonstrate that a drug is safe and effective in the treatment or prophylaxis of a disease. Clinical development is ideally a logical, step-wise procedure in which information from small, early studies is used to support and plan later, larger, more definitive studies. It is essential to identify characteristics of the investigational product in the early stages of development and to plan an appropriate development strategy based on this profile.

Clinical drug development is often described as consisting of four temporal phases (Phases I–IV). Phase I studies, often conducted with a simple formulation not intended for commercialization, evaluate the safety, clinical pharmacology, and clinical pharmacokinetics of a new drug. Phase II studies introduce the drug into the intended patient population and assess safety and efficacy in this population. Phase III studies are the pivotal, confirmatory studies of the product's safety and efficacy and are conducted using the final dosage form intended for commercialization. Phase IV, or postmarketing studies, offer insight into the drug's place in the therapeutic regimen (16).

Phase I

Phase I studies involve some combination of the evaluation of initial safety and tolerability, pharmacokinetics, pharmacodynamics, and an early measurement of drug activity. The initial clinical study is typically a single dose study conducted in normal healthy volunteers. The initial dose in the study is estimated from the nonclinical data and this dose is escalated until adverse events are seen. This study results in the determination of the maximally tolerated dose of the drug. Analysis of pharmacokinetic parameters and relation of blood levels to adverse events gives great insight for future studies.

Subsequent Phase I studies involve multiple doses for longer periods of time to assess longer term tolerance and accumulation of the drug or its metabolites. The data obtained from earlier studies are used to select the dose, dosing interval, and dosing duration for the later studies.

Phase I studies typically involve dozens of subjects. These are small, well-controlled studies with very close oversight by the investigator.

Phase II

After establishing the safety and kinetic properties of the investigational drug in Phase I, development moves into the intended patient population. Phase II studies are typically safety and efficacy studies conducted in the target patient population.

Early Phase II studies may look at the potential safety and efficacy of the product in its intended indication and the dose and dosing interval needed to have the desired effect while minimizing adverse events.

The goal in Phase II is to establish the lowest effective dose of the drug in the target indication. This is typically accomplished in a dose–response study, which looks at various doses and dosing regimens of a drug in the target patient population. These studies are designed to answer such questions as; is 5 mg twice a day as effective as 10 mg once a day or 10 mg twice a day? The desired outcome is to move into Phase III development with one dose and dosing regimen of the drug.

Phase II studies can involve several hundred patients and last several months or longer.

Phase III

Phase III studies are the pivotal safety and efficacy studies that confirm the therapeutic benefit of the drug product. Studies in Phase III are designed to confirm the evidence accumulated in Phase II that the drug is safe and effective for use in the intended indication and recipient population.

Phase III studies can be tested against a placebo control with the intent of showing superiority over placebo. Another type of study design is to show equivalence or noninferiority to an approved therapy. An equivalence trial is intended to show that the response to two or more treatments differs by an amount which is clinically unimportant. A noninferiority trial demonstrates that the response to the investigational product is not clinically inferior to a comparative agent.

While there are exceptions, the U.S. FDA typically requires two adequate and well-controlled Phase III studies whose results confirm each other in order to gain approval for marketing. One important aspect of worldwide development of a new drug is the FDA requirement for Phase III studies with a placebo arm for comparison while other Health Authorities, typically European, require Phase III studies with the current therapy of choice as the control arm in the trial. This requires the conduct of additional clinical testing to meet all requirements worldwide.

Phase III trials enroll hundreds to several thousands of patients. Depending on the indication the studies can last from several months to as long as several years.

Good Clinical Practices

Good clinical practice (GCP) is an international ethical and scientific quality standard for designing, conducting, recording, and reporting trials that involve the participation of human subjects. Compliance with this standard provides public assurance that the rights, safety, and well-being of trial subjects are protected, consistent with the ethical principles that have their origin in the Declaration of Helsinki. The rights, safety, and well-being of the trial subjects are the most important considerations in clinical study conduct and should prevail over interests of science and society.

A trial should be initiated and continued only if the anticipated benefits justify the risks. This means that adequate nonclinical and clinical information on an investigational product should be adequate to support the proposed clinical trial. Each investigator involved in the study should be qualified by education, training, and experience to perform his or her respective study related tasks and to provide appropriate medical care to the subjects enrolled in the study.

In addition to FDA review of a protocol as part of an IND, an Institutional Review Board (IRB) is also required to review and approve a protocol prior to study initiation. The IRB review is intended to safeguard the rights, safety, and well-being of all trial subjects. An IRB is composed of members who collectively have the qualifications and experience to review and evaluate the science, medical aspects, and ethics of the proposed trial. Each clinical site must have approval from its own IRB.

Freely given informed consent should be obtained from every subject prior to clinical trial participation. Informed consent of a subject includes informing the subject that the trial involves research, their participation is voluntary, and they may refuse to participate or withdraw at any time without penalty or loss of benefits. The subject is also informed of the purpose of the trial and the probability of being assigned to each treatment in the trial, the trial procedures to be followed, and the subject's responsibilities in the trial. The subject is informed of any reasonably foreseeable risks and potential benefits of the study and any alternative courses of treatment other than participation in the study.

Regulatory Issues Specific to Intraocular Drug Delivery

Diseases of the posterior segment of the eye include a number of disorders with severe visual disability and a lack of effective therapy. Age-related macular degeneration, diabetic retinopathy, macular edema, and retinal degenerations like retinitis pigmentosa are all examples of diseases with an unmet medical need. Although many diseases of the anterior segment of the eye can be effectively treated with topical application of medications, it is more difficult to deliver therapeutic levels of drugs to the back of the eye with topical administration. Since many of the diseases affecting the retina affect older patients, side effects may limit the systemic administration of drugs. Most agree that local drug delivery to the back of the eye is desirable for the treatment of retinal diseases; however, the development and regulatory approval of a medication in a sustained-release drug delivery system presents a number of challenges.

A number of the drug delivery systems deliver medications for a very long period of time; sometimes over a number of years. Rather than launching into large, expensive, and resource-consuming trials with implants that deliver the drug for many years, it is often prudent to prove that the drug is effective over a shorter period of time. Although an implant can be filled with drug and deliver the compound for several years, it may make sense to start studies with implants that last for less time. Similarly, this will also demonstrate that the drug is active when administered to a specific location. For example, just because a drug works when delivered systemically does not mean that it will work equally as well if the drug is delivered into the vitreous, even if similar intravitreal levels are achieved with both systemic and intravitreal drug delivery. Some drugs may have a systemic effect contributing to their efficacy. For other drugs, drug levels at the level of the retinal pigment epithelium (RPE) may be more important than intravitreal drug levels.

Pharmacokinetic studies are important in the development of local ocular drug delivery. Although a benefit of ocular drug delivery is the ability to achieve higher intraocular drug concentrations by avoiding the blood-retinal and blood–aqueous barriers. However, regardless of the route of administration, whether intraocular, topical, periocular, or systemic, drug levels in ocular tissues will vary depending on clearance from the aqueous and vitreous, tear turnover, absorption across such barriers as the cornea, RPE, and whether the compound concentrates in tissues like the lens, ciliary body, iris, and RPE. It is also important to determine which tissue level is most critical for the drugs activity. For some retinal diseases like proliferative vitreoretinopathy, vitreous levels may be important. For other diseases like macular degeneration, RPE or choroid levels may be most crucial.

It is critical to demonstrate consistent release rates. Regulatory agencies have suggested that release rates should be within ±10% of that specified. These release rates can be checked in vitro; however, some in vivo data confirming the release rates are desirable, since the human vitreous has unique characteristics that can affect drug release from many drug delivery devices. If one embarked on a clinical development plan, demonstrated preclinical safety, and clinical safety and efficacy with an implant later shown to release drug outside of the specifications, the initial studies could be invalid.

Before a drug delivery system is tested with an active drug, regulatory agencies require evidence to support the safety of the implant alone. Studies showing compatibility of the drug delivery system should be performed.

There have been examples of toxicity resulting from the sterilization of drug delivery systems. Sterilization can lead to changes in the implant materials or the release of residual products which can induce intraocular inflammation or other adverse events. Any changes in the manufacturing or sterilization procedures of drug delivery systems should be thoroughly tested before use in humans.

There has been debate on whether placebo implants should be mandated in clinical trials. Sham procedures for coronary artery bypass surgery have been employed in clinical trials as the appropriate control group. It is known that preparation for surgery, pre- and postoperative evaluation, and the psychological effects of surgery may introduce bias into a treatment group. Currently, the FDA has required at least two doses of drugs in intravitreal implants for initial clinical trials. A placebo implant has not been required. Sham procedures have been used in clinical trials using ocular drug delivery. A study arm with a low dose of drug in the delivery device may be accepted as an alternative control in some studies.

Drug–Device Combination Products

Combination products of a drug and a device offer special challenges to companies. A product of this type is typically developed by a company that possesses an expertise in either the development of drugs or devices, not both. Working in the new area, i.e., drug development for a device company, offers significant challenges.

A combination product can be a device that contains a drug product or a drug product that relies on a device for administration. The FDA will make a determination if the product will be regulated by the Center for Devices and Radiological Health (CDRH) or the Center for Drug Evaluation and Research (CDER). The FDA makes this determination based on the properties of the product (17–19). A drug that is delivered to the retina via an implantable device would be regulated as a drug since the intended outcome of therapy is dependent on the pharmacologic action of the drug. The implant is used solely to deliver the drug to the back of the eye.

A syringe that contains heparin to prevent clotting would be regulated as a device, since the activity of the product is dependent on the syringe. The drug, heparin, is present in the device to improve its action.

Device Development Requirements

Medical devices in the United States are regulated by the CDRH within the FDA. CDRH classifies medical devices into Classes I, II, and III based on the risk with the use of the device. Regulatory control increases from Class I to Class III.

Class I devices are subject to the least regulatory control. They present minimal potential for harm to the user and are often simpler in design than Class II or Class III devices. Most Class I devices are exempt from premarket notification and good manufacturing practice regulations. They are subject to general controls that include manufacturing under a quality assurance program, suitability for their intended use, adequately packaged and properly labeled, and have establishment registration and device listing forms on file with the FDA. Examples of Class I devices include elastic bandages, examination gloves, and hand-held surgical instruments.

Class II devices are those for which general controls alone are insufficient to assure safety and effectiveness, and existing methods are available to provide such assurances. In addition to complying with general controls, Class II devices are also subject to special controls. Special controls may include special labeling requirements, mandatory performance standards, and postmarketing surveillance. Examples of Class II devices include powered wheelchairs, infusion pumps, and surgical drapes.

Class III is the most stringent regulatory category for devices. Class III devices are those for which insufficient information exists to assure safety and effectiveness solely through general or special controls. Class III devices are usually those that support or sustain human life, are of substantial importance in preventing impairment of human health, or which present a potential, unreasonable risk of illness or injury. Examples of Class III devices are replacement heart valves, silicone-filled breast implants, and implanted cerebella stimulators.

CDRH approves medical devices through the premarket notification and premarket approval processes. Most marketed devices are approved by the FDA via submission of a Premarket Notification or 510(k). A 510(k) notification is required for Class I devices that are not exempt from notification, all Class II devices, and certain Class III devices. A 510(k) is a premarketing submission demonstrating that the device to be marketed is substantially equivalent to, or as safe and effective as, a legally marketed device that is not subject to premarket approval. The legally marketed, comparator device is termed the predicate device.

Applicants compare their 510(k) device to one or more similar devices currently on the market and make and support their substantial equivalency claims. A device is shown to be substantially equivalent if it has the same intended use as the predicate device, and, has the same technological characteristics as the predicate device, or, has different technological characteristics that do not raise new questions of safety and effectiveness. The FDA approves a 510(k) product by determining that the applicant has demonstrated substantial equivalence to the predicate device.

The Premarket Approval (PMA) process is more involved and requires the submission of clinical data to support claims made for the device. The PMA is reviewed and an actual approval of the device is granted by the FDA. PMA approval is required in order to market most Class III devices.

The PMA is a scientific, regulatory documentation to the FDA to demonstrate the safety and effectiveness of the Class III device. It contains the technical data on the design and manufacture of the device, nonclinical testing of the device, and clinical data showing the device is safe and effective for its intended use.

The clinical data submitted in a PMA is generated under an Investigational Device Exemption (IDE). An IDE contains information on previous clinical studies with the device, design, manufacture, and control of the device, the investigators who will conduct the study. The FDA must approve the IDE prior to the start of the clinical study and make a determination on the approvability of an IDE within 30 days of receipt (19–22).

FDA Issues—CDER vs. CDRH

Regulatory oversight of products that combine a drug and a device require coordination within the FDA divisions responsible for each aspect of the product. This causes increased difficulty for the sponsor company in determining who is primarily responsible for the review of their application. The sponsor finds themselves in a position of encouraging the two Centers' reviewers within the FDA to communicate and share information on their review and the status of their review. Reviews that involve coordination between FDA Review Divisions and reviewers who do not usually work together can add significant time to the FDA review and approval process.

The FDA has established a Request for Designation process that allows a Sponsor company to request the FDA to designate the lead Review Division for the product early in the development process. Thereafter, communication with the FDA on the product will go primarily to the lead Center; however, it is important to assure that reviewers from both Centers are involved in the development process and all concerns and comments are incorporated into the product development strategy.

Often a device company will work closely with CDRH staff to develop and submit a combination device–drug product, only to find out during the application review that upon consultation by the CDRH reviewer with CDER, new issues are brought up that could have been incorporated into the clinical study design. This points out the importance of early communication with all involved parties at FDA during product development. Assuring that representatives from both Review Divisions are present at FDA–sponsor meetings allows for identification and discussion of issues early in the process.

REFERENCES

1. Mathieu M. New Drug Development: A Regulatory Overview, 6th ed. Waltham, MA: Parexel International Corporation, 2000.
2. Guarino RA. New Drug Approval Process, 3rd ed. New York, NY: Marcel Dekker Inc., 2000.
3. FDA guidance for industry, content and format of investigational new drug applications (INDs) for Phase 1 studies of drugs, including well-characterized, therapeutic, biotechnology-derived products, November 1995.
4. ICH Harmonized Tripartite Guidance. Guideline for good clinical practice, E6, May 1, 1996.
5. ICH Harmonized Tripartite Guidance. Maintenance of the ICH guideline on non-clinical safety studies for the conduct of human clinical trials for pharmaceuticals, M3(M), November 9, 2000.
6. ICH Harmonized Tripartite Guidance. Safety pharmacology studies for human pharmaceuticals, S7A, November 8, 2000.
7. ICH Draft Consensus Guideline. Safety pharmacology studies for assessing the potential for delayed ventricular repolarization (QT interval prolongation) by human pharmaceuticals, S7B, February 7, 2002.
8. ICH Harmonized Tripartite Guideline. Note for guidance on toxicokinetics: the assessment of systemic exposure in toxicity studies, S3A, October 27, 1994.
9. ICH Harmonized Tripartite Guideline. Pharmacokinetics: guidance for repeated dose tissue distribution studies, S3B, October 27, 1994.
10. ICH Harmonized Tripartite Guideline, Guidance on specific aspects of regulatory genotoxicity tests for pharmaceuticals, S2A, July 19, 1995.
11. ICH Harmonized Tripartite Guideline. Genotoxicity: a standard battery for genotoxicity testing of pharmaceuticals, S2B, July 16, 1997.

12. ICH Harmonized Tripartite Guideline. Duration of chronic toxicity testing in animals (rodent and non rodent toxicity testing, S4, September 2, 1998.
13. ICH Harmonized Tripartite Guideline. Detection of toxicity to reproduction for medicinal products, S5A, June 24, 1993.
14. ICH Harmonized Tripartite Guideline. Maintenance of the ICH guideline on toxicity to male fertility. An addendum to the ICH Tripartite Guideline on detection of toxicity to reproduction for medicinal products, S5B(M), November 9, 2000.
15. ICH Harmonized Tripartite Guideline. Stability testing of new drug substances and products, Q1A(R2), February 6, 2003.
16. ICH Harmonized Tripartite Guideline. General considerations for clinical trials, E8, July 17, 1997.
17. FDA Manual of Standard Operating Procedures and Policies. Intercenter Consultative/Collaborative Review Process, February 14, 2003.
18. Assignment of agency component for review of premarket applications, Federal Register, Vol. 68, No. 120, June 23, 2003.
19. Center for Devices and Radiological Health. Device advice classification page, June 10, 2003.
20. Center for Devices and Radiological Health. Premarket notification 510(k): regulatory requirements for medical devices, August 1995.
21. The New 510(k) Paradigm. Alternative approaches to demonstrating substantial equivalence in premarket notifications. Final guidance, March 20, 1998.
22. Center for Devices and Radiological Health. Premarket approval manual, January 1998.

5

Antiangiogenic Agents: Intravitreal Injection

Sophie J. Bakri and Peter K. Kaiser
The Cole Eye Institute, Cleveland Clinic Foundation, Cleveland, Ohio, U.S.A.

INTRODUCTION

Ocular neovascularization is one of the major causes of blindness in many common ocular diseases. For example, age-related macular degeneration (AMD) is the leading cause of blindness in patients over the age of 65 years, with the neovascular (exudative) form accounting for more than 80% of the cases with severe visual loss (1,2). Similarly, diabetic retinopathy is the leading cause of visual loss in patients under the age of 55 years, with visual loss occurring due to macular edema or ischemia, vitreous hemorrhage, and vitreoretinal traction from the new blood vessels (2). Neovascularization of the iris and angle structures resulting in neovascular glaucoma occurs in several ocular conditions including diabetic retinopathy, central and branch retinal vein occlusions, and ocular tumors. A blind, painful eye secondary to neovascular glaucoma is the single most common cause of enucleation in North America (3). Visual loss from retinopathy of prematurity in preterm infants occurs due to retinal neovascularization and secondary vitreoretinal traction and retinal detachment. Visual impairment from this disease is estimated to affect 3400 infants and to blind 650 infants annually in the United States (4). Neovascularization (angiogenesis) is the hallmark of all these visually debilitating diseases. Thus, it is no surprise that agents that can block neovascularization are under active investigation.

Current, proven treatment regimens for ocular neovascularization include ocular photodynamic therapy (PDT) and laser photocoagulation to either directly treat the choroidal neovascular membrane (CNV) as in AMD, or ablate the ischemic retina sparing the macula and nonischemic areas as in diabetic retinopathy, vein occlusions, retinopathy of prematurity, and anterior segment neovascularization. Laser treatment is inherently destructive and creates a permanent scotoma at the site of retinal ablation. Since a large proportion of CNV lesions in AMD are subfoveal, direct laser ablation would lead to a permanent and immediate loss of central vision. Verteporfin (Visudyne^{TM}, Novartis Ophthalmics AG) ocular PDT is a relatively new treatment modality that has been used to selectively treat subfoveal CNVs sparing the overlying retina and surrounding choriocapillaris. However, the recurrence rate is high and over 90% of patients require retreatment after three months, and not all lesions benefit from treatment.

Antiangiogenic therapy relies on a different approach to treat neovascular diseases. It directly targets the angiogenic cascade that is thought to be initiated by several growth factors such as vascular endothelial growth factor (VEGF), platelet-derived growth factor, transforming growth factor, and basic fibroblast growth factor (bFGF). Angiogenesis is a balanced process between activators of neovascularization and inhibitors. If one side of the scale gets tipped, abnormal angiogenesis can occur. The advantage of antiangiogenic therapy is its potential to preserve the function of retinal tissue while directly targeting the neovascular complexes. Moreover, since many of these treatments work on different aspects of the angiogenesis cascade, the possibility exists for synergy with combination treatment.

INTRAVITREAL INJECTION

There are numerous administration routes for antiangiogenic compounds including systemic, topical, periorbital, and intraocular. Systemic administration of antiangiogenic compounds rarely delivers useful levels of the drug to the eye. Moreover, the risk of systemic angiogenesis blockade is problematic since numerous organ systems rely on angiogenesis to repair tissue, especially the cardiovascular system. One can easily understand the risk of blocking coronary remodeling in an elderly patient with AMD. In addition, for systemic medications to reach satisfactory intraocular levels, very high systemic levels are often required leading to unacceptable side effects. This is especially true for cytotoxic agents used to treat ocular inflammatory disease where the side effects of the medications often limit their usefulness. Topical application of medication is sufficient for anterior segment disorders, but usually does not deliver adequate retinal levels of medication due to several factors. These include the following: drops are eliminated from the precorneal area within 90 seconds; the corneal barrier allows only about 1% of nonhydrophilic drugs to be absorbed across the cornea; drugs are eliminated by aqueous outflow; and the drug is metabolized when it enters the eye. Periocular administration, in particular subtenon injections, is used to circumvent some of these problems. Steroids are routinely administered by subtenon injection for posterior segment diseases, but since the medication has to diffuse across the sclera and choroid, intraocular levels of medication are variable, difficult to quantify, and difficult to adjust. In addition, almost 90% of the medication is systemically absorbed. Thus, intraocular delivery is the best way to circumvent the blood–retinal barrier and to deliver adequate retinal drug levels. There are multiple methods to perform intraocular drug delivery; however, in this chapter we will concentrate on the use of intravitreal injections to deliver medications to the posterior segment.

TECHNIQUE FOR INTRAVITREAL INJECTION

Intravitreal injections are simple to perform and can be done in an office setting. A typical approach is as follows: topical anesthetic drops (e.g., 0.5% proparacaine hydrochloride) are instilled onto the ocular surface. Topical lidocaine 4% or proparacaine hydrochloride 0.5% is applied to the injection site using cotton tip pledgets. The conjunctiva, lids, and lashes are disinfected with 5% to 10% povidone iodine with at least a five minute observation period to allow the antibacterial properties to be fully effective. In patients with preexisting ocular disease such as blepharitis, one can consider pretreatment with a topical antibiotic drop for a few days prior to the injection (*Note*: this was required in the Genentech and Eyetech clinical trials—see chapter 16). An eyelid speculum is then placed; additional anesthesia in the form of 2% to 4% subconjunctival lidocaine administration can be utilized, but is rarely required. The intravitreal agent is then drawn into a 1 cc syringe with a 27-, 29-, or 30-gauge needle after first cleansing the top of the container with an alcohol swab. (*Note*: pegaptanib sodium is supplied in a unit dose syringe with a 27-gauge needle.) Care is taken to ensure there are no air bubbles in the syringe by inverting it prior to injection. A mark is placed 3–4 mm (depending on phakic status) posterior to the limbus, usually in the inferior or inferotemporal quadrant. The needle is then introduced into the midvitreous cavity, aiming posteriorly and slightly inferiorly, but the needle is not introduced all the way to the hub. Using a single, continuous maneuver, the drug is injected slowly into the eye. The needle is removed simultaneously with the application of a cotton tip pledget over the entry site. The optic nerve head is then examined for arterial pulsation, and indirect ophthalmoscopy is performed to ensure correct placement of the medication and to evaluate the retina. In general, an anterior chamber paracentesis is rarely necessary unless the intraocular pressure is markedly elevated or the volume injected is more than 0.1 mL. Finally, a drop of topical antibiotic solution is administered. Some physicians instruct the patient to take the topical antibiotic solution four times a day for three to seven days following injection and to sleep on his or her back for the next few days. Other clinicians do not give antibiotics on the days following the injection. Follow-up is variably scheduled for one to seven days postinjection.

VEGF

VEGF is a heparin-binding, homodimeric, peptide mitogen with narrow target cell specificity whose activity is limited to endothelial cells derived from small and large blood vessels (5). There are five isoforms of VEGF that arise from alternate mRNA splicing of a single gene; however, $VEGF_{165}$ is the predominant isoform and most abundant. VEGF was originally called vascular permeability factor and is a potent cause of vascular leakage in the retina; this vasopermeability is hypothesized to enhance angiogenesis by allowing translocation of plasma proteins. It is a critical rate-limiting step in the development of ocular neovascularization. In addition, it functions as a survival factor for newly formed blood vessels. VEGF is mainly upregulated by hypoxia and other factors. It is present in surgically excised choroidal neovascularization (6,7) and in the aqueous and vitreous humor in eyes with proliferative retinal vascular disorders (8). Primate eyes injected with intravitreal VEGF develop dilated, tortuous retinal vessels that leak fluorescein similar to that seen in diabetic retinopathy. The severity of the retinopathy correlates with the number of VEGF injections (9). VEGF, therefore, represents an ideal target of antiangiogenic therapy. Since VEGF inhibitors do not exist in nature, compounds must be manufactured. Inhibition of VEGF can be achieved by blocking its receptors or the molecule itself. In this chapter, we will discuss two intravitreal anti-VEGF molecule products: ranibizumab (Lucentis™, Genentech) and pegaptanib sodium (Macugen™, Eyetech Pharmaceuticals), as well as anecortave acetate (RETAANE™, Alcon Pharmaceuticals), delivered via the posterior juxtascleral route.

Ranibizumab (Lucentis, Genentech)

Ranibizumab is a humanized, antigen-binding fragment (Fab) of a second-generation, recombinant mouse monoclonal antibody directed toward VEGF. It consists of two parts: a nonbinding human sequence (humanized), making it less antigenic in humans, and a high-affinity binding epitope (Fab fragment) derived from the mouse, which serves to bind the antigen (10). Ranibizumab, with a molecular weight of 48 kDa, is a much smaller molecule than the full-length RhuMab VEGF (Avastin™, bevacizumab, Genentech). RhuMab VEGF, with a molecular weight of 148 kDa, is FDA approved for the treatment of colorectal cancer, and is in early clinical testing for the treatment of CNV via the intravitreal route. This size difference is very important since unlike RhuMAb VEGF, which does not penetrate through the retina, ranibizumab has been shown to completely penetrate the retina and enter the subretinal space after intravitreal injection (10,11). The ability of ranibizumab to penetrate the retina is likely related to the internal limiting lamina pores that only allow molecules smaller than roughly 50 kDa to pass through the retina (12). Ranibizumab has high specificity and affinity for all the soluble human isoforms of VEGF. Moreover, it has a higher affinity for binding VEGF than RhuMab VEGF. It is produced via a plasmid, containing the appropriate gene sequence, inserted into an *Escherichia coli* expression vector that undergoes large-scale fermentation. This is drained, the supernatant collected and purified to produce the active drug.

Preclinical Studies

Safety. Animal studies have shown that ranibizumab is a safe agent for intravitreal injection. In cynomolgus monkeys, intravitreal injections of 500 µg of ranibizumab at two-week intervals in a laser-induced CNV model (13) or in normal monkey eyes did not show any significant adverse effects (14). However, mild side effects of the injections were seen. All eyes treated with ranibizumab developed acute anterior chamber inflammation within 24 hours of the first intravitreal injection (13). In contrast, eyes injected with vehicle alone showed minimal or no inflammation. The inflammation resolved within one week, and the inflammatory response was less pronounced after subsequent intravitreal ranibizumab injections. Also, animal studies have shown that ranibizumab has no effect on electroretinography, including visually evoked potentials.

Efficacy. In preclinical animal studies intravitreal ranibizumab injections can prevent CNV and possibly have a beneficial effect on the treatment of established CNV in monkey eyes (13). Ryan produced, with argon green laser photocoagulation, a primate model of CNV (15). Krzystolik et al. injected 500 µg of ranibizumab intravitreally into one eye while the other eye received intravitreal ranibizumab vehicle on days 0 and 14, and then produced CNV in both eyes, as described by Ryan, on day 21. On day 28, each eye received another injection of the same substance. Analysis of the CNV lesions at days 35 and 42 showed a reduction in CNV leakage defined as the likelihood of reaching grade 4 fluorescein leakage ("clinically significant" hyperfluorescence in the early or midtransit stages with late leakage) in the ranibizumab-treated eyes compared to the vehicle-treated eyes ($p < 0.001$). On days 42 and 56, the vehicle-treated eyes received a crossover injection of intravitreal ranibizumab to assess its effect on mature CNV membranes. The number of lesions with grade 4 fluorescein leakage decreased in this group over time ($p = 0.01$) after day 42, indicating a significant beneficial effect on reducing leakage from mature CNV. However, it was pointed out that spontaneous regression may have played a role at the later time points since previous studies have shown decreased angiographic leakage of untreated CNV lesions as early as two to three weeks after laser induction, with a mean of thirteen weeks (13,15).

Pharmacodynamics. Since ranibizumab is delivered via an intravitreal injection, studies were undertaken to determine if the drug could cross the neural retina and access the subretinal space where the CNV lesions are located. A study in rhesus monkeys demonstrated that $25\,\mu g$ in $50\,\mu L$ of Fab antibody fragment diffused through the neural retina to the retinal pigment epithelial layer after one hour and persisted in this location for up to seven days (10). The half-life in the vitreous was 3.2 days. These data are consistent with the results of a pharmacokinetic study done by a noninvasive fluorophotometric method that showed that fluorescein-labeled ranibizumab disappeared from the vitreous with a mean terminal half-life of 2.9 days and a mean residence time of 4.2 days (11).

Since VEGF plays an important role in other parts of the body, especially the cardiovascular system, systemic exposure was evaluated in preclinical studies. In monkey experiments, systemic exposure to ranibizumab was low, with plasma concentrations of the Fab antibody remaining below the limit of quantitation ($<7.8\,ng/mL$) (10). Levels of plasma ranibizumab are highest on day one after injection and decrease rapidly by day seven (13). The ranibizumab antigen assay showed that the average detectable drug level in the vitreous was $32\,ng/mL$ after the first injection and increased with subsequent injections (13). Thus, intravitreal injection of ranibizumab does not appear to lead to significant systemic levels.

Pegaptanib Sodium (Macugen, Eyetech Pharmaceuticals)

The anti-VEGF pegylated aptamer pegaptanib sodium (Macugen®, formerly NX1838; Eyetech Pharmaceuticals) is a polyethylene glycol conjugated oligonucleotide with high specificity and affinity for the major soluble human VEGF isoform, $VEGF_{165}$. Pegylation decreases the clearance of the drug from the vitreous following intravitreal injection. Aptamers are chemically synthesized short strands of RNA or DNA (oligonucleotides) designed to bind to specific molecular targets based on their three-dimensional structure, and are made using SELEX technology (systematic evolution of ligands by exponential enrichment). Pegaptanib sodium is an aptamer composed of 28 nucleotide bases that avidly binds and inactivates $VEGF_{165}$. It is ~50 kDa in size and thus is small enough to diffuse across the internal limiting membrane and retina into the subretinal space (12).

Preclinical Studies

Safety. A three-month, multiple-dose pharmacokinetic and toxicology study was conducted in 24 rhesus monkeys (16). Pegaptanib sodium was administrated to both eyes as intravitreal injections every two weeks for three months for a total of six injections. A control group (Group 1) received phosphate-buffered saline vehicle alone. Group 2 received four doses of 0.10 mg per eye followed by two doses of 1.0 mg per eye. Groups 3 and 4 received 0.25 and 0.50 mg per eye, respectively. No animal in any dose group died or became moribund. No pegaptanib sodium–related effects were observed in any of the monkeys as measured by the following parameters: clinical signs, food consumption, body weight gain, hematology, clinical chemistry, urinalysis, direct ophthalmologic examination, fundus and slit lamp examination, intraocular pressure, electroretinograms, electrocardiograms, blood pressure, gross necropsy, and microscopic examination of tissues and organs. Thus, intravitreal administration of pegaptanib sodium was not associated with any significant systemic effects. Pegaptanib sodium did not activate complement, nor did it elicit the production of immunoglobulin-G (IgG)-directed antibodies. Pharmacokinetic data from this study are presented later. No safety issues from preclinical studies were identified that would preclude the intravitreal administration of pegaptanib sodium in clinical trials or warrant special precautions in the conduct of these trails.

Efficacy.   Several preclinical animal models were used to examine the efficacy of pegaptanib sodium (17) on ocular neovascularization. The cutaneous vascular permeability assay (Miles assay) showed that VEGF-induced leakage of the Evans Blue indicator dye from the intradermal vasculature of guinea pigs was almost completely inhibited by the coadministration of pegaptanib sodium at concentrations as low as 100 nm (17). The corneal angiogenesis assay demonstrated that systemic treatment with pegaptanib sodium results in 65% inhibition of VEGF-dependent angiogenesis in rat corneas when compared with phosphate-buffered saline solution (17). In a prematurity model of mice retinopathy, there was an 80% reduction in retinal neovasculature compared with the untreated control at both the 10 and 3 mg/kg doses (17). Finally, treatment of mice with 10 mg/kg of pegaptanib sodium once daily inhibited A673 rhabdomyosarcoma tumor growth by 74% at day 16 of treatment compared with the control (17).

Pharmacodynamics.   The pharmacokinetics of intravitreal pegaptanib sodium has been evaluated in several preclinical models. Examination of the plasma and vitreous humor concentration data following intravitreal administration of pegaptanib sodium in rhesus monkeys (16) and rabbits (17) indicates that the systemic and local pharmacokinetics of pegaptanib sodium are linear and are predictable based on dose, over the dose ranges tested. The study design for the rhesus monkeys was described earlier. Eighteen New Zealand rabbits were administered a bilateral intravitreal injection of 0.5 mg pegaptanib sodium per eye in a volume of 40 μL per eye (17). Vitreous humor and ethylenediaminetetraacetic acid plasma samples in monkeys and rabbits were collected over a 28-day period (one sample per eye at one time point) and stored frozen until assayed.

Pegaptanib sodium was eliminated from the eye through systemic circulation with a terminal half-life from the vitreous of three to five days in both monkeys and rabbits. The plasma terminal half-life mimicked the vitreous humor half-life, indicative of "flip-flop" kinetics whereby the rate-limiting step that determines the systemic pegaptanib sodium concentration is the exit of the drug from the eye. From these observations one can estimate the vitreous humor terminal half-life in patients that would approximate the plasma terminal half-life.

A key finding of the study in rhesus monkeys was that after residing in the vitreous humor for 28 days, pegaptanib sodium was fully capable of binding to $VEGF_{165}$. In the rabbit, initial vitreous concentrations were 350 μg/mL and these concentrations decreased to 1.7 μg/mL by day 28 after intravitreal injection of 0.5 mg per eye. Substantial concentrations of the drug in the rabbit and rhesus monkey, well above the K_D for VEGF (200 pM), were present in the vitreous 28 days after a single intravitreal injection. These data suggest that a dosing frequency of every six weeks is an appropriate regimen. The pharmacokinetic results in both rhesus monkeys and rabbits were consistent with a highly stable aptamer that undergoes a slow release from the vitreous into the systemic circulation. Once in the systemic circulation, pegaptanib sodium is cleared by a first-order elimination process that occurs at a faster rate than the exit out of the eye. These pharmacokinetic properties ensure that the vitreous humor concentrations exceed by several hundred to thousand times the concentrations that are seen in the plasma throughout the dosing interval.

Intravitreal Triamcinolone Acetonide

Corticosteroids have antiangiogenic, antifibrotic, and antipermeability properties. The principle effects of steroids are stabilization of the blood–retinal barrier, resorption of exudation, and downregulation of inflammatory stimuli. Antiangiogenesis is a secondary effect felt to be mediated primarily by upregulation of extracellular matrix protein plasminogen activator inhibitor-1 (PAI-1) in vascular endothelial cells (18). This inhibits activation of plasmin and alters extracellular matrix degradation.

Experimentally, corticosteroids have been shown to reduce inflammatory mediators including interleukin 5, interleukin 6, interleukin 8, prostaglandins, interferon-gamma, and tumor necrosis factor (19–21), decrease levels of VEGF (22,23), and improve blood–retinal barrier function (see Chapter 2) (24). Wilson reported that in rabbit eyes, intravitreal triamcinolone successfully reduced blood–retinal barrier breakdown (as quantified by both gadolinium-enhanced magnetic resonance imaging and fluorescein angiography) induced by photocoagulation, whereas posterior subtenon triamcinolone did not (24). Several known corticosteroid mechanisms of action of could explain blood–retinal barrier stabilization; corticosteroids may stabilize cell and lysosomal membranes (25), reduce the release (25) or synthesis (26) of prostaglandins, inhibit cellular proliferation (27), block macrophage recruitment in response to macrophage inhibitory factor, inhibit phagocytosis by mature macrophages, and decrease polymorphonuclear infiltration into injured tissues (28). However, the doses of intravitreal steroid used clinically may be much greater than those necessary to activate corticosteroid receptors, and the mechanism of action of intravitreal steroids may not be due to the pharmacological actions described previously. Data from the National Acute Spinal Cord Injury Study (29) indicate that high-dose intravenous methylprednisolone is beneficial in reducing the morbidity of acute spinal cord trauma. The proposed mechanism for this effect is reduction of tissue edema, resulting in increased vascular perfusion. This is thought to be due to inhibition of lipid peroxidation and hydrolysis that damages microvascular and neuronal membranes after injury. In addition, membrane stabilization is postulated to reduce tissue necrosis. Triamcinolone, in particular, has been shown to have an antiangiogenic effect. It inhibits bFGF-induced migration and tube formation in choroidal microvascular endothelial cells and downregulates metalloproteinase-2 (30), decreases permeability, downregulates intercellular adhesion molecule-1 (ICAM-1) expression in vitro (31), and decreases MHC-II antigen expression (32).

The use of intravitreal corticosteroids was first popularized by Machemer in 1979 (33) in an effort to halt cellular proliferation after retinal detachment surgery, and Graham (34), McCuen (35), Tano (36), and others have studied its use in both animal models and humans. In contrast to other corticosteroids with short half-lives following intravitreal injection, triamcinolone acetonide is an effective and well-tolerated (35,37) agent for intravitreal injection in conditions such as uveitis (38,39), macular edema secondary to ocular trauma or retinal vascular disease (40), proliferative diabetic retinopathy (41), intraocular proliferation such as proliferative vitreoretinopathy (42), and choroidal neovascularization from AMD (43,44).

Jonas et al. (45) found no significant effect of intravitreal triamcinolone on blood glucose in a series of diabetic patients treated with intravitreal triamcinolone after pars plana vitrectomy for proliferative diabetic retinopathy.

Preclinical Corticosteroid Studies

Safety. A single pure intravitreal triamcinolone injection is well tolerated in rabbit eyes (35). Electroretinographic data showed no significant differences between treated and control eyes and both light and electron microscopy were normal in both groups. Hida and associates (37) investigated the vehicles of six commercially available depot corticosteroids in rabbit eyes and found no effect on the retina and lens with the vehicle in Kenalog™ (commercially available triamcinolone acetonide) at levels two times higher than in the marketed drug. However, preservatives present in the vehicle for Kenalog including benzyl alcohol were shown, in the same report, to have toxic effects on the retina in other steroid preparations. The use of a preservative-free triamcinolone for macular edema has been reported (46). This formulation has a shelf-life of 45 days and can be obtained from a compounding pharmacy. Other pharmaceutical companies are evaluating a purified, preservative-free, single-use triamcinolone acetonide formulation. They are being tested in the current National Eye Institute–sponsored clinical trials evaluating intraocular steroids for macular edema.

Efficacy. Corticosteroids have an inhibitory effect on the growth of fibroblasts (47,48). Triamcinolone acetonide inhibits experimental intraocular proliferation in rabbits (36). Intravitreal injection of 1 mg of triamcinolone significantly reduced both retinal neovascularization and retinal detachment in an experimentally induced rabbit model (36). A 4-mg intravitreal triamcinolone injection inhibited preretinal and optic nerve head neovascularization in a pig model of iatrogenic branch vein occlusion; all untreated eyes developed neovascularization by six weeks (49). Intravitreal triamcinolone is also a potent inhibitor of laser-induced CNV in a rat model; however, this animal model may not be ideal since laser-induced CNV may be caused by a traumatic repair process or inflammatory response and may be more susceptible to steroids than neovascularization in human disease states (50). In addition, the intravitreal triamcinolone acetonide was administered at the time of laser treatment; thus, the treatment may only inhibit new vessel formation and not existing neovascularization.

Penfold and associates found that triamcinolone acetonide significantly decreased MHC-II expression consistent with immunocytochemical observations that revealed condensed microglial morphology (32). The modulation of subretinal edema and microglial morphology correlated with in vitro observations suggesting that downregulation of inflammatory markers and endothelial cell permeability are significant features of the triamcinolone acetonide mode of action. In another study (31), they investigated the capacity of triamcinolone to modulate the expression of adhesion molecules and permeability using a human epithelial cell line (ECV304) as a model of the outer blood–retinal barrier (BRB). They found that triamcinolone modulated transepithelial resistance of TER and ICAM-1 expression in vitro, suggesting that re-establishment of the BRB and downregulation of inflammatory markers are the principal effects of intravitreal triamcinolone in vivo. The results indicate that triamcinolone has the potential to influence cellular permeability, including the barrier function of the retinal pigment epithelium.

Pharmacodynamics. After a single triamcinolone acetonide injection (0.5 mg) in rabbits, corticosteroid was undetectable ophthalmoscopically by 41 days in

normal eyes, by 16.8 days in vitrectomized eyes, and by 6.5 days in eyes that had undergone combined vitrectomy and lensectomy (51). The colorimetric evaluation of the
ocular tissues for triamcinolone correlated well with the clinical disappearance of
the corticosteroid crystals. In another study (52), 0.4 mg of triamcinolone acetonide
was injected into the vitreous of rabbit eyes, and the vitreous was harvested at
intervals ranging from one hour to 46 days and analyzed by high-performance
liquid chromatography. This study found a shorter half-life of 1.6 days, and triamcinolone was visible in the vitreous for 23.3 days.

Indirect ophthalmoscopy of intravitreal crystals and measurement of intravitreal triamcinolone concentrations have been used to estimate the rate of triamcinolone elimination from rabbit vitreous humor (52). However, it is important to
know the pharmacokinetics of intravitreal triamcinolone acetonide in human eyes
with the number of diseases proposed for treatment with this medication. In patients
who have not clinically responded to intravitreal triamcinolone, or have relapsed, it
is not known whether response failure is due to insufficient drug levels, or whether
the patients are steroid nonresponders. Beer et al. (53) described the pharmacodynamics of intravitreal triamcinolone in human eyes. An aqueous humor sample
was obtained by anterior chamber paracentesis from five eyes at days 1, 3, 10, 17,
and 31 following injection of intravitreal triamcinolone. Intraocular triamcinolone
concentrations were measured using high-performance liquid chromatography and
population pharmacokinetic parameters were calculated using an iterative, non-
linear, weighted, least-squares regression computer program. Pharmacodynamic
data followed a two-compartment model. Peak aqueous humor concentrations
ranged from 2151 to 7202 ng/mL, half-lives from 76 to 635 hours, and the integral
of the area under the concentration–time curve (AUC_{0-t}) from 231 to 1911 ng hr/mL.
Following a single intravitreal injection of triamcinolone, the mean elimination
half-life was 18.6 days in nonvitrectomized patients. The half-life in a patient
who had undergone a vitrectomy was shorter at 3.2 days. Following intravitreal
injection, measurable concentrations of triamcinolone would be expected to last
for ~3 months (93 ± 28 days) in the absence of a vitrectomy (53).

Following a single intravitreal injection of triamcinolone acetonide, it has been
shown that one can deliver a concentration of thousands of nanograms of triamcinolone to the vitreous cavity (53). Drug concentrations rapidly decrease and are
followed by a subsequent prolonged elimination rate. It has long been recognized
clinically that a decisive initial amount of immunosuppression followed by a relatively rapid but sustained taper is often the most effective strategy in treating uveitis.
The pharmacodynamics results in this study mimic typical dosing regimens for treating uveitis, and are therefore consistent with the clinical results of the effectiveness of
intravitreal triamcinolone administration for posterior segment inflammation.

Intravitreal triamcinolone acetonide use has increased recently following case
series of successful treatment of edema due to various etiologies as well as choroidal
neovascularization. It has been advocated for the treatment of macular edema
associated with diabetes (54), uveitis (38,39), and central retinal vein occlusion (55).
In addition, it has been proposed as adjunctive treatment for proliferative diabetic
retinopathy (41) and exudative AMD (43,44).

Anecortave Acetate (RETAANE, Alcon Pharmaceuticals)

Anecortave acetate (RETAANE®, Alcon Pharmaceuticals) is an angiostatic agent
given as a posterior juxtascleral depot. The drug is under clinical evaluation for the

treatment of neovascular AMD, for the prevention of neovascular AMD in patients with high-risk non-neovascular AMD, and neovascular AMD in the fellow eye (see chap. 16).

Anecortave acetate is one of a new class of steroids that inhibits angiogenesis, yet has little glucocorticoid (anti-inflammatory) or mineralocorticoid (salt retaining) activity, and was introduced in 1985 (56). The formula of anecortave acetate is 4,9(11)-pregnadien-17,21-diol-3,20-dione-21-acetate. It is a synthetic analog of cortisol acetate with specific and irreversible chemical modifications made to its original structure. Removal of the 11-beta hydroxyl and the addition of a new double bond at the C9–11 position resulted in a novel angiostatic cortisene that has not exhibited typical glucocorticoid receptor-mediated bioactivity. These modifications also resulted in an apparent elimination of anti-inflammatory activity typical of the initial cortisol molecule.

Angiostatic steroids have since proven effective in inhibiting angiogenesis in a variety of systems, including chick chorioallantoic membrane (57,58), rat mammary carcinoma (59), rabbit cornea (60,61), rat cornea (62), and mouse intraocular tumors (62–76). Anecortave acetate inhibits angiogenesis further downstream from VEGF, and therefore has the potential to inhibit angiogenesis driven by multiple stimuli (64). It has been demonstrated that angiostatic steroids exert their inhibitory effect on endothelial cell growth in vitro by increasing the synthesis of plasminogen-activator inhibitor-1 (PAI-1) (65). This induction of PAI-1 then inhibits u-PA activity, which is essential for the invasive aspect of angiogenesis—the breakdown of vascular endothelial basement membrane and extracellular matrix. Therefore, the result of steroid-induced suppression of PA function is that endothelial cells cannot proliferate and migrate toward an angiogenic stimulus to participate in new blood vessel formation. There is evidence that angiostatic steroids may operate by the same mechanism in vivo (66).

Posterior Juxtascleral Administration Technique

Anecortave acetate is a white depot suspension preparation, available in a 15-mg dose (0.5 mL of 30 mg/mL) and a 30-mg dose (0.5 mL of 60 mg/mL). It is administered by the posterior juxtascleral route, using a specifically designed protocol and cannula. After topical anesthetic drops, and applying a pledget soaked in 4% lidocaine to the superotemporal quadrant, a small incision is made in that quadrant 8 mm posterior to the limbus. Using blunt Westcott scissors, a dissection through conjunctiva and Tenon's capsule is made so that bare sclera is visualized. Using a 56° posterior juxtascleral cannula, 0.5 mL of anecortave acetate is injected. At the same time, a counter-pressure device, consisting of a modified sponge on an applicator, is held on the conjunctiva surrounding the cannula, as posterior as possible, to prevent reflux. There have been several recent modifications to improve the posterior juxtascleral administration of anecortave acetate. These include use of a 1-cm^3 tuberculin syringe, a radial incision, slower infusion time (around 10 seconds), and slower cannula withdrawal (10 seconds). These measures have decreased the incidence of reflux, improving the contact of the medication with the sclera.

ACKNOWLEDGMENTS

Research grant support (PKK): Allergan, Alcon, Novartis, QLT, Eyetech, Genentech, SIRNA.

REFERENCES

1. Ferris FL III, Fine SL, Hyman L. Age-related macular degeneration and blindness due to neovascular maculopathy. Arch Ophthalmol 1984; 102:1640–1642.
2. Klein R, Klein BE, Jensen SC, Meuer SM. The five-year incidence and progression of age-related maculopathy: the Beaver Dam Eye Study. Ophthalmology 1997; 104:7–21.
3. Wand M, Dueker DK, Aiello LM, Grant WM. Effects of panretinal photocoagulation on rubeosis iridis, angle neovascularization, and neovascular glaucoma. Am J Ophthalmol 1978; 86:332–339.
4. Penn JS, Tolman BL, Lowery LA. Variable oxygen exposure causes preretinal neovascularization in the newborn rat. Invest Ophthalmol Vis Sci 1993; 34:576–585.
5. Ferrara N, Davis-Smyth T. The biology of vascular endothelial growth factor. Endocr Rev 1997; 18:4–25.
6. Kvanta A, Algvere PV, Berglin L, Seregard S. Subfoveal fibrovascular membranes in age-related macular degeneration express vascular endothelial growth factor. Invest Ophthalmol Vis Sci 1996; 37:1929–1934.
7. Lopez PF, Sippy BD, Lambert HM, Thach AB, Hinton DR. Transdifferentiated retinal pigment epithelial cells are immunoreactive for vascular endothelial growth factor in surgically excised age-related macular degeneration-related choroidal neovascular membranes. Invest Ophthalmol Vis Sci 1996; 37:855–868.
8. Aiello LP, Avery RL, Arrigg PG, et al. Vascular endothelial growth factor in ocular fluid of patients with diabetic retinopathy and other retinal disorders. N Engl J Med 1994; 331:1480–1487.
9. Tolentino MJ, Miller JW, Gragoudas ES, et al. Intravitreous injections of vascular endothelial growth factor produce retinal ischemia and microangiopathy in an adult primate. Ophthalmology 1996; 103:1820–1828.
10. Mordenti J, Cuthbertson RA, Ferrara N, et al. Comparisons of the intraocular tissue distribution, pharmacokinetics, and safety of 125I-labeled full-length and Fab antibodies in rhesus monkeys following intravitreal administration. Toxicol Pathol 1999; 27: 536–544.
11. Gaudreault J, Escandon E, Maruoka M, Reich M, Li D, Hsei D. Vitreal pharmacokinetics of rhuFab V2 in rabbits using a non-invasive method. Association for Research in Vision and Ophthalmology, 2002 (Abstract 2801).
12. Bunt-Milam AH, Saari JC, Klock IB, Garwin GG. Zonulae adherentes pore size in the external limiting membrane of the rabbit retina. Invest Ophthalmol Vis Sci 1985; 26(10):1377–1380.
13. Krzystolik MG, Afshari MA, Adamis AP, et al. Prevention of experimental choroidal neovascularization with intravitreal anti-vascular endothelial growth factor antibody fragment. Arch Ophthalmol 2002; 120:338–346.
14. O'Neill CA, Christian B, Murphy CJ, et al. Safety evaluation of intravitreal administration of Lucentis™ in cynomolgus monkeys for 3 months. Invest Ophthalmol Vis Sci 2000; 41:S142 (Abstract 732).
15. Ryan SJ. Subretinal neovascularization. Natural history of an experimental model. Arch Ophthalmol 1982; 100:1804–1809.
16. Drolet DW, Nelson J, Tucker CE, et al. Pharmacokinetics and safety of an anti-vascular endothelial growth factor aptamer (NX1838) following injection into the vitreous humor of rhesus monkeys. Pharm Res 2000; 17:1503–1510.
17. Eyetech Study Group. Preclinical and phase 1A clinical evaluation of an anti-VEGF pegylated aptamer (EYE001) for the treatment of exudative age-related macular degeneration. Retina 2002; 22:143–152.
18. Folkman J, Klagsbrun M. Angiogenic factors. Science 1987; 235(4787):442–447.
19. Kang BS, Chung EY, Yun YP, et al. Inhibitory effects of anti-inflammatory drugs on interleukin-6 bioactivity. Biol Pharm Bull 2001; 24(6):701–703.

20. Umland SP, Nahrebne DK, Razac S, et al. The inhibitory effects of topically active glucocorticoids on IL-4, IL-5, and interferon-gamma production by cultured primary CD4+ T cells. J Allergy Clin Immunol 1997; 100:511–519.
21. Floman N, Zor U. Mechanism of steroid action in ocular inflammation: inhibition of prostaglandin production. Invest Ophthalmol 1977; 16:69–73.
22. Bandi N, Kompella UB. Budesonide reduces vascular endothelial growth factor secretion and expression in airway (Calu-1) and alveolar (A549) epithelial cells. Eur J Pharmacol 2001; 425:109–116.
23. Fischer S, Renz D, Schaper W, Karliczek GF. In vitro effects of dexamethasone on hypoxia-induced hyperpermeability and expression of vascular endothelial growth factor. Eur J Pharmacol 2001; 411(3):231–243.
24. Wilson CA, Berkowitz BA, Sato Y, et al. Treatment with intravitreal steroid reduces blood-retinal barrier breakdown due to retinal photocoagulation. Arch Ophthalmol 1992; 110:1155–1159.
25. Naveh N, Weissman C. Prolonged corticosteroid treatment exerts transient inhibitory effect on prostaglandin E2 release from rabbits' eyes. Prostaglandins Leukot Essent Fatty Acids 1991; 42(2):101–105.
26. Lewis GD, Campbell WB, Johnson AR. Inhibition of prostaglandin synthesis by glucocorticoids in human endothelial cells. Endocrinology 1986; 119(1):62–69.
27. Heffernan JT, Futterman S, Kalina RE. Dexamethasone inhibition of experimental endothelial cell proliferation in retinal venules. Invest Ophthalmol Vis Sci 1978; 17(6):565–568.
28. Bhattacherjee P, Williams RN, Eakins KE. A comparison of the ocular anti-inflammatory activity of steroidal and nonsteroidal compounds in the rat. Invest Ophthalmol Vis Sci 1983; 24(8):1143–1146.
29. Bracken MB, Shepard MJ, Hellenbrand KG, et al. Methylprednisolone and neurological function 1 year after spinal cord injury. Results of the National Acute Spinal Cord Injury Study. J Neurosurg 1985; 63(5):704–713.
30. Wang YS, Friedrichs U, Eichler W, Hoffmann S, Wiedemann P. Inhibitory effects of triamcinolone acetonide on bFGF-induced migration and tube formation in choroidal microvascular endothelial cells. Graefes Arch Clin Exp Ophthalmol 2002; 240(1):42–48.
31. Penfold PL, Wen L, Madigan MC, Gillies MC, King NJ, Provis JM. Triamcinolone acetonide modulates permeability and intercellular adhesion molecule-1 (ICAM-1) expression of the ECV304 cell line: implications for macular degeneration. Clin Exp Immunol 2000; 121(3):458–465.
32. Penfold PL, Wong JG, Gyory J, Billson FA. Effects of triamcinolone acetonide on microglial morphology and quantitative expression of MHC-II in exudative age-related macular degeneration. Clin Experiment Ophthalmol 2001; 29(3):188–192.
33. Machemer R, Sugita G, Tano Y. Treatment of intraocular proliferations with intravitreal steroids. Trans Am Ophthalmol Soc 1979; 77:171–180.
34. Graham RO, Peyman GA. Intravitreal injection of dexamethasone. Treatment of experimentally induced endophthalmitis. Arch Ophthalmol 1974; 92(2):149–154.
35. McCuen BW II, Bessler M, Tano Y, Chandler D, Machemer R. The lack of toxicity of intravitreally administered triamcinolone acetonide. Am J Ophthalmol 1981; 91(6):785–788.
36. Tano Y, Chandler D, Machemer R. Treatment of intraocular proliferation with intravitreal injection of triamcinolone acetonide. Am J Ophthalmol 1980; 90(6):810–816.
37. Hida T, Chandler D, Arena JE, Machemer R. Experimental and clinical observations of the intraocular toxicity of commercial corticosteroid preparations. Am J Ophthalmol 1986; 101(2):190–195.
38. Antcliff RJ, Spalton DJ, Stanford MR, Graham EM, Ffytche TJ, Marshall J. Intravitreal triamcinolone for uveitic cystoid macular edema: an optical coherence tomography study. Ophthalmology 2001; 108(4):765–772.

39. Young S, Larkin G, Branley M, Lightman S. Safety and efficacy of intravitreal triamcinolone for cystoid macular oedema in uveitis. Clin Experiment Ophthalmol 2001; 29(1):2–6.

40. Martidis A, Duker JS, Greenberg PB, et al. Intravitreal triamcinolone for refractory diabetic macular edema. Ophthalmology 2002; 109:920–927.

41. Jonas JB, Hayler JK, Sofker A, Panda-Jonas S. Intravitreal injection of crystalline cortisone as adjunctive treatment of proliferative diabetic retinopathy. Am J Ophthalmol 2001; 131(4):468–471.

42. Jonas JB, Hayler JK, Panda-Jonas S. Intravitreal injection of crystalline cortisone as adjunctive treatment of proliferative vitreoretinopathy. Br J Ophthalmol 2000; 84(9):1064–1067.

43. Danis RP, Ciulla TA, Pratt LM, Anliker W. Intravitreal triamcinolone acetonide in exudative age-related macular degeneration. Retina 2000; 20(3):244–250.

44. Challa JK, Gillies MC, Penfold PL, Gyory JF, Hunyor AB, Billson FA. Exudative macular degeneration and intravitreal triamcinolone: 18 month follow up. Aust N Z J Ophthalmol 1998; 26(4):277–281.

45. Jonas JB, Hayler JK, Sofker A, Panda-Jonas S. Intravitreal injection of crystalline cortisone as adjunctive treatment of proliferative diabetic retinopathy. Am J Ophthalmol 2001; 131(4):468–471.

46. Bakri SJ, Shah A, Falk NS, Beer PM. Preservative-free intravitreal triamcinolone for macular edema. Eye 2005; 19(6):686–688.

47. Ruhmann AG, Berliner DL. Influence of steroids on fibroblasts. II. The fibroblast as an assay system for topical anti-inflammatory potency of corticosteroids. J Invest Dermatol 1967; 49(2):123–130.

48. Pratt WB, Aronow L. The effect of glucocorticoids on protein and nucleic acid synthesis in mouse fibroblasts growing in vitro. J Biol Chem 1966; 241(22):5244–5250.

49. Danis RP, Bingaman DP, Yang Y, Ladd B. Inhibition of preretinal and optic nerve head neovascularization in pigs by intravitreal triamcinolone acetonide. Ophthalmology 1996; 103(12):2099–2104.

50. Ciulla TA, Criswell MH, Danis RP, Hill TE. Intravitreal triamcinolone acetonide inhibits choroidal neovascularization in a laser-treated rat model. Arch Ophthalmol 2001; 119(3):399–404.

51. Schindler RH, Chandler D, Thresher R, Machemer R. The clearance of intravitreal triamcinolone acetonide. Am J Ophthalmol 1982; 93(4):415–417.

52. Scholes GN, O'Brien WJ, Abrams GW, Kubicek MF. Clearance of triamcinolone from vitreous. Arch Ophthalmol 1985; 103(10):1567–1569.

53. Beer PM, Bakri SJ, Singh RJ, Liu W, Peters GB III, Miller M. Intraocular concentration and pharmacokinetics of triamcinolone acetonide after a single intravitreal injection. Ophthalmology 2003; 110(4):681–686.

54. Martidis A, Duker JS, Greenberg PB, et al. Intravitreal triamcinolone for refractory diabetic macular edema. Ophthalmology 2002; 109:920–927.

55. Ip MS, Gottlieb JL, Kahana A, et al. Intravitreal triamcinolone for the treatment of macular edema associated with central retinal vein occlusion. Arch Ophthalmol 2004; 122(8):1131–1136.

56. Crum R, Szabo S, Folkman J. A new class of steroids inhibits angiogenesis in the presence of heparin or a heparin fragment. Science 1985; 230:1375–1378.

57. McNatt LG, Lane D, Clark AF. Angiostatic activity and metabolism of cortisol in the chorioallantoic membrane. J Steroid Biochem Mol Biol 1992; 42:687–693.

58. McNatt LG, Weimer L, Yanni J, Clark AF. Angiostatic activity of steroids in the chick embryo CAM and rabbit corneal models of neovascularization. J Ocular Pharm Therapeut 1999; 15:413–423.

59. Oikawa T, Hiragun A, Yoshida Y, et al. Angiogenic activity of rat mammary carcinomas induced by 7,12-dimethylbenz[a]anthracene and its inhibition by medroxyprogesterone

acetate: possible involvement of antiangiogenic action of medroxyprogesterone acetate in its tumor growth inhibition. Cancer Lett 1988; 43:85–92.

60. Li WW, Casey R, Gonzalez EM, Folkman J. Angiostatic steroids potentiated by sulfated cyclodextrins inhibit corneal neovascularization. Invest Ophthalmol Vis Sci 1991; 32:2898–2905.

61. BenEzra D, Griffin BW, Maftzir G, et al. Topical formulations of novel angiostatic steroids inhibit rabbit corneal neovascularization. Invest Ophthalmol Vis Sci 1997; 38:1954–1962.

62. Proia AD, Hirakata A, McInnes JS, et al. The effect of angiostatic steroids and b-cyclo-dextrin tetradecasulfate on corneal neovascularization in the rat. Exp Eye Res 1993; 57:693–698.

63. Clark AF, Mellon J, Li X-Y, et al. Inhibition of intraocular tumor growth by topical application of the angiostatic steroid anecortave acetate. Invest Ophthalmol Vis Sci 1999; 40:2156–2162.

64. Casey R, Li WW. Factors controlling ocular angiogenesis. Am J Ophthalmol 1997; 124(4):521–529.

65. Blei F, Wilson EL, Mignatti P, Rifkin DB. Mechanism of action of angiostatic steroids: suppression of plasminogen activator activity via stimulation of plasminogen activator inhibitor synthesis. J Cell Physiol 1993; 155:568–578.

66. Penn JS, Rajaratnam VS, Collier RJ, Clark AF. The effect of an angiostatic steroid on neovascularization in a rat model of retinopathy of prematurity. Invest Ophthalmol Vis Sci 2001; 42(1):283–290.

6

Intravitreal Antimicrobials

Travis A. Meredith
Department of Ophthalmology, University of North Carolina, Chapel Hill, North Carolina, U.S.A.

INTRODUCTION

Intravitreal injection of antimicrobials has become the mainstay of treatment of intraocular infections (1,2). Early in the development of antimicrobials, studies were done on the intraocular use of penicillin and sulfa drugs by various authors (3–6). For a number of years, intravitreal antibiotic injection was considered controversial, but antimicrobials are now used not only for treatment but also for prophylaxis against infection by placing them in the infusion fluid during vitrectomy (7), injecting them into the vitreous cavity after vitrectomy, or by injecting them into the lens capsule where they have access to the anterior vitreous after cataract surgery. The studies on intravitreal antimicrobial pharmacokinetics are fundamental in assessing their utility for clinical therapy. Typical studies of antimicrobial clearance from the eye after intravitreal injection (Table 1) are made after the single bolus of drug is given, followed by sampling of the concentrations in a noninflamed phakic eye. Intraocular surgery, inflammation, and multiple intravenous dosing regimens change the pharmacokinetics of antimicrobials in the vitreous cavity significantly (8,10,17, 21,22). Results in noninflamed phakic eyes may therefore have little relevance in many clinical situations.

BASIC PHARMACOKINETICS

Once injected into the eye, antibiotics diffuse through the vitreous cavity and are eliminated by either a posterior or an anterior route (23–25). Several factors govern the intravitreal concentration of an antibiotic at any given time. The initial concentration is a result of the extended distribution and the initial dose. Subsequently, the volume of distribution, the dose of the initial injection, and the rate of elimination govern the concentration of the drug at a given time (26). Two parameters characterize the elimination phase of the drug: (i) the elimination half-life and (ii) the apparent volume of distribution.

The volume of distribution is a measure of the extent of physical distribution of the drug, but corresponds to the physical volume only rarely. Volume of distribution may be lower or many times higher than the actual physical volume, depending on

Table 1 Concentration (µg/ml) and Half-Life After Intravitreal Injection

Drug	Model	Dose	Phakic			Aphakic			Aphakic vitrectomized			References
			24 hours	48 hours	$T_{1/2}$ (hr)	24 hours	48 hours	$T_{1/2}$ (hr)	24 hours	48 hours	$T_{1/2}$ (hr)	
Carbenicillin disodium	Monkey, normal	1 mg	50[a]	11[a]	10[b]							8
>Cefazolin sodium	Monkey, normal	1 mg	350[a]	30[a]	7[c]							8
	Rabbit, infected	2.25 mg	8.7	3.15								9
	Rabbit, control	2.25 mg	147	8.97	6.5	183.3	17.7	8.3	25.7	3.7	6	10
	Rabbit, inflamed	2.25 mg	340	57.4	10.4	242.5	31.9	9.0	33.4	5.7	6.7	10
Ceftriaxone sodium	Rabbit, normal	2 mg	200	45								11
	Monkey, normal	2 mg	434	59								12
Vancomycin	Rabbit, normal	2 mg	>100	>100					90[d]	16[d]		13
	Rabbit, phakic	2 mg									25.5	14
	Rabbit, inflamed	2 mg									31.5	14
	Rabbit, infected	2 mg									56	15
Amikacin	Rabbit, normal	250 µg	19	7								16
	Rabbit, control	400 µg	100.9	55.6	25.5	25.3	15.3	14.3	15.5	3.0	7.9	17
	Rabbit, inflamed	400 µg	97.6	31.4	15.5	7.6	1.5	7.4	7.2	1.4	7.7	17
Gentamicin	Human, infected	100 µg	10.5				4.6					18
	Monkey, normal	100 µg										8
	Cat, normal	400 µg	50	6.5	34							19
	Cat, infected	400 µg	11									19
	Rabbit, control	71 µg/mL[a]	75	40	32	25	7	12				18
	Rabbit, infected	40–65 µg/mL[a]	35	18	19	20	6	14				18

[a]Estimated from graphic representation.
[b]With probenecid $T_{1/2}$ = 20 hours.
[c]With probenecid $T_{1/2}$ = 30 hours.
[d]Without intact capsule.
Source: Modified from Ref. 20.

many factors. The half-life $(T_{1/2})$ is the period of time required for the drug concentration to fall by one half. Elimination of a drug from any compartment is usually a first-order process; by definition the rate of elimination is proportional to the amount of drug present. The elimination of drug when plotted on a semilog scale is usually a linear function.

The rate constant k is of first order with a dimension of time (-1). The elimination constant defines the fractional rate of drug removal and is equivalent to the rate of elimination divided by the amount of drug in the compartment. Given this definition, the half-life is thus $T_{1/2} = 0.693$ divided by k.

Clearance is the parameter that relates the drug concentration in a cavity to the rate of its elimination. Clearance multiplied by concentration equals the rate of elimination. Units of clearance are expressed in volume per unit of time. The half-life is then expressed as $T_{1/2} = (0.693 \times$ volume of distribution$)/$clearance. The half-life and elimination constant both reflect (not control) the volume of distribution and the clearance of drugs.

INTRAVITREAL INJECTIONS

Once injected into the vitreous cavity, drugs diffuse rapidly through the vitreous, although this may take several hours (24,25). There is little resistance to diffusion of drugs within the vitreous cavity due to low average concentration of collagen in the vitreous gel. Molecular action is thought to be more important than fluid flow. Maurice has calculated that therapeutic concentration of drug at the retinal surface may be achieved within about three hours of injecting 100 times the therapeutic concentration into the central vitreous cavity (24,25). Furthermore, eye movements enhance drug mixing in the vitreous cavity. When vitreous has been removed, the injected drug may settle quickly onto the most dependent areas of retina and thereby increase the potential toxicity (27). Other models suggest that both vitreous diffusivity and retinal permeability are important factors that can be quantitated (28,29).

After drug injection, larger molecules are thought to be removed by an anterior route. Drugs flow through the vitreous cavity, around the lens, and enter into the anterior chamber where they then exit through the trabecular meshwork into the canal of Schlemm. The lens itself is thought to have negligible contribution to drug elimination but may provide a physical barrier to movement of the drug anteriorly. Maurice has calculated that if elimination is entirely through the anterior chamber, the amount of drug lost in one hour of the vitreous body equals $k_v \times c_v \times v_v = fc_a$ where k_v is the fraction lost every hour, c_v is the average concentration of the vitreous body, f is the volume of the aqueous flow in one hour, v_v is the volume in the vitreous body, and c_a is the concentration of drug in the aqueous. Vancomycin, sulfacetamide sodium, aminoglycosides, streptomycin sulfate, and newer betalactam antimicrobials are thought to be eliminated anteriorly (9,23,25,30).

Posterior elimination is thought to be the predominant route for clindamycin, dexamethasone, and first- and second-generation cephalosporins (23). There is a barrier posteriorly that is normally impermeable to materials of high molecular weight, although active transport out of the vitreous by the retina occurs for numerous substances. The barrier is thought to be shared between the retinal capillaries and the retinal pigment epithelium. Both posterior and anterior routes of egress may come into play for drugs such as ceftriaxone sodium. Because of the wide surface area for absorption and the contribution of active transport, posteriorly excreted drugs often have shorter half-lives than do anteriorly excreted ones (23). In the case of betalactam antibiotics, competitive inhibition and metabolic inhibition have been demonstrated, suggesting properties of saturation kinetics. There may be a correspondence between vitreous half-life of drugs and renal tubular excretion of drugs in humans (24,30). The half-life of intravitreal carbenicillin is prolonged from 5 to 13 hours by probenicid administration in the rabbit and from 10 to 20 hours in the monkey (8). Probenicid prolongs the half-life of cefazolin in the monkey from 7 to 30 hours (8).

Inflammation has been noted to change the half-lives of intravitreal drugs, presumably through interference with active transport. The half-life of cefazolin in phakic rabbit eyes was increased by inflammation from 6.5 to 10.4 hours (10). The anterior removal route may become more important as the posterior route is blocked by metabolic or competitive inhibition. In the monkey treated with probenicid the half-life of cefazolin, normally excreted posteriorly, is essentially the same life as the half-life of gentamicin, an anteriorly secreted drug (8). The effects of inflammation may be complex, however, since the permeability in posterior structures may be increased, partially compensating for the effect of decrease in active transport. Aphakic and aphakic-vitrectomized eyes when injected with cefazolin have no difference in half-life when comparing control to the inflamed eyes (10).

Surgical alteration of the eye also alters the half-life. The presence of the lens appears to slow removal of drug from the eye leading Maurice to suggest that it is a bottleneck in the elimination process (24,25). In studies of intravitreal amikacin, removal of the lens in rabbits decreases the half-life from 25.5 to 14.3 hours and decreases the half-life of intravitreal gentamicin from 32 to 12 hours (17,18). Removal of the lens and vitreous significantly shortens half-life for multiple antimicrobials. Meredith and associates studied amikacin and demonstrated that the half-life was reduced from 14.3 to 7.9 hours after removal of the lens and vitreous. Martin et al. (22) found that the cefazolin half-life was reduced from 8.3 to 6.0 hours when comparing clearance of intravitreal cefazolin from phakic eyes versus aphakic-vitrectomized eyes (22). The half-life of intravitreal amphotericin was reduced from 4.7 days in aphakic eyes to 1.4 days in eyes in which the lens and vitreous were removed in a study by Doft et al. (31).

The half-life of anteriorly excreted drugs is also decreased by inflammation, although the mechanism is not clear. Studies of amikacin demonstrated that in phakic eyes the half-life was diminished from 25.5 to 15.5 hours by inducing inflammation, while the aphakic eye half-life was reduced from 14.3 to 7.4 hours (17). Studies of gentamicin showed inflammation decreased the half-life from 32 to 19 hours (18). It is possible that inflammation increases posterior permeability and thereby allows these drugs to be eliminated by both anterior and posterior routes, accounting for the observed increase in the rate of elimination.

Simultaneous and continuing administration of intravenous or parenteral antimicrobials after intravitreal injection can theoretically favorably influence the amount of drug retained in the vitreous cavity. The intravitreal concentration of drug results from the amount given by intravitreal injection plus the inflow of drug through the aqueous or posterior structures, minus the amount lost from the eye through egress across the trabecular meshwork, across retinal structures by passive diffusion, or eliminated by active transport.

The entry of drug into the vitreous cavity is predominantly limited by permeability issues. If a drug is maintained within the vascular compartment for a sufficient period of time, it should reach distribution equilibrium with the concentration in

Table 2 Vitreous Volume and Aqueous Flow

Species	Vitreous volume (mL)	Aqueous flow (μL/min)
Rabbit	1.4–1.7	3.6
Cat	2.4	13.0
Dog	3.2	
Monkey	3.0–4.0	3.0
Man	3.9–5.0	2.5–3.0

Source: From Ref. 32.

the vitreous cavity equal to that in the plasma. Such equality is not always observed because of such factors as active transport and pH gradients across cell membranes.

The vitreous cavity volume affects the half-life of injected intravitreal antibiotics. Maurice has postulated that because of the increase in vitreous cavity volume in the human as compared to the rabbit, the time of the diffusion to the retinal surface is expected to be greater and the half-life of a drug in a human may be 1.7 times longer than in the rabbit vitreous cavity (24,25). Most data for actual concentrations of antimicrobial in the scientific literature do not correct for the difference in vitreous cavity volume (Table 2). The vitreous volume in the human eye is roughly 4 mL. Thus, on a concentration basis, a given amount of drug injected in the human eye will achieve an initial concentration of only $\sim$35% of the concentration in the rabbit eye.

THERAPEUTIC CONCENTRATION RANGE

The therapeutic window or therapeutic concentration range is a span of drug concentration over which therapy is effective without undue toxicity (26). The therapeutic concentration range for most drugs in plasma is narrow and the upper and lower limits differ by only a factor of two or three. The upper limit of the concentration range may be established by toxicity or in some cases by diminished effectiveness of the drug at higher concentrations. Toxicity may be totally unrelated to the therapeutic effect or may be an extension of the drug's pharmacologic properties. When the therapeutic range is narrow, it is more difficult to maintain values within the proper range. With intravitreal injections not only is the concentration within the vitreous cavity an issue but some studies have suggested that the concentration of the injected dose may be a factor in causing toxicity. Initial concentrations much higher than might be seen in other tissues may be achieved with an intravitreal dose. Initial injection of a milligram of vancomycin, for example, produces an initial concentration of $\sim$250 μg/mL whereas the minimum inhibitory concentration for most microorganisms being targeted is between 1 and 5 μg/mL. Since drug concentration begins to diminish soon after the injection has been performed, high initial doses are sometimes chosen in order to prolong the therapeutic effect. It should be noted, however, that doubling the dose of a drug will essentially only prolong its effect for one half-life.

TOXICITY

There is no standard definition of intravitreal injection toxicity. Histopathologic criteria are most frequently used in animal models. Various authors evaluate changes in the retina or retina pigment epithelium usually after a single injection of intra-vitreal antimicrobials. The rabbit has been chosen as the animal for toxicity studies for reasons of convenience and expense. The rabbit retina is merangiotic with less vasculature when compared to the holangiotic retina of higher-level primates. This difference in structure caused vascular infarction as a toxic reaction to aminoglyco-sides to go unrecognized until it occurred in humans. These changes are now thought to be the most common type of aminoglycoside toxicity as well as potentially the most damaging to vision.

Electroretinographic toxicity criteria have also been employed but concurrent controls in which a placebo is injected in the fellow eye simultaneously are important since surgical invasion of the eye alone can reduce the electroretinographic response (33). The antimicrobial itself, the vehicle, or the preservatives may all be sources of intraocular toxicity. Osmolality or pH have been suggested by Marmor as potential causes for iatrogenic tissue damage to the retina (34). Injection of low volumes of drug probably minimizes this risk.

Peyman et al. (35) hypothesized that the injection of antibiotic into an eye from which vitreous has been removed might increase the risk of toxicity. Lim demon-strated that the settling of aminoglycosides onto the retinal surface in vitrectomized eye did in fact predispose the eyes to higher local doses on the retinal surface and therefore to higher risk of toxicity (27).

Aminoglycosides have been the most widely studied class of antibiotics for their intraocular toxic potential. In the rabbit, Zachary and Foster (36) demon-strated dose-related damage to the outer retina with marked disruption of the outer nuclear layer and impressive loss of outer segments with gentamicin doses exceeding 0.2 mg (27). Retinal pigment epithelial mottling, pigment clumping, and individual areas of depigmentation were noted ophthalmoscopically. At doses of 0.4–0.5 mg, the electroretinogram did not provide any results. Electron microscopy specimens, after gentamicin intravitreal injections, were studied by Talamo et al. (37), who noted lamellar lysosomal inclusions similar to drug-induced lipid storage problems. They felt this suggested the retinal pigment epithelium (RPE) as a primary site for toxicity of gentamicin.

Conway and Campochiaro (38) identified the syndrome of macular vessel infarction in human eyes after gentamicin was injected intraocularly. A subsequent study by retinal specialists identified multiple similar cases secondary to both ami-kacin and gentamicin (39). In a monkey model, Conway injected 1000 µg of genta-micin (40). This resulted in a clinical picture of macular infarction characterized three days later by cotton-wool spots and intraretinal hemorrhages. There was strik-ing inner retinal layer damage observed by electron microscopy with less severe changes in the outer layers. Vascular changes could not be identified despite the clinical picture of obstruction. They suggested that the neurotoxic changes lead to shutdown of local blood flow, perhaps through granulocytic plugging.

Five aminoglycosides were studied by D'Amico et al. (41) after intravitreal injection, characterizing them using the clinical picture, histopathology, and electron microscopy. As noted by Talamo, they found the first abnormality produced was lysosomal overloading the RPE with lamellar lipid material. Toxic reactions in the outer retina were noted at doses of 1500 µg of amikacin and 400 µg of gentamicin. Macrophages with storage lysosomes were noted in the subretinal space along with disorganization of photoreceptor outer segments and focal necrosis of the retinal pigment epithelium. Later in the course of study, focal disappearance of photo-receptors and RPE with reactive gliosis was noted. Full-thickness retinal necrosis with corresponding destruction of the photoreceptor–RPE complex was noted by doubling these doses. They found that the relative antibiotic toxicities in this model were: gentamicin > netilmicin sulfate = tobramycin > amikacin = kanamycin sulfate.

Little work has been done to evaluate toxicity of repeated intraocular antibiotic doses but there is a suggestion that the risk for complications may be increased. Antibiotic toxicity is often related to peak dose and with multiple intraocular injections these high–peak doses are repeated. In a study of 1 mg of vancomycin along with 400 μg of amikacin or 100 μg of gentamicin, Oum et al. (42) found no toxicity after one injection (41). When two injections of each antibiotic were spaced 48 hours apart, three of the six amikacin-injected eyes and five of the six gentamicin-injected eyes demonstrated changes at multiple levels. Abnormalities of the retinal pigment epithelium, mild loss of photoreceptor interdigitation, and disorganization of the outer segments of the photoreceptors were documented. When a third injection was given 48 hours later, multiple white dots appeared ophthalmoscopically throughout the retina in almost 50% of the eyes that were given gentamicin and in over 20% of the eyes that received amikacin; these findings were not seen in the control eyes. Initially, retinal pigment epithelial disturbance was noted followed by photoreceptor outer segment disorganization. Focal disorganization of the retinal pigment epithelium was accompanied by hyperpigmentation and hypopigmentation.

CHARACTERISTICS OF SELECTED ANTIMICROBIALS

Vancomycin

Vancomycin is derived from *Streptomyces orientalis* and is a bacteriocidal agent. As it covers almost all the gram-positive organisms which are pathogens for exogenous endophthalmitis, vancomycin is perhaps the most commonly administered intravitreal antibiotic. Its coverage is limited to gram-positive organisms with no gram-negative effect, and therapeutic ranges are thought to be 1–5 μg/mL. Vancomycin inhibits synthesis and assembly of bacterial cell walls and also interferes with RNA synthesis. Dosages of 1 or 2 mg are used in the human eye without identifiable toxicity.

The half-life of vancomycin following intraocular injection is long, and ranges from 25.5 to 56 hours (14,15). Removal of the lens dramatically decreases the half-life to 9.8 hours (14). In inflamed eyes the half-life is slightly longer in the phakic eye and is slightly shorter in a model of infection (14,15).

In two studies, samples were obtained from the vitreous cavity after intraocular injection. Ferencz (43) identified a range of 25–182 μg 44–72 hours after injection. Gan (44) identified average levels of 10.3 μg/mL three days after injection. Rough estimates indicate this is consistent with the half-life of ~24 hours as suggested by data in the rabbit (44).

Betalactams

Betalactam antibiotics resemble penicillin with the exception of replacement of a five-member thiazolidine ring. Various modifications of these antimicrobials have been made to alter their penetration, pharmacokinetics, and range of efficacy.

The mechanisms of action of betalactams are complex. Betalactams are bactericidal and are known to inactivate specific targets on the inner aspects of bacterial cell membranes, thus making them inactive. Betalactams are time-dependent antimicrobials that have been successively developed into four generations. The first-generation agents are active against gram-positive cocci but not methicillin-resistant streptococci or penicillin-resistant *S. pneumoniae*. Second-generation agents are more potent against *Escherichia coli*, *Klebsiella*, and certain proteus species. The third-generation agents are the most potent against gram-negative organisms and are divided based on their efficacy against pseudomonas. They do not have the same degree of activity against gram-positive organisms as do first- and second-generation agents. Ceftazadime is the most widely used drug in the treatment of endophthalmitis but is being replaced by a fourth-generation agent, Cefapime. Ceftazadime has been shown to be safe in doses up to 2.25 mg after injection into the monkey vitreous cavity (45). Its spectrum of coverage against gram-negative organisms is similar to that of the aminoglycosides.

Clearance of first-generation cephalosporins after intravitreal injection suggests a posterior clearance route (23). Newer agents may be removed by the anterior route or by the combined posterior and anterior route (30). In phakic eyes, the half-life of cefazolin in the monkey and the rabbit is 6.5–7 hours (22,45). The half-life is decreased by inflammation, presumably by interference with active transport across posterior structures. In studies of third-generation agents (ceftizoxime and ceftriaxone), there was increased drug half-lives in infected rabbit eyes as compared to controls (30). Ceftazadime has a half-life of 13.8–20 hours in the rabbit, but the half-life is dramatically lowered by removal of the lens and vitreous (46).

Aminoglycosides

Aminoglycosides are semisynthetic analogs of *Actinomycetes* fungi products. They were initially popular for intraocular injection to treat endophthalmitis but have fallen into some disfavor because of intraocular vascular complications. Aminoglycosides are active against aerobic gram-negative bacilli and in higher concentrations are bactericidal against gram-positive pathogens including staphylococcal species. Aminoglycosides inhibit protein synthesis, although this action may not completely explain their effects. The recommended doses for intravitreal injection are gentamicin 0.1 mg and amikacin 0.2–0.4 mg. Vascular occlusive changes in the macula have been reported on occasion even at these dose ranges, however. A more complete view of toxicity studies in animals was discussed earlier (47,48).

Amikacin is eliminated anteriorly and has a half-life similar to that of vancomycin at 25.5 hours in the normal phakic eye (49). Inflammation decreases the half-life and removal of the lens significantly reduces the half-life to 14.3 hours in the rabbit although the effect is not quite as marked as the changes for vancomycin. Amikacin is a concentration-dependent antibiotic with a postantibiotic effect; that is, there is prolonged activity even after its concentrations have fallen below therapeutic levels.

Fluoroquinolones

Fluoroquinolones are becoming increasingly popular in ophthalmology because of their broad spectrum of coverage. DNA synthesis in susceptible bacterial cells is inhibited while mammalian cells are spared. Ciprofloxacin was initially thought to be effective against most strains of ocular pathogens including both gram-positive and gram-negative bacteria. Risk-resistant strains to ciprofloxacin appeared and it has more recently been replaced by second-generation agents such as ofloxacin and levofloxacin. Ciprofloxacin has been of little use for intraocular injection. It has been shown to be toxic to the rabbit retina in doses of 250 µg in one study, although another suggested that 500 µg was a safe dose (48,50,51). The half-life is only 2.2 hours in the normal phakic eye and 1.0 hour in aphakic eyes, suggesting rapid posterior elimination (52). Recent studies have suggested that injections of 500 µg of ofloxacin were tolerated in the rabbit (53).

Antifungals

Amphotericin B is the most frequently chosen antimicrobial for treatment of intra-ocular fungal infections. The usual intraocular dose of amphotericin B is 0.005 µg. There is a long half-life of 4.7 days in the phakic rabbit eye, but there is a significant reduction to 1.4 days by removal of the vitreous and the lens (31). Gupta studied injection of fluconazole in the rabbit demonstrating a very short half-life of only 3.08 hours (54). The half-life can be prolonged to 23.4 hours by entrapping the fluco-nazole in liposomes. In a study of efficacy in a rabbit model of *Candida* endophthal-mitis, 200-µg injections of plain fluconazole was associated with sterility of the vitreous cavity in 75% of the treated animals at day 16 after treatment. Lysosomal trapped fluconazole was not as efficacious in sterilizing the vitreous cavity (55).

REFERENCES

1. Baum J, Peyman GA, Barza M. Intravitreal administration of antibiotic in the treatment of bacterial endophthalmitis. III. Consensus. Surv Ophthalmol 1982; 26:204–206.
2. Endophthalmitis Vitrectomy Study Group. Results of the Endophthalmitis Vitrectomy Study: a randomized trial of immediate vitrectomy and of intravenous antibiotics for the treatment of postoperative bacterial endophthalmitis. Arch Ophthalmol 1995; 113:1479–1496.
3. Mann I. The intraocular use of penicillin. Br J Ophthalmol 1946; 30:134–136.
4. Von Sallman L, Meyer K, DiGrandi J. Experimental study on penicillin treatment of ectogenous infection of vitreous. Arch Ophthalmol 1944; 32:179–189.
5. Leopold IH, Scheie HG. Studies with microcrystalline sulfathiazole. Arch Ophthalmol 1943; 29:811.
6. Leopold IH. Intravitreal penetration of penicillin and penicillin therapy of infections of the vitreous. Arch Ophthalmol 1945; 33:211–216.
7. Liang C, Peyman GA, Sonmez M, Molinari LC. Experimental prophylaxis of *Staphylo-coccus aureus* endophthalmitis after vitrectomy: the use of antibiotics in irrigating solu-tion. Retina 1999; 19:223–229.
8. Barza M, Kane A, Baum J. Pharmacokinetics of intravitreal carbenicillin, cefazolin, and gentamicin in rhesus monkeys. Invest Ophthalmol Vis Sci 1983; 24:1602–1606.
9. Fisher JP, Civiletto SE, Forster RK. Toxicity, efficacy, and clearance of intravitreally injected cefazolin. Arch Ophthalmol 1982; 100:650–652.
10. Ficker LA, Meredith TA, Gardner SK, Wilson LA. Cefazolin levels after intravitreal injection: effects of inflammation and surgery. Invest Ophthalmol Vis Sci 1990; 31:502–505.
11. Shockley RK, Jay WM, Friberg TR, Aziz AM, Rissing JP, Aziz MZ. Intravitreal cef-triaxone in a rabbit model. Arch Ophthalmol 1984; 102:1236–1238.
12. Jay WM, Aziz MZ, Rissing JP, Shockley RK. Pharmacokinetic analysis of intravitreal ceftriaxone in monkeys. Arch Ophthalmol 1985; 103:121–123.
13. Pflugfelder SC, Hernandez E, Fleisler SJ, Alvarez J, Pflugfelder M, Forster RK. Intra-vitreal vancomycin. Arch Ophthalmol 1987; 105:831–837.
14. Aguilar HE, Meredith TA, El-Massry A, et al. Vancomycin levels after intravitreal injec-tion. Retina 1995; 15:428–432.

15. Park SS, Vallar RV, Hong CH, von Gunten S, Ruoff K, D'Amico DJ. Intravitreal dexamethasone effect on intravitreal vancomycin elimination in endophthalmitis. Arch Ophthalmol 1999; 117:1058–1062.

16. Nelson P, Peyman GA, Bennett TO. BB-K8: a new aminoglycoside for intravitreal injection in bacterial endophthalmitis. Am J Ophthalmol 1974; 78:82–89.

17. Meredith TA, Mandell BA, Aguilar EA, El-Massry A, Sawant A, Gardner S. Amikacin levels after intravitreal injection: effects of inflammation and surgery. Invest Ophthalmol Vis Sci 1992;33/34:747.

18. Cobo LM, Forster RK. The clearance of intravitreal gentamicin. Am J Ophthalmol 1981; 92:59–62.

19. Ben-Nun J, Joyce DA, Cooper RL. Pharmacokinetics of intravitreal injection: assessment of a gentamicin model by ocular dialysis. Invest Ophthalmol Vis Sci 1989; 30:1055–1061.

20. Gardner S. Treatment of bacterial endophthalmitis III. Ocul Ther Manage 1991; 2:14–19.

21. Barza M. Factors affecting the intraocular penetration of antibiotics. The influence of route, inflammation, animal species and tissue pigmentation. Scan J Infect Dis 1978; 14(suppl):151–19.

22. Martin DF, Ficker LA, Aguilar HA, Gardner SK, Wilson LA, Meredith TA. Vitreous cefazolin levels after intravenous injection: effects of inflammation, repeated antibiotic doses, and surgery. Arch Ophthalmol 1990; 108:411–414.

23. Barza M. Antibacterial agents in the treatment of ocular infections. Infect Clin N Am 1989; 3:533–551.

24. Maurice DM. Injection of drugs into the vitreous body. In: Leopold IJ, Burns RP, eds. Symposium on Ocular Therapy. New York: John Wiley, 1976:59–72.

25. Maurice DM, Mishima S. Ocular pharmacokinetics. In: Sears ML, ed. Pharmacology of the Eye. New York: Springer-Verlag, 1984:19–116.

26. Rowland M, Tozer TN. Clinical Pharmacokinetics. Lea and Febiger. Philadelphia, PA: Saunders, 1989.

27. Lim JI, Anderson CT, Hutchinson A, Buggage RR, Grossniklaus HE. The role of gravity in gentamicin-induced toxic effects in a rabbit model. Arch Ophthalmol 1994; 112: 1363–1367.

28. Friedrich S, Saville B, Cheng YL. Drug distribution in the vitreous humor of the human eye: the effects of aphakia and changes in retinal permeability and vitreous diffusivity. J Ocul Pharmacol Ther 1997; 13:445–459.

29. Tojo KJ, Ohtori A. Pharmacokinetic model of intravitreal drug injection. Math Biosci 1994; 123:59–75.

30. Barza M, Lynch E, Baum JL. Pharmacokinetics of newer cephalosporins after subconjunctival and intravitreal injection in rabbits. Arch Ophthalmol 1993; 111:121–125.

31. Doft BH, Weiskopf J, Nillson-Ehle I, Wingard L. Amphotericin clearance in vitrectomized versus non-vitrectomized eyes. Ophthalmology 1985; 92:1601–1605.

32. Barza, M. Animal models in evaluation of chemotherapy of ocular infections. In: Zak O, Sande MA, eds. Experimental Models in Antimicrobial Chemotherapy. London: Harcourt Brace Jovanovich, 1986:187–211.

33. Meredith TA, Lindsey DT, Edelhauser HF, Goldman AI. Electroretinographic studies following vitrectomy and intraocular silicone oil injection. Br J Ophthalmol 1985; 69:254–260.

34. Marmor MF. Retinal detachment from hyperosmotic intravitreal injection. Invest Ophthalmol Vis Sci 1979; 18:1237–1244.

35. Peyman GA, Vastine DW, Raichand M. Postoperative endophthalmitis: experimental aspects and their clinical application. Ophthalmology 1978; 85:374–385.

36. Zachary IG, Forster RK. Experimental intravitreal gentamicin. Am J Ophthalmol 1976; 82:604–711.

37. Talamo JH, D'Amico DJ, Hanninen LA. The influence of aphakia and vitrectomy on experimental retinal toxicity of aminoglucoside antibiotics. Am J Ophthalmol 1985; 100:840–847.
38. Conway BP, Campochiaro PA. Macular infarction after endophthalmitis treated with vitrectomy and intravitreal gentamicin. Arch Ophthalmol 1986; 104:367–371.
39. Campochiaro PA, Conway BP. Aminoglycoside toxicity: a survey of retinal specialists. Arch Ophthalmol 1991; 109:946–950.
40. Conway BP, Tabatabay CA, Campochiaro PA, D'Amico DJ, Hanninen LA, Kenyon KR. Gentamicin toxicity in the primate retina. Arch Ophthalmol 1989; 107:107–112.
41. D'Amico DJ, Caspers-Velu L, Libert J, et al. Comparative toxicity of intravitreal aminoglycoside antibiotics. Am J Ophthalmol 1985; 100:264–275.
42. Oum BS, D'Amico DJ, Wong KW. Intravitreal antibiotic therapy with vancomycin and aminoglycoside. An experimental study of combination and repetitive injections. Arch Ophthalmol 1989; 107:1055–1060.
43. Ferencz JR, Assia EI, Diamantstein L, Rubinstein E. Vancomycin concentration in the vitreous after intravenous and intravitreal administration for postoperative endophthalmitis. Arch Ophthalmol 1999; 117:1023–1027.
44. Gan IM, van Dissel JT, Beekhuis WH, Swart W, van Meurs JC. Intravitreal vancomycin and gentamicin concentrations in patients with postoperative endophthalmitis. Br J Ophthalmol 2001; 85:1289–1293.
45. Campochiaro PA, Green WR. Toxicity of intravitreal ceftazidime in primate retina. Arch Ophthalmol 1992; 110:1625–1629.
46. Meredith TA. Antimicrobial pharmacokinetics in endophthalmitis treatment; studies of ceftazidime. Trans Am Ophth Soc 1993; 91:653–699.
47. Campochiaro PA, Lin JI, Group AS. Aminoglycoside toxicity in the treatment of endophthalmitis. Arch Ophthalmol 1994; 112:48–53.
48. Galloway G, Ramsay A, Jordan K, Vivian A. Macular infarction after intravitreal amikacin: mounting evidence against amikacin. Br J Ophthalmol 2002; 86:359–360.
49. Mandell BA, Meredith TA, Aguilar E, El-Massry A, Sawant A, Gardner S. Effects of inflammation and surgery on amikacin levels in the vitreous cavity. Am J Ophthalmol 1993; 115:770–774.
50. Stevens SX, Fouraker BD, Jensen HG. Intraocular safety of ciprofloxacin. Arch Ophthalmol 1991; 109:1737–1743.
51. Wiechens B, Grammer JB, Johannsen U, Pleyer U, Hedderich J, Duncker GI. Experimental intravitreal application of ciprofloxacin in rabbits. Ophthalmologica 1999; 213:120–128.
52. Pearson PA, Hainsworth DP, Ashton P. Clearance and distribution of ciprofloxacin following intravitreal injection. Retina 1993; 13:326–330.
53. Wiechens B, Neumann D, Grammer JB, Pleyer U, Hedderich J, Duncker GI. Retinal toxicity of liposome-incorporated and free ofloxacin after intravitreal injection in rabbit eyes. Int Ophthalmol 1998; 22:133–143.
54. Gupta SK, Velpandian T, Dhingra N, Jaiswal J. Intravitreal pharmacokinetics of plain and liposome-entrapped fluconazole in rabbit eyes. J Ocul Pharmacol Ther 2000; 16:511–518.
55. Gupta SK, Dhingra N, Velpandian T, Jaiswal J. Efficacy of fluconazole and liposome entrapped fluconazole for C. albicans induced experimental mycotic endophthalmitis in rabbit eyes. Acta Ophthalmol Scand 2000; 78:448–450.

7

Pharmacologic Retinal Reattachment with INS37217 (Denufosol Tetrasodium), a Nucleotide P2Y$_2$ Receptor Agonist

Ward M. Peterson
Department of Biology, Inspire Pharmaceuticals, Durham, North Carolina, U.S.A.

DESCRIPTION OF DRUG DELIVERY SYSTEM

INS37217 (denufosol tetrasodium) is an investigational new drug administered as an intravitreous injection into the midvitreal cavity of the affected eye at a volume of 0.05 mL using standard intravitreous injection procedures. At the present stage of clinical development, INS37217 is administered using a sterile, appropriately sized tuberculin syringe attached to a sterile 30-gauge needle. INS37217 is formulated in a physiological phosphate buffered saline (PBS) adjusted to a pH of ~7.3 and a tonicity of ~290 mOsm. The final formulated drug product is a true aqueous solution that is clear, colorless, and particulate free. An aqueous formulation of INS37217 in a smaller prefilled syringe with affixed needle capable of holding and administering a small dose volume is under consideration for later development.

SPECTRUM OF DISEASES

The ability to reattach the retina in the treatment of retinal detachment is contingent on removing the pathological accumulation of subretinal fluid between the retina and the underlying retinal pigment epithelium (RPE). Although most of the fluid can be removed using surgical techniques, it is ultimately the contribution of the RPE fluid "pump" machinery that enables complete subretinal fluid reabsorption and full anatomical reattachment. A number of ion channels and transporters, cell-surface receptors, extracellular signaling molecules, and intracellular signaling pathways have been shown to influence ion and fluid transport across the RPE (1). One such cell-surface receptor is the G protein-coupled P2Y$_2$ receptor, which is normally activated by extracellular adenosine 5′-triphosphate (ATP) and uridine 5′-triphosphate (UTP) and functionally expressed at the RPE apical membrane (2). Previous in vitro work has demonstrated that activation of this receptor by ATP, UTP, or a novel synthetic, hydrolysis-resistant agonist INS37217 (Fig. 1),

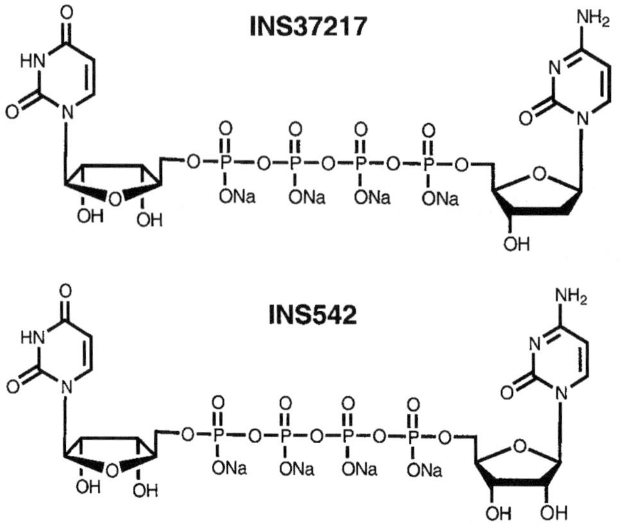

Figure 1 Structure of INS37217 and INS542. INS37217 is a deoxcytidine–uridine dinucleotide [P^1-(Uridine 5′)-P^4-(2′-deoxycytidine 5′)tetraphosphate, tetrasodium salt] and INS542 is a cytidine–uridine dinucleotide [P^1-(Uridine 5′)-P^4-(cytidine 5′)tetraphosphate, tetrasodium salt]. The only difference between the two molecules is the absence of the OH-group in the 2′ position of the cytidine ribose in INS37217. Both molecules have nearly identical activity in a variety of pharmacological and metabolism assays.

stimulates ion transport and fluid absorption (i.e., apical-to-basolateral or subretinal-to-choroidal direction) across RPE monolayers (2,3). Intravitreous injection of INS37217 was also shown to stimulate subretinal fluid reabsorption in experimental models of induced retinal detachment in vivo (3,4). Thus, INS37217 may provide therapeutic benefit in the treatment of retinal detachment by stimulating reabsorption of subretinal fluid. This chapter reviews the background, clinical rationale, preclinical findings, and drug delivery issues relating to the clinical development of INS37217, administered as an intravitreous injection for the treatment of retinal detachment.

Retinal detachments are generally categorized as rhegmatogenous, traction, or exudative, depending on the underlying primary mechanism that leads to subretinal fluid accumulation (5). Rhegmatogenous retinal detachments (RRD) are most common and occur as a result of single or multiple retinal breaks that permit vitreal fluid to enter into the subretinal space at a rate exceeding that of the fluid reabsorption rate of the RPE. Tractional retinal detachments, such as those secondary to proliferative diabetic retinopathy or retinopathy of prematurity, occur when vitreoretinal contractile forces pull the retina away from the RPE without necessarily creating a retinal tear. Exudative retinal detachments occur in the absence of evident retinal breaks or tractional forces and are thought to result in part from an imbalance of RPE transport that favors subretinal fluid accumulation. For example, a reversal in the direction of the RPE fluid pump from absorption to secretion is thought to lead to the progression of exudative retinal detachment (6,7).

Three surgical approaches are used to reattach the retina in the treatment of RRD: scleral buckle, pneumatic retinopexy, and vitrectomy. When used singly or in various combinations, these surgeries have been reported to be successful in achieving anatomical reattachment rates of 60–95% following initial operation and >95% following repeat operations (8,9). Depending on specific causes of the detachment and other factors (such as the ability of the patient to comply with the demands of surgical follow-up), the time to achieve retinal reattachment usually takes one or two days, but can on occasion take more than a week. When the retina is sufficiently flattened against the RPE, the retinal breaks are closed and repaired using cryotherapy or laser photocoagulation. Reattachment can be facilitated by additional procedures such as mechanically draining the subretinal fluid or injecting gas (such as sulfur hexafluoride, SF_6) into the vitreous (10,11). For example, a common adjunctive procedure used in scleral buckle surgery comprises posterior insertion of a small needle through the sclera, choroid, and RPE directly into the subretinal space to drain most of the extraneous fluid. In pneumatic retinopexy and pneumatic buckle surgeries, expanding gas is injected into the vitreous and the patient positions himself or herself postoperatively such that the surface tension of the gas acts as a tamponade to block vitreal fluid entry through the retinal break, and the gas buoyancy acts to flatten the retina against the RPE. Sometimes this effect is enhanced by having the patient perform specific head movements to help facilitate the gas bubble forcing fluid out from around the tear in a technique referred to as the "steam roller." These mechanical procedures have proven very useful to aid in the reattachment process but are associated with significant surgical risk, patient morbidity, and protracted periods of convalescence. For example, serious complications such as subretinal hemorrhage and retinal perforation are associated with the drainage procedure (9,12). Successful pneumatic retinopexy requires that the patient is able to comply with a rigorous postoperative period of precise head positioning that can last for a few days, and the gas bubble itself remains in the vitreous cavity for a few weeks, during which period the patient's mobility—such as air travel—is limited (13). Thus, pharmacological stimulation of subretinal fluid reabsorption by INS37217 may provide significant clinical benefits to patients by reducing the need for these invasive procedures. If sufficiently robust, INS37217 may also provide adequate reattachment in a subset of RRD patients such that the surgeon can repair the break with cryotherapy or laser photocoagulation without the need for reattachment surgery.

MECHANISM OF ACTION

Some of the known or expected mechanisms of action of INS37217 on RPE physiology are summarized in Figure 2 (3). Previous in vitro work on freshly isolated RPE monolayers has shown that binding of INS37217 (or UTP) to the P2Y$_2$ receptor at the apical membrane stimulates active ion transport, which provides the major osmotic driving force for fluid absorption across the RPE (2,3). Chloride is the predominant ion-mediating active fluid transport across the RPE (1). Net apical-to-basolateral transport of Cl$^-$ occurs as a result of polarized distribution of specific ion channels and transporter proteins at both membranes. Chloride enters the apical membrane via Na$^+$, K$^+$, and Cl$^-$ cotransporter proteins and exits the basolateral membrane via Cl$^-$ channel proteins. INS37217 and other P2Y$_2$ receptor agonists have been shown to stimulate an increase in cytosolic Ca^{2+}, which in turn increases basolateral membrane Cl$^-$ conductance and decreases apical membrane K$^+$ conductance (2). This is expected to result in net absorption of Cl$^-$, which along with a counterion (most likely Na$^+$ across the paracellular pathway) drives osmotically coupled fluid transport in the apical-to-basolateral direction.

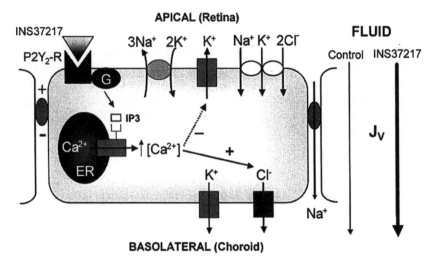

Figure 2 A diagram summarizing the known and expected effects of INS37217 on RPE ion and fluid transport. Binding of P2Y$_2$ receptor (P2Y$_2$-R) by INS37217 at the apical membrane activates heterotrimeric G proteins and generates intracellular inositol 1,4,5 trisphosphate (IP3), which releases Ca^{2+} from intracellular endoplasmic reticulum (ER) stores. Elevation of cytosolic Ca^{2+} in turn leads to an increase in basolateral membrane Cl$^-$ conductance, a decrease in apical membrane K$^+$ conductance, and stimulation of net apical-to-basolateral fluid absorption.

ANIMAL MODELS OF DISEASE USED

The effects of INS37217 on subretinal fluid reabsorption were evaluated in mice, rats, and rabbits by injecting saline solution into the subretinal space using a small needle (32 gauge or less), which produces a non-RRD because the induced retinotomy appears to seal itself immediately (3). This model of induced retinal detachment was chosen because RRDs occur too infrequently in animals to be useful for preclinical proof-of-concept studies.

Retinal detachments were induced in Long–Evans rats by inserting a guidance needle behind the limbus and into the vitreous, and then inserting a smaller flat-tip needle directly into the barrel of the guidance needle. The flat-tip needle was attached to a Hamilton syringe containing modified PBS solution, of which ~3 μL was injected directly into the subretinal space to create the detachment. (For more details on the surgical procedure, see Ref. 3.) The modified PBS solution was formulated to contain an empirically chosen balance of ions and pH that allowed the induced "subretinal blebs" to remain relatively constant in size, at least for an initial 24-hour period. A masked investigator used indirect ophthalmoscopy techniques adapted for rat eyes to evaluate the extent of the induced detachments, which initially comprised 20–30% of the total retinal surface area. Following the creation of a retinal detachment, a subsequent intravitreous injection (3 μL) of PBS, with or without INS37217, was given to evaluate their effects of subretinal fluid reabsorption.

The effects of INS37217 on subretinal fluid reabsorption in rabbits were made in a similar manner (4). As with the rat studies, a single non-RRD was produced in New Zealand white rabbits by injecting modified PBS (~50 μL administration volume) solution in the subretinal space, which resulted in a detachment that was less than 10% of the retinal surface area of the rabbit eye. Immediately following the creation of the subretinal bleb, an intravitreous injection (50 μL) of saline alone or saline containing INS37217 was administered into the vitreous directly above the detachment. Masked investigators viewed the fundus to quantify the extent of retinal detachment and reattachment by using the nearby optic disk as a size marker.

Non-RRDs were induced in mouse eyes to evaluate the effects of subretinal delivery INS37217 on recovery of electroretinography (ERG) function following experimental retinal detachment and spontaneous reattachment. Subretinal injection was conducted using an anterior approach through the cornea. In brief, a 28-gauge beveled hypodermic needle was used to puncture the cornea, avoiding any contact with the lens, and a subretinal injection was conducted using a 33-gauge blunt needle and the transvitreal approach. One microliter of saline solution alone or saline solution containing INS37217 was then injected into the subretinal space. Extent of induced retinal detachment was estimated by adding fluorescent microbeads into the subretinally injected saline solution and monitoring the distribution of fluorescent signal histologically and in eyecup preparations.

RESULTS OF ANIMAL MODEL STUDIES

In the rat model of induced non-RRD, an intravitreous injection of PBS solution containing INS37217 into the vitreous was shown to significantly enhance subretinal fluid reabsorption when compared with vehicle (PBS) alone, and the effects of INS37217 were apparent even at one hour following administration (3), as shown in Figure 3. In contrast, intravitreous injection of PBS solution containing UTP

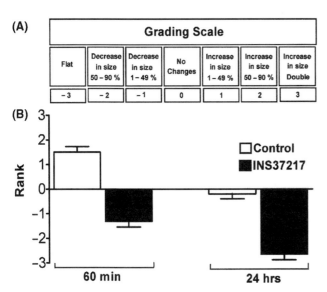

Figure 3 (A) Grading scale used to subjectively quantify the effects of INS37217 versus vehicle on retinal detachment in a rat model of induced nonrhegmatogenous retinal detachment. INS37217 (5 mM) was administered as a 3-μL intravitreous injection. INS37217-containing solution and vehicle solution were formulated to equal tonicity and pH in physiological saline. Subjective evaluation of retinal detachment and reattachment was conducted under investigator-masked conditions. (B) Mean placebo and INS37217 results from 12 experiments conducted by the same investigator. The mean ± SEM of the estimated rank for placebo and INS37217 experiments are plotted at 60 minutes and 24 hours. At both 60 minutes and 24 hours, the mean of the placebo and INS37217 data are significantly different ($P < 0.005$; two-tailed Mann–Whitney test).

did not stimulate subretinal fluid reabsorption, perhaps owing to UTP's metabolic instability and its increased likelihood for degradation by the retina (discussed later). These findings represent the first in a series of proof-of-concept findings for the use of intravitreously administered INS37217 to reabsorb extraneous subretinal fluid.

Confirming the effects seen in rats, rabbit intravitreous delivery of INS37217 was shown to significantly enhance subretinal fluid reabsorption in a dose-dependent manner when compared with vehicle control (Fig. 4). Optical coherence tomography (OCT) techniques were used to image retinal detachments in these rabbit studies and to provide an independent, qualitative confirmation of the topographic observations made by indirect ophthalmoscopy. Time-lapsed OCT images of subretinal blebs taken from an animal treated with INS37217 in one eye and modified PBS solution in the other eye revealed an initial dome-shaped elevation of the retina immediately following the creation of a subretinal bleb (Fig. 5). During the early post-operative

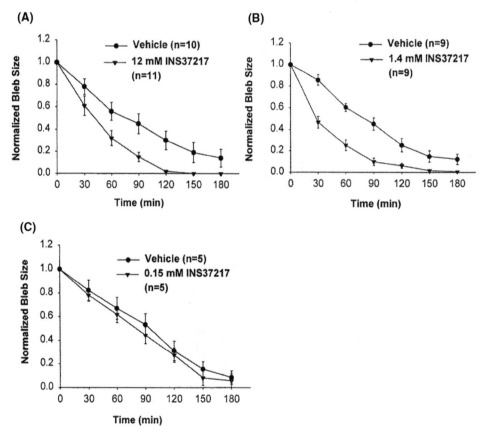

Figure 4   Effects of a single 50-μL intravitreous injection of INS37217-containing solution at concentrations of (**A**) 12 mM, (**B**) 1.4 mM, and (**C**) 0.15 mM versus vehicle on retinal reattachment in a rabbit model of induced nonrhegmatogenous retinal detachment. INS37217-containing solution and vehicle solution were formulated to equal tonicity and pH in physiological saline. Retinal detachment was first induced by injecting a ~50-μL volume solution of modified PBS using a 29-gauge needle into the subretinal space. This was immediately following by injection of INS37217 into the vitreous. Results show that INS37217 administered at 12 and 1.4 mM, but not at 0.15 mM, increased the rate of clearance of subretinal blebs when compared with vehicle control.

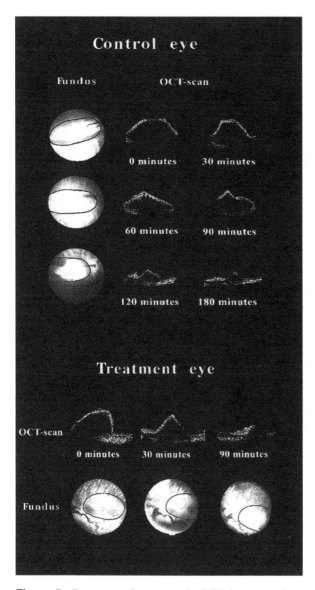

Figure 5 Representative grayscale OCT images and corresponding fundus photographs of induced retinal detachment taken from an animal injected intravitreously (50 μL) with 12 mM INS37217 ("treatment eye") and another animal treated with vehicle ("control eye"). The initial, elliptical border representing the visible contour of each subretinal bleb in the fundus images at baseline (pre-INS37217 or prevehicle treatment) is drawn in, and the border is drawn over the same corresponding areas in the follow-up images. In the control eye, fundus photographs were taken at baseline and at 60 and 120 minutes post-treatment, and OCT scans were taken every 30 minutes. Subretinal blebs in the control eye were initially dome-shaped and assumed a more concave contour during the post-treatment period. Subretinal fluid appeared to be largely reabsorbed by 180 minutes. In the treatment eye, fundus photographs and OCT images were taken at baseline and at 30 and 90 minutes post-treatment. The initial dome-shaped retinal detachment assumed a more triangular profile at 30 minutes postinjection, and by 90 minutes the subretinal bleb was no longer visible.

period the bleb lost the convex contour and the surface became irregular. Subretinal fluid appeared largely resolved by 90 minutes in the INS37217-treated eye and 180 minutes in the vehicle-treated eye, thus confirming observations made by indirect ophthalmoscopy. OCT imaging revealed the development of small retinal folds as subretinal fluid reabsorbed.

The effects of INS37217 on recovery of ERG function were evaluated in the mouse following experimental retinal detachment and spontaneous reattachment. Because of the small size of a mouse eye, a subretinal injection of 1-μL saline solution resulted in a relatively large retinal detachment. This was clearly demonstrated, for example, by adding fluorescent microbeads to the subretinally injected solution. A single 1-μL injection of saline solution containing fluorescent microbeads detached most of the mouse retina and distributed the microbeads to almost all of the subretinal space (14). It was noted that within 24 hours following a subretinal injection, grossly evident retinal reattachment accompanied by extensive retinal folding was observed. The retinal folding generally resolved within a week following the induced detachment, and histological evaluations revealed that the retina was reattached at this time. However, the time course of recovery of retinal function as determined by ERG responses dramatically lagged behind the time course for morphological reattachment, as was the case seen in a previous study in cats (14,15). For example, the recovery of dark-adapted a-wave ERG amplitudes in mice at 14 days following induced retinal detachment was only ~60% of contralateral, mock-surgery control eyes evaluated at the same time. (Mock-surgery eyes received all surgical manipulations except for the actual subretinal injection.) Subretinal injection of 1 μL saline solution containing 10 μM of INS37217 dramatically reduced the extent of retinal folding associated with induced detachments and significantly enhanced recovery of scotopic a- and b-wave amplitudes at 1 and 10 days postinjection, when compared with saline-injected controls (Fig. 6). Thus, INS37217 markedly improved post-reattachment ERG function in this model of induced retinal detachment.

The rat, rabbit, and mouse retinal detachment studies described previously strongly suggest, but do not directly demonstrate, that INS37217 stimulates active transport across the RPE in vivo. Therefore, additional studies using the noninvasive technique of differential vitreous fluorophotometry (DVF) were conducted with a similar $P2Y_2$ receptor agonist (INS542, Fig. 1) in rabbit eyes to demonstrate direct stimulation of RPE-active transport and to assess the duration of pharmacological action (16). Previous studies have shown that following systemic administration of fluorescein, both fluorescein (F) and its metabolite fluorescein glucuronide (FG) initially diffuse inwardly across the blood–retinal barrier (BRB) and accumulate in the vitreous. After two to three hours following systemic administration of F, vitreal F and FG are then transported outwardly back to the systemic circulation (17). Although both vitreal F and FG can passively diffuse outward across the BRB, the majority of the outward F movement and a smaller part of FG movement depend on an active transport mechanism in the RPE (18). Vitreal F and FG, both of which are differentially fluorescent, can be spectrally resolved and quantified using DVF techniques. Thus, the measurements of fluorescence from F and FG using DVF and calculations of the resultant F/FG ratios (at two or more hours following systemic administration of F) provide a measure of the outward active transport of F across the BRB at the level of the RPE (19). For example, an increase in active F transport across the RPE results in less F in the vitreous and thus a smaller F/FG ratio. Figure 7 shows that intravitreous injection of INS542 in intact rabbit eyes reduced F/FG ratios beginning as early as 30 minutes following administration

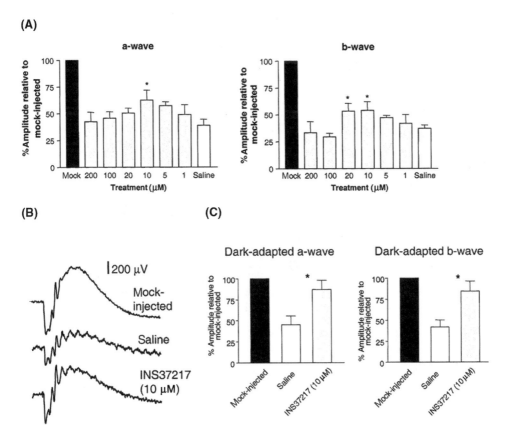

Figure 6 (**A**) Results summarizing the effects of subretinally administered INS37217 (1–200 µM), compared with vehicle (saline) and mock-injected controls, on a- and b-wave amplitudes measured from scotopic ERG recordings taken at one day following an induced nonrhegmatogenous retinal detachment in normal mice. In these experiments, INS37217 was directly added to the saline solution used for subretinal injection. Note that mock-injected eyes did not receive actual subretinal injections but otherwise received all other surgical manipulations as INS37217- and saline-control-treated eyes. The effects of INS37217 show a "bell-shaped" dose response with an optimal improvement of ERG function observed at the 10 µM dose. (**B**) Representative dark-adapted ERG waveforms from mock-, saline-, and INS37217-injected eyes recorded at 10 days following surgical treatment. (**C**) Results summarizing the amplitude of scotopic a- and b-wave ERG responses from mock-, saline-, and INS37217-injected eyes at 10 days following treatment.

and the pharmacological effect was evident for at least the initial 12 hours. These results therefore indicate that INS542 stimulates active transport of F across the RPE. Insofar as the active transport of F can be taken as a probe for active fluid and ion transport, these results further support the notion that the RPE is the direct in vivo INS542 (and INS37217) pharmacological target.

DRUG DELIVERY AND DISTRIBUTION

From a drug delivery perspective, localization of P2Y$_2$ receptors at the RPE apical membrane requires that INS37217 must be present in the subretinal space to bind to

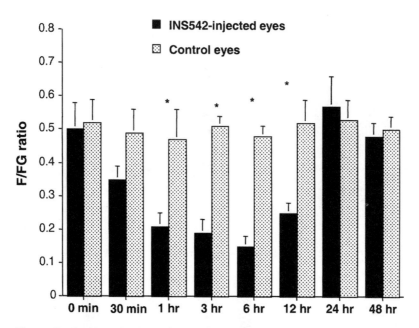

Figure 7 A comparison of F/FG ratios in treated rabbit eyes injected with 1.0 mM INS542 and contralateral, untreated eyes at baseline (0 min) and at 0.5, 1, 3, 6, 12, and 24 hours after vitreous injection of INS542. These rabbit eyes were intact insofar as no retinal detachments were induced in these studies. The F/FG ratios in INS542-treated eyes are significantly smaller than contralateral eye controls at the time points labeled with an asterisk ($P < 0.05$) (see text for details).

the target receptor. Thus, delivery of INS37217 to the site of action can feasibly be achieved using subretinal or intravitreous injection techniques in the clinic. Obvious practical difficulties are associated with delivering drugs via subretinal injection in the clinic, including both novelty and difficulty of approach and the clear potential exacerbation of detachment. Thus, intravitreous injection of a small volume (such as 0.10 mL or less) represents a much more reasonable approach for drug administration. INS37217 would need to remain intact as it diffuses across the retina to reach the apical membrane of the RPE. In RRD, the presence of single or multiple retinal tears or holes affords an additional passageway for compound diffusion into the subretinal space. ATP and UTP are highly labile compounds that are rapidly degraded by extracellular ectonucleotidases (20). INS37217 is a synthetic dinucleotide that is engineered with improved metabolic stability when compared with ATP and UTP (21). Previous work has shown that INS37217 is approximately four times more stable than UTP in retinal tissue (22).

To track the ocular biodistribution of ^{3}H-INS37217 and its radiolabeled metabolites, Dutch-belted rabbits were given a single intravitreous administration of ^{3}H-INS37217 and eyes were sectioned and processed for autoradiography for up to 48 hours postadministration (22). Figure 8 shows that the ^{3}H-signal distributed throughout the vitreous and retina within 15 minutes postinjection. Time-dependent signal localization was detected throughout the vitreous, retina, and ciliary body/iris during the 24 hour postadministration period. The radioactivity in the anterior and posterior chambers was sometimes absent at 15 minutes or 2 hours postdose, was at the highest level six hours postdose, and either decreased or was absent at 24 hours

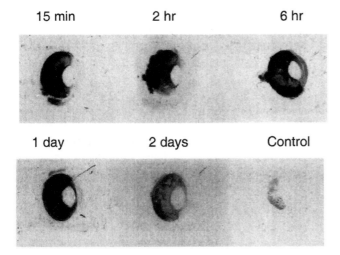

Figure 8 Representative autoradiographic images taken from cross sections of rabbit eyes injected intravitreously with radiolabeled INS37217 (^{3}H-INS37217 at 3 mg per eye) showing the distribution of radiolabeled signal in various ocular structures at the postinjection time points indicated. Radioactivity from INS37217 or its metabolites is distributed throughout the entire vitreous within 15 minutes and is largely absent by 48 hours.

postdose. The signal was only slightly above background levels at the 48 hour time point. No radioactivity was observed in the cornea, lens, choroid/sclera, and optic nerve of any eyes at any time point. Thus, the biodistribution results here for ^{3}H-INS37217 and its metabolites are in reasonable accordance with the pharmacodynamic data from the rabbit DVF studies described earlier.

CLINICAL STUDY

INS37217 is currently in clinical development for the treatment of RRD. Preliminary results of a Phase I clinical study on the tolerability and preliminary efficacy of INS37217 in 14 patients with RRD were presented in 2003 (23). The study was a randomized, placebo-controlled, double-masked, dose-escalation comparison of INS37217 to placebo (balanced saline solution). Three doses were evaluated in the study, 0.12, 0.24, and 0.48 mg. Both INS37217 and placebo were delivered as a single intravitreous injection (0.05 or 0.10 mL). The study consisted of two phases: the pharmacologic activity phase and the safety follow-on phase. The pharmacologic activity phase assessed the action of a single dose of INS37217 intravitreal injection versus placebo during the first 24 hours after dosing. The safety follow-on phase provided for monitoring of the subjects for one year to ensure no acute or chronic toxicities.

The purpose of this trial was to assess the tolerability of INS37217 when administered as a single intravitreal injection in subjects with RRD. Only patients with macula-on RRD were enrolled in the study. The secondary objective of this trial was to determine the pharmacologic activity of INS37217 by assessing its ability to clear extraneous fluid from the subretinal space and thereby facilitate retinal reattachment with a single injection. Subjects that responded positively to INS37217 received treatment for repairing the retinal tear, such as laser photocoagulation or cryopexy. Subjects that did not respond to treatment proceeded to rescue therapy of pneumatic retinopexy (PR). The effect of INS37217 or placebo on the extent of retinal detachment was evaluated using two independent, quantitative measures. One measure involved quantifying the extent of retinal detachment using fundus examination, and the second measure involved quantifying extent and height of retinal detachment using B-scan ultrasound images of the eye. Fundus and B-scan evaluations were conducted under conditions in which the identity of the drug versus placebo was masked.

INS37217 was well tolerated at all doses tested with no drug-related serious adverse events reported in the study. There was no evidence of systemic or ocular toxicity, endophthalmitis, or maculopathy associated with INS37217 treatment. When compared with placebo-treated eyes, INS37217-treated eyes showed a greater decrease in extent of retinal detachment, as observed using both direct fundus examination and B-scan ultrasound. One subject receiving 0.12 mg INS37217 did not require PR to reattach the retina and was treated with cyropexy to repair the tear. All other subjects required PR prior to repair of the retinal tear. Retinal reattachment was achieved in all subjects following PR therapy. Four cases of retinal redetachment in the study eye were observed at varying time points during the one-year observation period following the initial repair. The frequency of redetachment is consistent with the published literature of redetachment rates (13). Although all randomized subjects had a macula-on retinal detachment and were therefore at high risk for development of a macula-off detachment, none of the subjects progressed to a macula-off detachment. Further details of the results of this clinical study will be revealed at a later date. A larger Phase II clinical study took place in 2004 and 2005.

FUTURE HORIZONS

Surgeries to repair retinal detachment are generally successful in terms of achieving ophthalmoscopically evident anatomical reattachment. However, this anatomical reattachment frequently does not produce a commensurate full restoration in visual function. In macula-off detachments, successful reattachment resulted in only ~20% of patients achieving better than 20/50 visual acuity (24). Enhancing retinal reattachment or preventing the progression of a macula-on detachment to a macula-off detachment via pharmacological means may improve visual outcomes in RRD. There are additionally retinal conditions, such as central serous retinopathy, that cannot be treated with surgical approaches and also may be amenable for pharmaceutical intervention. No pharmaceutical agents are currently approved as part of standard treatment of retinal detachments, and the ability to define efficacy outcome measures in pivotal clinical trials may prove challenging because of the novelty of this proposed treatment modality. The following list provides a number of efficacy measures that are clinically meaningful and perhaps achievable with INS37217's pharmacological mechanism of action:

- Improve surgical outcomes in terms of reattachment rates and frequency.
- Eliminate the need for surgery in limited cases of RRD (such as those involving shallow detachments, pinhole tears, or detachments with negligible tractional component).
- Eliminate or reduce the need for adjunctive procedures in surgery (such as drainage or pneumatic procedures in scleral buckle surgery).

- Improve visual outcome following reattachment surgery by resolving persistent accumulation of subfoveal fluid (25).
- Treat other disorders of the retina associated with intraretinal or subretinal fluid build-up, including central serous retinopathy, central and branch retinal vein occlusion, and cystoid and diabetic macular edema.
- Improve the time course for retinal reattachment in macular translocation surgery (26,27).

Thus, there exist a variety of predominantly acute edematous retinal disorders that may be amenable for treatment with an intravitreous injection of INS37217. For treatment of chronic conditions such as cystoid or diabetic macular edema, alternative means of intravitreous delivery, such as intravitreous insert or implant or a sustained-release formulation, will likely be required.

ACKNOWLEDGMENTS

I wish to thank the following principal investigators, their scientific personnel, and institutions for supporting the preclinical and early clinical development of this project: Sheldon Miller, Ph.D. (University of California, Berkeley), Glenn Jaffe, M.D., and Cynthia Toth, M.D. (Duke University Eye Center), Muna Naash, Ph.D. (University of Oklahoma Health Sciences Center), Taiichi Hikichi, M.D. (Asahikawa Medical College), Paul Tornambe, M.D., and Lon Poliner, M.D. (Retina Consultants, San Diego), Greg Fox, M.D., and Brett King, O.D. (Retina Associates, Kansas City) and Michael Barricks, M.D. Thanks to Mark Vezina and Gianfranca Piccirilli at ClinTrial BioResearch, Ltd., for managing additional preclinical toxicology and biodistribution studies. Thanks, also, to Ramesh Krishnamoorthy, Amy Schaberg, and Robin Sylvester for reviewing and editing parts of this chapter, and Inspire Pharmaceuticals for supporting the development of this program.

REFERENCES

1. Hughes BA, Gallemore RP, Miller SS. Transport mechanisms in the retinal pigment epithelium. In: Marmor MF, Wolfensberger TJ, eds. The Retinal Pigment Epithelium. New York: Oxford University Press, 1998:103–134.
2. Peterson WM, Meggyesy C, Yu K, Miller SS. Extracellular ATP activates calcium signaling, ion, and fluid transport in retinal pigment epithelium. J Neurosci 1997; 17: 2324–2337.
3. Maminishkis A, Jalickee S, Blaug SA, et al. The P2Y$_2$ receptor agonist INS37217 stimulates RPE fluid transport in vitro and retinal reattachment in rat. Invest Ophthalmol Vis Sci 2002; 43:3555–3566.
4. Meyer CH, Hotta K, Peterson WM, Toth CA, Jaffe GJ. Effect of INS37217, a P2Y$_2$ receptor agonist, on experimental retinal detachment and electroretinogram in adult rabbits. Invest Ophthalmol Vis Sci 2002; 43:3567–3574.
5. Hay A, Lander MB. Types of pathogenetic mechanisms of retinal detachment. In: Glaser BM, ed. Retina. St. Louis: Mosby, 1994:1971–1977.
6. Anand R, Tasman WS. Non-rhegmatogenous retinal detachment. In: Glaser BM, ed. Retina. St. Louis: Mosby, 1994:2463–2488.
7. Marmor MF, Yao XY. Conditions necessary for the formation of serous detachment. Experimental evidence from the cat. Arch Ophthalmol 1994; 112:830–838.

8. Wilkinson CP, Rice TA. Results of retinal reattachment surgery. In: Craven L, ed. Michels Retinal Detachment. St. Louis: Mosby, 1997:935–972.
9. The repair of rhegmatogenous retinal detachments. American Academy of Ophthalmology. Ophthalmology 1990;97:1562–1572.
10. Wilkinson CP, Rice TA. Operative methods. In: Craven L, ed. Michels Retinal Detachment. St. Louis: Mosby, 1997:537–594.
11. Wilkinson CP, Rice TA. Alternative methods for retinal reattachment. In: Craven L, ed. Michels Retinal Detachment. St. Louis: Mosby, 1997:595–640.
12. Wilkinson CP, Rice TA. Complications of retinal detachment surgery and its treatment. In: Craven L, ed. Michels Retinal Detachment. St. Louis: Mosby, 1997:979–1080.
13. Tornambe PE. Pneumatic retinopexy: current status and future directions. Int Ophthalmol Clin 1992; 32:61–80.
14. Nour M, Quiambao A, Peterson WM, Al-Ubaidi MR, Naash MI. $P2Y_2$ receptor agonist INS37217 enhances functional recovery after detachment caused by subretinal injection in normal and RDS mice. Invest Ophthalmol Vis Sci 2003; 44:4505–4514.
15. Sakai T, Calderone JB, Lewis GP, Linberg KA, Fisher SK, Jacobs GH. Cone photoreceptor recovery after experimental detachment and reattachment: an immunocytochemical, morphological, and electrophysiological study. Invest Ophthalmol Vis Sci 2003; 44:416–425.
16. Takahashi J, Hikichi T, Mori F, Atsushi K, Yoshida A, Peterson WM. Effect of nucleotide $P2Y_2$ receptor agonists on outward active transport of fluorescein across normal blood-retina barrier in rabbit. Exp Eye Res 2004; 78:103–108.
17. Larsen M. Ocular fluorometry methodological improvements and clinical studies—with special reference to the blood-retina barrier permeability to fluorescein and fluorescein glucuronide. Acta Ophthalmol Suppl 1993; 211:1–52.
18. Koyano S, Araie M, Eguchi S. Movement of fluorescein and its glucuronide across retinal pigment epithelium-choroid. Invest Ophthalmol Vis Sci 1993; 34:531–538.
19. Takahashi J, Mori F, Hikichi T, Yoshida A. Effect of acetazolamide on outward permeability of blood-retina barrier using differential vitreous flyorophotometry. Curr Eye Res 2001; 23:166–170.
20. Zimmermann H, Braun N. Extracellular metabolism of nucleotides in the nervous system. J Auton Pharmacol 1996; 16:397–400.
21. Yerxa BR, Sabater JR, Davis CW, et al. Pharmacology of INS37217 [P(1)-(uridine 5′)-P(4)-(2′-deoxycytidine 5′)tetraphosphate, tetrasodium salt], a next-generation $P2Y_2$ receptor agonist for the treatment of cystic fibrosis. J Pharmacol Exp Ther 2002; 302:871–880.
22. Peterson WM, Venturini AE, Cowlen MS, Vezina M, Piccirilli G. Metabolism, ocular distribution, and ERG effects of INS37217, a novel $P2Y_2$ receptor agonist, for the treatment of retinal detachment. Association for Research in Vision and Ophthalmology, Fort Lauderdale, FL, April 29–May 4, 2001.
23. Tornambe PE, Fox GM, Poliner LS, et al. A double-masked, randomized, placebo-controlled, dose-escalating study of a single intravitreous injection of INS37217 in subjects with retinal detachment. Association for Research in Vision and Ophthalmology, Fort Lauderdale, FL, May 4–9, 2003.
24. Burton TC. Recovery of visual acuity after retinal detachment involving the macula. Trans Am Ophthalmol Soc 1982; 80:475–497.
25. Hagimura N, Iida T, Suto K, Kishi S. Persistent foveal retinal detachment after successful rhegmatogenous retinal detachment surgery. Am J Ophthalmol 2002; 133:516–520.
26. Machemer R. Macular translocation. Am J Ophthalmol 1998; 125:698–700.
27. Akduman L, Karavellas MP, MacDonald JC, Olk RJ, Freeman WR. Macular translocation with retinotomy and retinal rotation for exudative age-related macular degeneration. Retina 1999; 19:418–423.

8

Cell-Based Delivery Systems: Development of Encapsulated Cell Technology for Ophthalmic Applications

Weng Tao
Neurotech USA, Lincoln, Rhode Island, U.S.A.

Rong Wen and Alan Laties
Department of Ophthalmology, University of Pennsylvania School of Medicine, Philadelphia, Pennsylvania, U.S.A.

Gustavo D. Aguirre
James A. Baker Institute for Animal Health, College of Veterinary Medicine, Cornell University, Ithaca, New York, U.S.A.

DESCRIPTION OF ENCAPSULATED CELL TECHNOLOGY

Encapsulated cell technology (ECT) was developed to treat diseases of the central nervous system (CNS) (1–10) and the eye (11). ECT implants consist of living cells encapsulated within a semipermeable polymer membrane and supportive matrices. The encapsulated cells are genetically engineered to produce a specific therapeutic substance to target a specific disease or condition. Once surgically implanted into the CNS or eye, the semipermeable polymer membrane has two main functions: it allows the outward passage of the therapeutic product while protecting the encapsulated cells from rejection by the patient's immune system. It also permits ready access to oxygen and nutrients (Fig. 1).

The ability to deliver biologically active molecules directly to the target site is a major hurdle to their use in the treatment of CNS and eye diseases. The blood–retina barrier (BRB) prevents the penetration of most molecules to the neurosensory retina, in the same way the blood–brain barrier (BBB) hinders access to the CNS. ECT offers the potential for controlled, continuous, long-term delivery of therapeutics, including a wide variety of novel proteins and other compounds, directly to the retina, bypassing the BRB. In addition, the implants can be retrieved, providing an added level of safety. Therefore, ECT has promising applications to major types of ocular disorders such as retinal degeneration, ocular inflammation, and angiogenesis.

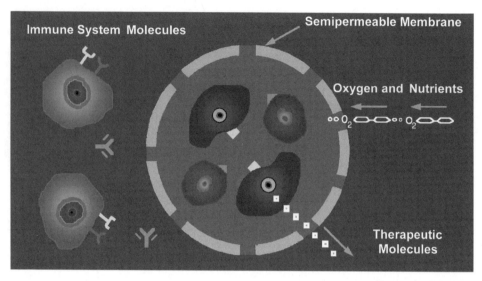

Figure 1 Diagram of a cross-section view of an ECT implant. *Abbreviation*: ECT, encapsulated cell technology.

An intraocular implantable encapsulated cell unit prototype for chronic delivery of therapeutic agents has been developed to treat ophthalmic disorders (Fig. 2) (11). The implant consists of genetically modified cells packaged in a hollow, semipermeable membrane. The hollow fiber membrane (HFM) prevents immune molecules, e.g., antibodies and host immune cells, from entering the implant, while allowing nutrients and therapeutic molecules to diffuse freely across the membrane. The encapsulated cells continuously secrete therapeutic agents (Fig. 2A), and derive nourishment from the host milieu. The ECT capsule is implanted through a small pars plana incision and anchored to the sclera by a small titanium wire loop (Fig. 2B). The active intravitreal portion of the implant measures ∼1 mm in diameter and 10 mm in length. It is fixed outside the visual axis.

Advances in molecular biology over the last two decades have led to the discovery of potent proteins such as cytokines and neurotrophic factors. The potential therapeutic value of these molecules is impressive; however, realization of this potential has been slow. Effective delivery of these molecules to the target sites, particularly the CNS and the eye, has proven to be a formidable task, due to the barrier properties of the brain and eye. Despite promising results in short-term animal studies, few, if any, proteins have become successful therapeutics for human CNS or eye disorders. A clinical trial sponsored by Regeneron of systemically administered ciliary neurotrophic factor (CNTF, a 24-kDa member of the interleukin-6 cytokine family) for amyotrophic lateral sclerosis is a good example. In this trial, despite high systemic doses, CNTF was undetectable in the CNS and there was no therapeutic benefit. In addition, the high peripheral CNTF levels were associated with major side effects, such as fever, fatigue, and blood chemistry changes that are consistent with activation of the acute-phase response (12,13). One reason for these disappointing results may be difficulty in achieving adequate concentrations of drug at the appropriate site; systemic administration may simply not be an effective way to treat CNS or ocular disorders. A continuous and site-specific delivery system may optimize the pharmacokinetics of these potential therapeutic agents in these two areas.

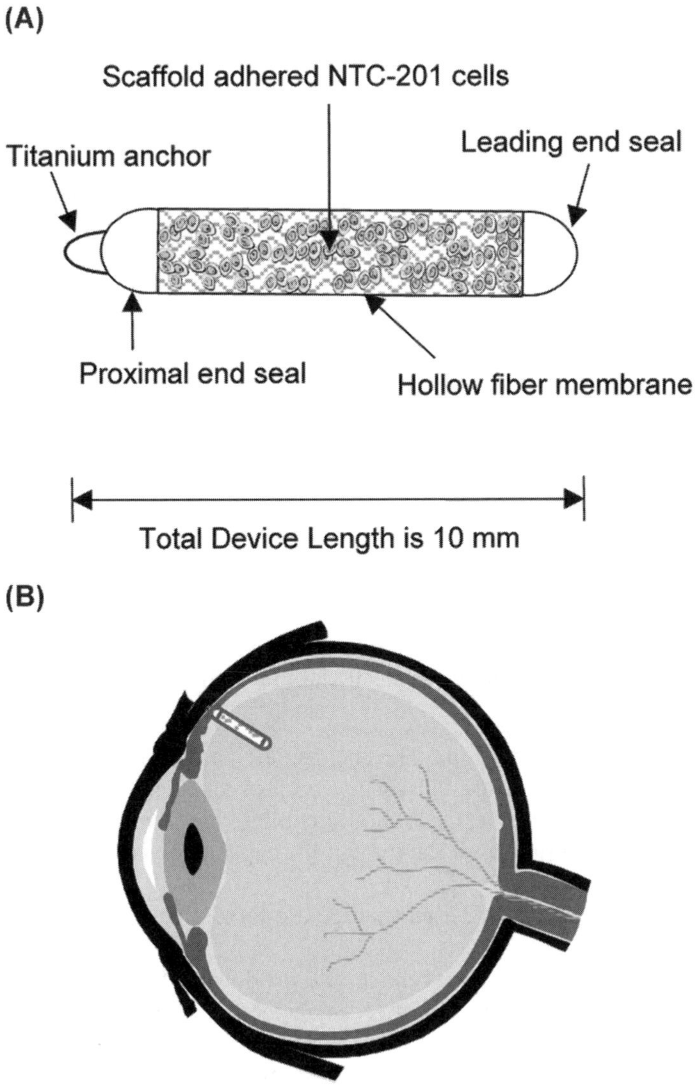

(A)

Scaffold adhered NTC-201 cells

Titanium anchor

Leading end seal

Proximal end seal

Hollow fiber membrane

Total Device Length is 10 mm

(B)

Figure 2 Neurotech's proprietary encapsulated cell therapy. Encapsulated cell implants consist of living cells encapsulated within semipermeable polymer membranes and supportive matrices: (**A**) longitudinal view of a cell-containing implant; (**B**) intraocular placement of an encapsulated cell implant. *Source*: From Ref. 11.

ECT provides an alternative to the conventional means of administration. It is particularly attractive for the following reasons: (i) it potentially allows any therapeutic agent to be engineered into the cells and therefore has a broad range of applications; (ii) in at least one system the mammalian cell produced protein factor, freshly synthesized and released within the target site in situ, is more potent than the purified recombinant factors (14) thereby reducing the dose requirement; (iii) for proteins delivered directly into the cerebrospinal fluid (CSF) or eye the limited CNS and eye volume of distribution, the presence of the BBB and BRB, and the low dose requirement minimize potential systemic toxicity associated with the protein; and (iv) the cell-containing capsule can be retrieved.

Selection of a Platform Cell Line for ECT

To successfully develop ECT-based systems, it is essential to identify the specific parameters that determine the survival, output, and immunological behavior of cells in an
encapsulated environment. These parameters can then be used to screen and develop
appropriate cells that show good survival and stable function when encapsulated
within a capsule. Methods were developed to identify hardy platform cell lines for
a wide range of encapsulated cell therapies. The focus of the strategy was to identify
human cell lines that show long-term survival in an encapsulated environment; that
can be genetically engineered to secrete protein factors; and that are nontumorigenic.

The treatment of neurological and ocular diseases is particularly attractive for
ECT. The CSF, brain parenchyma, and vitreous gel of the eye are all potential sites
for implantation of encapsulated cells releasing therapeutic factors. Transplantation
of encapsulated cells releasing therapeutic factors can bypass either the BBB or the
BRB. However, long-term survival of encapsulated cells in these environments
is challenged by the potential of cell overgrowth and the stressful environmental
factors such as low O_2 and poor nutrient flux.

Selection Criteria

The translation of success from short-term animal models to a practical treatment
for chronic human diseases depends on long-term cell viability in the capsule in vivo.
Ideally, the encapsulated cells will:

1. Be hardy under stringent conditions. The encapsulated cells should be both
   viable and functional in the avascular tissue cavities such as in the CNS or
   the vitreous cavity environment. Cells should exhibit >80% viability for a
   period of more than one month in the implant or capsule in vivo to ensure
   long-term delivery.
2. Be genetically modifiable. The desired therapeutic factors need to be engineered into the cells.
3. Have a relatively long life span. The cells should produce sufficient progeny
   so they can be tissue banked, characterized, engineered, safety tested, and
   clinical lot manufactured.
4. Deliver an appropriate quantity of a useful biological product to ensure
   treatment effectiveness.
5. Have no or a low level of host immune reaction to ensure graft longevity.
   Preferably cells should be of human origin to increase compatibility
   between the encapsulated cells and the host.
6. Be nontumorigenic to ensure safety to the host, in case of implant leakage.

Selection Strategies

A three-step hardy cell screen was designed to identify potential platform cell lines.
The schematic representation of the screening process is presented in Figure 3.

1. *Rapid in vitro screening for hardy cells that are transfectable.* The two
   critical characteristics that cells must possess in order to be useful for
   long-term delivery of therapeutic agents are (i) viability under stringent
   conditions and (ii) transfectability (so that desired therapeutic factors
   can be genetically engineered into the cells). Because of the number of both

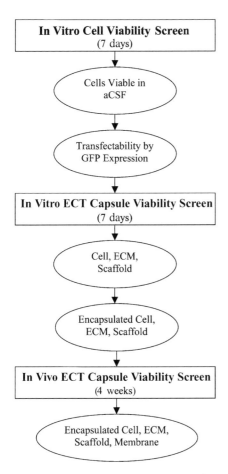

Figure 3 Schematic representation of the three-step hardy cell screening process. The cells that pass the in vitro viability and transfectability screen proceed to the in vitro ECT capsule viability screen, and the optimal combination of the cell–ECM–scaffold that passed the in vitro capsule viability screen proceed to the in vivo ECT capsule viability screen. *Abbreviations*: aCSF, artificial CSF; ECM, extracellular matrix.

xenogeneic and human cell lines available, a screening method was developed to rapidly and efficiently assess viability and transfectability. Viability was screened using increasingly challenging conditions from optimal (DMEM F12 + 10% FBS) to artificial aqueous humor (aAH) or artificial CSF (aCSF). Transfectability was assessed using a green fluorescent protein (GFP) expression vector transfected using a variety of transfection techniques. Approximately 25–30 currently available cell lines were rapidly screened at this level and the most promising cells were selected and carried through further stages (described next).

2. *In vitro ECT capsule viability screen.* For "hardy cell" candidates that passed the initial cell screen and transfectability screen, the cells were encapsulated and their viability was evaluated using different combinations of extracellular matrix (ECM) and scaffold. The encapsulated performance was examined under optimal tissue culture conditions or under stringent tissue culture conditions (such as artificial CSF). The best

combination of cell–ECM–scaffold that passed the stringent in vitro capsule screen was further evaluated in vivo.

3. *In vivo capsule viability screen.* The intrinsic behavior of the cells, diffusion
   of nutrients and cofactors, and immunogenicity all affect the viability
   and function of the cells in capsules. Multiple capsule configurations and
   cell–matrix–membrane combinations were tested. The combination of
   cell–ECM–scaffold was encapsulated with different membranes and capsules
   were implanted into CNS and ocular sites (such as rat ventricle, rabbit eye,
   dog eye, pig eye, and sheep intrathecal space). The optimal cell–matrix–
   membrane combinations that show longevity and functional stability in vivo
   were chosen for further development.

Following the three-step screening process, NTC-200 was identified as the
best platform cell line for ECT. NTC-200 cells are retinal pigment epithelial cells
derived from a human donor. In combination with the HFM and polyethylene
terephthalate (PET) yarn, NTC-200 cells demonstrated the best in vitro and in
vivo viabilities. NTC-200 cells can also be genetically modifiable to secrete a
desired factor, such as CNTF. The modified cells being designated NTC-201.
The cells have a long lasting life span and are of human origin. The encapsulated
NTC-200 cells have good in vivo viability (>80% viable after one month in vivo).
NTC-200 cells can deliver sufficient quantity of growth factor to achieve efficacy,
trigger no or a low level of host immune reaction, and are nontumorigenic in
nude mice.

SPECTRUM OF DISEASES FOR WHICH THIS DELIVERY SYSTEM MIGHT BE APPROPRIATE

ECT is ideally suited for treating CNS and eye diseases for which there are currently
no effective therapies.

Diseases of the CNS

The efficacy of ECT-based therapeutic factor delivery has been consistently demonstrated in animal models of neurodegenerative diseases of the CNS, including CNTF
in the rodent and primate models of Huntington's disease, glial cell line–derived
neurotrophic factor in the rat model of Parkinson's disease, and nerve growth factor
in rodent and primate models of Alzheimer's disease (4–10). Furthermore, previous
studies have shown that mammalian cell–derived growth factor, synthesized de novo,
is more potent than purified, *Escherichia coli*–derived growth factor (3,14). NsGene,
a Danish biotech company and a sublicensee of Neurotech, is actively pursuing CNS
applications using ECT.

Diseases of the Eye

Neurotech USA is developing ECT for ophthalmic applications, primarily due to
the many unmet medical needs in the field of ophthalmology. Although many topical
pharmaceutical agents such as antibiotics and anti-inflammatory agents are available
for the eye, few treatments, if any, are available for the common causes of blindness
that affect millions of people worldwide. Many of these devastating diseases are

associated with the degenerating retina. Although previous studies have shown promise of growth factors in reducing or halting the pathogenesis of retinal degeneration, unfortunately, progress has been slow in this field due to a number of challenges. First, the therapeutic agents that have shown promise cannot pass through the BRB. Second, repeated intraocular injection is not practical due to the chronic nature of these diseases. Third, an effective delivery system is not yet available.

Although there are a wide variety of eye diseases, there are three main clinical manifestations that represent targets for therapy: photoreceptor degeneration in the neural retina, vascular proliferation, and inflammation. Several proteins show powerful neurotrophic, antiangiogenic, and anti-inflammatory properties. These proteins have the potential to significantly slow or halt retinal diseases. The lack of effective concentration at the target site, and the adverse effects associated with frequent intraocular injections are current challenges to administering these therapeutic proteins.

ECT-Based Delivery of Neurotrophic Factors for the
Treatment of Retinitis Pigmentosa

NT-501 is an ECT-CNTF product that consists of encapsulated cells that secrete recombinant human CNTF. After implantation, CNTF is released from the cells constitutively into the vitreous gel. The NT-501 implant is manufactured to be sterile, nonpyrogenic, and retrievable. The current implant or capsule is about 1.1 cm in length (including titanium loop) and will be placed well outside the visual axis in the human eye. This same implant and implant size has been used in preclinical toxicity and efficacy evaluation studies in dogs, pigs, and rabbits. The therapeutic intent of intraocular CNTF delivery is to reduce or arrest the progressive loss of photoreceptors, which is characteristic of retinitis pigmentosa (RP) and related retinopathies.

RP is a group of incurable retinal degenerative diseases that have a complex molecular etiology. Approximately 100,000 Americans suffer from RP. More than 100 RP-inducing mutations have been identified in several genes including: rhodopsin, the rod visual pigment; peripherin, a membrane structure protein; and PDEB, the beta subunit of rod cyclic GMP (cGMP) phosphodiesterase. However, the genotype is unknown for the majority of patients. Despite this genetic heterogeneity, there tends to be a common pattern of visual loss in patients with RP. Typically, patients experience disturbances in night vision early in life because of rod photoreceptor degeneration. The remaining cone photoreceptors become their mainstay of vision, but over the years and decades, the cones slowly degenerate, leading to blindness. These two phases of degeneration in the visual life of a patient with RP may involve different underlying pathogenic mechanisms. Regardless of the initial causative defects, however, the end result is photoreceptor degeneration. This common pathogenesis pathway provides a target for therapeutic intervention.

There are many naturally occurring and genetically engineered animal models of RP. Many studies have demonstrated the promise of growth factors, neurotrophic factors, and cytokines as therapeutics for RP in short-term animal models. Among them CNTF is reported to be the most effective in reducing retinal degeneration (15). Unfortunately, the local adverse effects associated with the intraocular administration of these factors at relatively high levels, their short half-life following intravitreal administration, and the existence of the BRB, which precludes useful systemic administration of these agents for treatment of RP, have prevented their further clinical development and therapeutic practicality for RP patients. To circumvent these CNTF delivery problems, the NT-501 implant has been developed.

ECT Neurotrophic Factors for the Treatment of Glaucoma

In addition to RP, ECT could be applied to neurodegeneration in glaucoma. This group of diseases is characterized by a progressive degeneration within the neural retina eventually leading to blindness. In a number of in vitro systems, several neurotrophic factors protect various types of retinal neurons, including retinal ganglion cells and photoreceptors. These factors have been shown in animal models to have protective effects on the neural retina, as well as on other parts of the CNS, despite the complicated nature of the underlying molecular mechanisms that trigger degeneration in different diseases. Preserving photoreceptors and ganglion cells by slowing the degenerative process would have enormous therapeutic benefit, even if the underlying pathophysiology of the disease was not corrected. The delivery of neurotrophic factors from encapsulated cells in the eye may significantly delay the loss of visual function associated with diseases in which degeneration plays a role. One of the most promising molecules that has shown significant efficacy in these models is CNTF, and the efficacy of local administration of this agent has been demonstrated in several large animal models (16–18). Also see Chapter 3.

ECT Antiangiogenic Factors for the Treatment of Age-Related Macular Degeneration and Diabetic Retinopathy

Retinal vascular proliferation can occur in a number of different sites within the eye, and plays a role in many ocular diseases. In age-related macular degeneration, angiogenesis of the choroidal vasculature can cause leakage of fluid and bleeding into the retina, subretinal, subretinal pigmented epithelium, and subneurosensory spaces, leading ultimately to loss of neural retinal elements. In diabetic retinopathy, neovascularization in the optic nerve and the retina leads to hemorrhaging within the vitreous cavity and subsequent retinal detachments from traction. In addition, angiogenesis within the iris (called rubeosis iridis) causes neovascular glaucoma. Delivery of antiangiogenic factors either alone or in combination with neurotrophic factors could significantly impact the progression of these diseases. Some antiangiogenic factors that are believed to be promising for the treatment of vasoproliferative diseases are inhibitors of vascular endothelial growth factor (VEGF), soluble receptors for VEGF, endostatin, angiostatin, and pigment epithelium-derived factor. These are all testable using existing well-established animal models that mimic these human diseases. Also see Chapter 5.

ECT Anti-Inflammatory Factors for the Treatment of Uveitis

Uveitis is a general term used to describe a group of syndromes that have, as a common feature, inflammation of the uveal tract. Experimental autoimmune uveoretinitis can be induced in many animals by a subretinal injection of bovine serum albumin or fibroblasts, although most studies have been done in rabbits. A number of nonsteroidal, anti-inflammatory proteins are known which may be valuable for long-term therapy of uveitis if delivered from encapsulated cells. Some anti-inflammatory factors that are believed to be promising for the treatment of uveitis diseases are inhibitors of inflammatory cytokines, such as antibodies or soluble receptors.

ANIMAL MODELS USED TO INVESTIGATE THE APPLICABILITY OF THIS DELIVERY SYSTEM FOR THE DISEASES MENTIONED

Although ECT can be applied to treat a number of human ocular diseases, RP was chosen for the following reasons: (i) in many instances the cause and pathogenesis of the disease are well defined (19); (ii) animal models that are molecular homologs of the human disease are available (20–26); and (iii) a number of neurotrophic factors have shown protective effects against photoreceptor degeneration (15,27–29).

Prior studies have demonstrated the promise of growth factors, neurotrophic factors, and cytokines given by intravitreal injection in short-term animal experiments as potential therapy for RP (15,27,28,30). Among these, ciliary neurotrophic factor (CNTF) is effective (15). Unfortunately, the chronic nature of the disease and the adverse effects associated with repeated short-duration intraocular administration mitigates the use of CNTF by this delivery method.

Two animal models, the S334ter-3 transgenic rat model and the rod–cone degeneration 1 (*rcd1*) mutant dog model, were used to investigate the therapeutic efficacy and safety of prolonged intraocular CNTF delivery via either unencapsulated or encapsulated cells.

Transgenic Rat Model for Retinal Degeneration

Heterozygous S334ter-3 rats carrying the rhodopsin mutation S334ter were produced by mating homozygous breeders (kindly provided by Dr. M. M. LaVail, University of California, San Francisco, CA) with wild-type Sprague–Dawley rats. In this model, photoreceptor degeneration begins soon after birth (P8), and degeneration continues rapidly. By P20 only one layer of photoreceptor remains (25). This model was used to assess the effect of unencapsulated CNTF secreting cells. At P9, approximately 10^5 NTC-201 cells (in vitro CNTF output at 100 ng/million cells/day) in 2-μL phosphate buffered saline (PBS) were injected into the vitreous of the left eye of S334ter-3 rats ($n = 6$) using a 32-gauge needle. Control animals were injected with untransfected parental cells ($n = 6$). Contralateral eyes were not treated. For the CNTF bolus injections, 1 μg CNTF in 1 μL of PBS was injected into the vitreous at P9. Eyes were collected at P20, and processed for histologic evaluation. Plastic embedded sections of 1-μm thickness stained with toluidine blue were examined by light microscopy.

rcd1 Dog Model for Retinal Degeneration

The efficacy of intravitreal CNTF was investigated using NT-501 implant in the *rcd1* dog model. The *rcd1* affected dogs carrying a mutation of the gene encoding the β-subunit of the rod cGMP phosphodiesterase (PDEB) were provided by the Retinal Disease Studies Facility (Kennett Square, PA), which is a resource maintained by the NEI/NIH (EY-06855) and the Foundation Fighting Blindness. The pathogenesis of the *rcd1* dog model has been characterized (26,31). In this model, photoreceptor degeneration begins 3.5 weeks after birth and continues for a year, with 50% photoreceptor loss at seven weeks of age and additional 50% loss at 14 weeks. For each dog, the left eye received an NT-501 implant and the right eye was untreated (control). The study duration was seven weeks. At the end of the study, the animals were sacrificed and implant were explanted and evaluated for CNTF output by enzyme-linked immunosorbent assay (ELISA) and for cell viability by histological analysis. The eyes were enucleated and fixed in Bouin's solution, embedded in paraffin, and sectioned at 6 μm. Vertical sections through optic nerve and pupil were stained with hemotoxylin and eosin and examined by light microscopy. The ECT implants were fixed in 4% paraformaldehyde and processed for glycidylmethacrylate (GMA) embedding, sectioned, and stained for histological evaluation. Each section was reviewed to determine cell density and cell viability.

PHARMACOKINETIC AND PHARMACODYNAMIC STUDIES USING THE DELIVERY SYSTEM

To evaluate the pharmacokinetics of CNTF in the vitreous humor when delivered via intraocular implantation of NT-501 as well as the long-term function of the NT-501 after implantation, capsules were implanted into rabbit eyes. At different time points after implantation, capsules were explanted and vitreous samples harvested. The CNTF output from the explanted capsules and CNTF levels in vitreous samples were determined by ELISA.

RESULTS OF ANIMAL MODEL STUDIES

Efficacy

Protective Effect of NTC-201 in a Transgenic Rat Model for RP

To investigate whether CNTF delivered via mammalian cells was effective in photoreceptor protection, a short-term study using unencapsulated NTC-201 cells was carried out in a rapid retinal degeneration, transgenic rat RP model, S334ter-3 (25). The S334ter-3 transgenic rats were treated with either NTC-200 (parental cell line, $n = 6$) or NTC-201 (CNTF-secreting cell line, $n = 6$) via intravitreal injection into one eye on postnatal day (P) 9 when retinal degeneration has already begun. The contralateral eye was not treated. The experiment was terminated on P20, and the eyes processed for histologic evaluation. In untreated eyes of S334ter-3 transgenic rats, severe photoreceptor degeneration was observed by P20, and examination of the outer nuclear layer (ONL) showed only one row of nuclei remaining (Fig. 4A). The NTC-201-injected eyes had five to six rows of nuclei in the ONL (Fig. 4C), while in the control eyes that were injected with nontransfected cells (NTC-200), only one to two rows of nuclei remained (Fig. 4B). Furthermore, no evidence of retinal inflammation was observed in any of the treated or control eyes. In animals treated with a single intravitreal injection of purified human recombinant CNTF, the ONL had two to three rows of nuclei (Fig. 5). These results clearly demonstrate that continuous delivery of CNTF via mammalian cells protected against retinal degeneration in this model.

NT-501 Implant Protects Photoreceptors in the rcd1 Dog RP Model

To evaluate the effect of CNTF delivered via an NT-501 implant, an experiment was conducted using the *rcd1* dog model for RP (26,31). The *rcd1* dogs carry a mutation on the PDE6B gene and the retinal degeneration of this model is well characterized. To evaluate photoreceptor protection, NT-501 capsules secreting 1–2 ng/day of CNTF were surgically implanted into one eye of each *rcd1* dog ($n = 2$) at seven weeks of age, a point at which 40% to 50% of photoreceptors have already been lost due to degeneration (i.e., five to six layers of ONL remain). This is three weeks after weaning for the dogs and the earliest time point that the surgical procedure can be performed without disruption of the retina (the eyes of younger dogs would be too small to accommodate a 1-cm-long implant). The contralateral eye was not treated. Capsules were explanted at 14 weeks of age after which time, if untreated, an additional 50% of photoreceptors are expected to be lost, leaving only two to three layers of ONL remaining (26,32). After explantation, the eyes were processed for histologic evaluation and the explanted capsules were assayed for CNTF output and cell

(A)

(B)

(C)

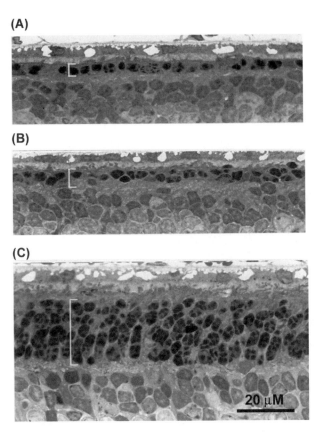

Figure 4 Retinal photomicrographs of transgenic rats carrying the rhodopsin mutation S334ter: (**A**) S334ter untreated eye; (**B**) NTC-200 parental cell treated eye; and (**C**) NTC-201 cell treated eye. The cells were injected on P9 and the experiment was terminated on P20. Brackets denote ONL. *Abbreviation*: ONL, outer nuclei layer. *Source*: From Ref. 11.

viability. As expected, the ONL in untreated eyes was about two to three layers thick. In contrast, the ONL in the NT-501-treated eyes still had five to six layers remaining, similar to the number of nuclei rows present at the time when the treatment began (Table 1 and Fig. 6). Moreover, the protection of photoreceptors was evenly distributed throughout the retina (Table 1) and not localized near the implant site. The observed protection is statistically significant ($p < 0.0001$). Again, there were no apparent adverse effects existing in the retina. All explanted capsules contained viable cells (Fig. 7).

Photoreceptor Protection in the rcd1 Dog by NT-501 Implant Is Dose Dependent

To determine the minimum effective dose and the optimal therapeutic dose of CNTF, a dose-ranging study was conducted. Thirty-one *rcd1* dogs were included in this study. Capsules that released different levels of CNTF were implanted into one eye of small groups of *rcd1* dogs at seven weeks of age. The contralateral eye was not treated. The level of implant CNTF output (ng/day) was defined as follows: <0.1 ($n = 4$), 0.2–1 ($n = 8$), 1–2 ($n = 7$), 2–4 ($n = 9$), 5–15 ($n = 3$). Again the capsules were explanted at 14 weeks of age and assayed for CNTF output and viability, and

(A)

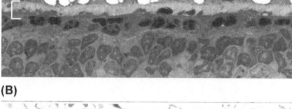

(B)

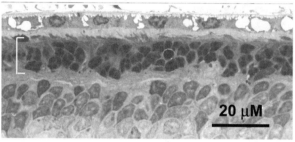

20 µM

Figure 5 Retinal photomicrographs of transgenic rats carrying the rhodopsin mutation S334ter: (**A**) S334ter untreated eye and (**B**) CNTF treated eye by bolus intravitreal injection. The CNTF was injected on P9 and the experiment was terminated on P20. Brackets denote ONL. *Abbreviations*: CNTF, ciliary neurotropic factor; ONL, outer nuclei layer. *Source*: From Ref. 11.

the eyes were processed for histologic evaluation. As can be seen in Figure 8, NT-501 implant significantly protected photoreceptors in the *rcd1* mutant dog model from degeneration in a dose-dependent manner. Complete protection was achieved at the highest dose (5–15 ng/day of CNTF), and minimal, but statistically significant,

Table 1 Protective Effect of NT-501 on Photoreceptors of *rcd1* Dogs

Animal[#]	Retina area[*]	Photoreceptor ONL		
		ECT-CNTF treated	Not treated	*p*-value
1485	S1	5.5	3.0	
	S2	5.8	2.7	
	S3	6.0	4.0	
	I1	4.0	2.5	
	I2	3.8	2.3	
	I3	4.2	2.7	
	Average	4.8 ± 0.23	2.9 ± 0.15	<0.0001
1489	S1	5.3	3.5	
	S2	7.5	3.3	
	S3	7.5	4.0	
	I1	5.5	3.7	
	I2	4.5	3.2	
	I3	5.3	2.7	
	Average	5.9 ± 0.0.29	3.4 ± 0.14	<0.0001

Abbreviations: ONL, outer nuclei layer; ECT, encapsulated cell technology; CNTF, ciliary neurotrophic factor.
Source: From Ref. 11.

(A)

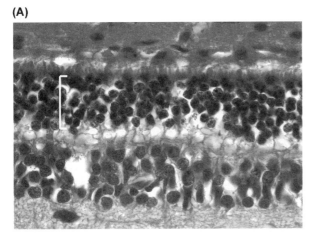

(B)

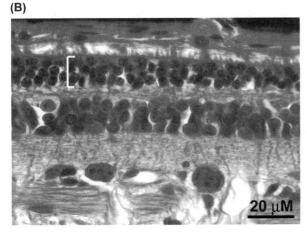

Figure 6 Retinal photomicrographs of *rcd1* dog model of retinitis pigmentosa. Comparison of ONLs in NT-501 (**A**) treated versus (**B**) nontreated eyes. The capsule was implanted into one eye at 7 weeks of age and explanted at 14 weeks of age. The contralateral eye was not treated. *Abbreviation*: ONL, outer nuclei layer. *Source*: From Ref. 11.

protection was observed at levels as low as 0.2–1 ng/day of CNTF. CNTF delivered below 0.1 ng/day had no protective effect, indicating that the observed protective effect was due to the presence of CNTF and not the ECT implant itself.

Histological evaluation indicated that all implants contained healthy, viable cells throughout. No cellular evidence of an immune reaction, inflammation, or damage to the retina was observed. Clinical and histological examination of the eye and focal areas of opacity of the lens were observed in some animals, which in most cases were located adjacent to the placement site of the implant. CNTF dosage received by the animal could not be correlated with the incidence or severity of these lens changes.

Pharmacokinetics

The explanted NT-501 capsules produced a consistent amount of CNTF up to 12 months in vivo. CNTF was readily detectable in the vitreous. The results are

(A)   **(B)**

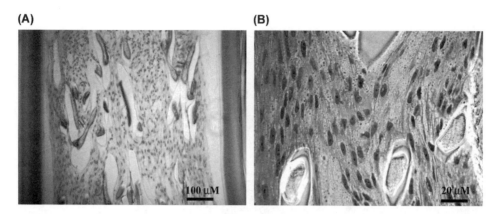

Figure 7 Representative implant cell viability after seven weeks in vivo. GMA embedded sections, 4-μm thick, stained with hemotoxylin and eosin, were examined under light microscope under the following magnifications: (**A**) low power and (**B**) high power. Sections of PET yarn scaffold along with cells were shown. *Abbreviations*: GMA, glycidyl methe arylate; PET yarn, polyethylene terephthalate (PET) yarn. *Source*: From Ref. 11.

presented in Figure 9. The data from these rabbit pharmacokinetic and long-term implant function studies suggest that the NT-501 capsule delivers CNTF into the vitreous for periods of at least one year after implantation. Histologic evaluation indicated that all capsules contained healthy, viable cells (33).

TECHNIQUES FOR IMPLANTING OR PLACING THE IMPLANT IN HUMANS

For clinical trials, the ECT capsule will be implanted in a manner similar to that described for fluocinolone acetonide intravitreal implant (see Chapter 14).

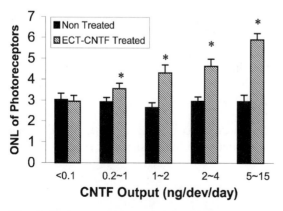

Figure 8 Dose response protection of photoreceptors in *rcd1* dog model of retinitis pigmentosa. Comparison of ONLs in NT-501 treated eyes with the nontreated eyes (mean ± SEM); $^*p < 0.05$. *Abbreviation*: ONL, outer nuclei layer. *Source*: From Ref. 11.

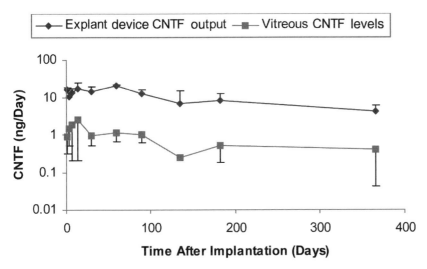

Figure 9 Time course of CNTF output (after explanting) and vitreous CNTF levels after NT-501 intraocular implantation in rabbits. *Abbreviation*: CNTF, ciliary neurotropic factor. *Source*: From Ref. 33.

NT-501 Implant Surgery

An implant will be removed from its package by manually securing the notched cap that closes the inner baffle of the package, twisting the cap counterclockwise, and gently pulling the cap from the baffle opening, revealing an implant supported by a titanium clip. The implant is examined for any gross defects. A single throw of double-armed Prolene will be placed through the suture anchor on the implant. The implant is then released from the titanium clip. The implant is then inserted through a 2-mm scleral incision, 3.75 mm posterior to the limbus in the inferonasal quadrant. The implant is secured to the sclera, and the wound closed, as described for the fluocinolone acetonide intravitreal implant (see Chapter 14).

NT-501 Explant Surgery

The patient is prepared in a manner similar to that used for implantation. Under microscopic visualization, a limbal conjunctiva peritomy will be performed at the site of the previous implant. Bipolar wetfield cautery is applied as needed to control bleeding and the previously placed prolene sutures will be identified. The two lateral prolene sutures will be removed. A super-sharp blade is used to create a 1.5-mm sclerotomy on either side of the anchoring 9–0 prolene suture. Bleeding from the pars plana sclerotomy is controlled with a 23-gauge tapered bipolar cautery. Care is taken to ensure the scleral wound lips are not cauterized. The sclerotomy edges are gently spread and a microforceps is used to grasp the anchoring loop on the implant. The sclerotomy may have to be enlarged to adequately visualize the implant. Once grasped, the implant is steadied while the sclerotomy is completed (joining adjacent sclerotomies) using the super-sharp blade. The anchoring prolene knot will be transected by this maneuver but the suture should remain attached to the implant. An attempt is then made to gently remove the implant using the microforceps. If resistance is encountered, the sclerotomy will be inspected and adhesions will be transected using the super-sharp blade or microscissors. Once the implant has been removed the sclerotomy is closed with 9–0 nylon suture and the retina is inspected with the indirect ophthalmoscope for tears or bleeding. The intraocular pressure is restored to a normal level, and the conjunctiva is closed with 7–0 vicryl suture. At the end of surgery, 100 mg cefazolin is administered by subconjunctiva injection.

FUTURE HORIZONS

Preclinical development of ECT has demonstrated the therapeutic efficacy, long-term delivery, and relative safety in the animal eyes. Validation of the ECT technology in human eyes will be a critical step. If safety and consistent delivery of ECT are demonstrated in clinical trials, ECT could potentially serve as a delivery system for a number of ophthalmic diseases for which currently no effective therapies are available.

Higher-output cell lines that provide adequate dosage with fewer cells have been developed. A shorter implant, 6 mm in length, that incorporates these cells is currently under development. It is desirable to make the implant as small as possible.

REFERENCES

1. Aebischer P, Buchser E, Joseph JM, et al. Transplantation in humans of encapsulated xenogeneic cells without immunosuppression. A preliminary report. Transplantation 1994; 58:1275–1277.
2. Aebischer P, Schluep M, Deglon N, et al. Intrathecal delivery of CNTF using encapsulated genetically modified xenogeneic cells in amyotrophic lateral sclerosis patients. Nat Med 1996; 2:696–699.
3. Lindner MD, Kearns CE, Winn SR, Frydel B, Emerich DF. Effects of intraventricular encapsulated hNGF-secreting fibroblasts in aged rats. Cell Transplant 1996; 5: 205–223.
4. Emerich DF, Lindner MD, Winn SR, Chen EY, Frydel BR, Kordower JH. Implants of encapsulated human CNTF-producing fibroblasts prevent behavioral deficits and striatal degeneration in a rodent model of Huntington's disease. J Neurosci 1996; 16:5168–5181.
5. Emerich DF, Winn SR, Hantraye PM, et al. Protective effect of encapsulated cells producing neurotrophic factor CNTF in a monkey model of Huntington's disease. Nature 1997; 386:395–399.
6. Emerich DF, Plone M, Francis J, Frydel BR, Winn SR, Lindner MD. Alleviation of behavioral deficits in aged rodents following implantation of encapsulated GDNF-producing fibroblasts. Brain Res 1996; 736:99–110.
7. Tseng JL, Baetge EE, Zurn AD, Aebischer P. GDNF reduces drug-induced rotational behavior after medial forebrain bundle transection by a mechanism not involving striatal dopamine. J Neurosci 1997; 17:325–333.
8. Emerich DF, Hammang JP, Baetge EE, Winn SR. Implantation of polymer-encapsulated human nerve growth factor- secreting fibroblasts attenuates the behavioral and neuropathological consequences of quinolinic acid injections into rodent striatum. Exp Neurol 1994; 130:141–150.
9. Emerich DF, Winn SR, Harper J, Hammang JP, Baetge EE, Kordower JH. Implants of polymer-encapsulated human NGF-secreting cells in the nonhuman primate: rescue and sprouting of degenerating cholinergic basal forebrain neurons. J Comp Neurol 1994; 349:148–164.
10. Kordower JH, Winn SR, Liu YT, et al. The aged monkey basal forebrain: rescue and sprouting of axotomized basal forebrain neurons after grafts of encapsulated cells secreting human nerve growth factor. Proc Natl Acad Sci USA 1994; 91:10,898–10,902.

11. Tao W, Wen R, Goddard MB, et al. Encapsulated cell-based delivery of CNTF reduces photoreceptor degeneration in animal models of retinitis pigmentosa. Invest Ophthalmol Vis Sci 2002; 43:3292–3298.

12. Cedarbaum JM, Chapman C, Charatan M, et al. The pharmacokinetics of subcutaneously administered recombinant human ciliary neurotrophic factor (rHCNTF) in patients with amyotrophic lateral sclerosis: relation to parameters of the acute-phase response. The ALS CNTF treatment study (ACTS) phase I–II study group. Clin Neuropharmacol 1995; 18:500–514.

13. Cedarbaum JM. A phase I study of recombinant human ciliary neurotrophic factor (rHCNTF) in patients with amyotrophic lateral sclerosis. The ALS CNTF treatment study (ACTS) phase I–II study group. Clin Neuropharmacol 1995; 18:515–532.

14. Hoane MR, Puri KD, Xu L, et al. Mammalian-cell-produced neurturin (NTN) is more potent than purified escherichia coli-produced NTN. Exp Neurol 2000; 162:189–193.

15. LaVail MM, Yasumura D, Matthes MT, et al. Protection of mouse photoreceptors by survival factors in retinal degenerations. Invest Ophthalmol Vis Sci 1998; 39:592–602.

16. Van Adel BA, Kostic C, Deglon N, Ball AK, Arsenijevic Y. Delivery of ciliary neurotrophic factor via lentiviral-mediated transfer protects axotomized retinal ganglion cells for an extended period of time. Hum Gene Ther 2003; 14:103–115.

17. Watanabe M, Fukuda Y. Survival and axonal regeneration of retinal ganglion cells in adult cats. Prog Retin Eye Res 2002; 21:529–553.

18. Agarwal N, Martin E, Krishnamoorthy RR, et al. Levobetaxolol-induced up-regulation of retinal bFGF and CNTF mRNAs and preservation of retinal function against a photic-induced retinopathy. Exp Eye Res 2002; 74:445–453.

19. Bedell MA, Largaespada DA, Jenkins NA, Copeland NG. Mouse models of human disease. Part II: recent progress and future directions. Genes Dev 1997; 11:11–43.

20. LaVail MM, Sidman M, Rausin R, Sidman RL. Discrimination of light intensity by rats with inherited retinal degeneration: a behavioral and cytological study. Vision Res 1974; 14:693–702.

21. Bowes C, Li T, Danciger M, Baxter LC, Applebury ML, Farber DB. Retinal degeneration in the rd mouse is caused by a defect in the beta subunit of rod cGMP-phosphodiesterase. Nature 1990; 347:677–680.

22. Pittler SJ, Baehr W, Wasmuth JJ, et al. Molecular characterization of human and bovine rod photoreceptor cGMP phosphodiesterase alpha-subunit and chromosomal localization of the human gene. Genomics 1990; 6:272–283.

23. Suber ML, Pittler SJ, Qin N, et al. Irish setter dogs affected with rod/cone dysplasia contain a nonsense mutation in the rod cGMP phosphodiesterase beta-subunit gene. Proc Natl Acad Sci USA 1993; 90:3968–3972.

24. McLaughlin ME, Ehrhart TL, Berson EL, Dryja TP. Mutation spectrum of the gene encoding the beta subunit of rod phosphodiesterase among patients with autosomal recessive retinitis pigmentosa. Proc Natl Acad Sci USA 1995; 92:3249–3253.

25. Liu C, Li Y, Peng M, Laties AM, Wen R. Activation of caspase-3 in the retina of transgenic rats with the rhodopsin mutation s334ter during photoreceptor degeneration. J Neurosci 1999; 19:4778–4785.

26. Aguirre GF, Farber D, Lolley R, et al. Retinal degenerations in the dog III abnormal cyclic nucleotide metabolism in rod-cone dysplasia. Exp Eye Res 1982; 35:625–642.

27. Faktorovich EG, Steinberg RH, Yasumura D, Matthes MT, LaVail MM. Photoreceptor degeneration in inherited retinal dystrophy delayed by basic fibroblast growth factor. Nature 1990; 347:83–86.

28. LaVail MM, Unoki K, Yasumura D, Matthes MT, Yancopoulos GD, Steinberg RH. Multiple growth factors, cytokines, and neurotrophins rescue photoreceptors from the damaging effects of constant light. Proc Natl Acad Sci USA 1992; 89:11249–11253.

29. Steinberg RH. Survival factors in retinal degenerations. Curr Opin Neurobiol 1994; 4:515–524.

30. Unoki K, LaVail MM. Protection of the rat retina from ischemic injury by brain-derived neurotrophic factor, ciliary neurotrophic factor, and basic fibroblast growth factor. Invest Ophthalmol Vis Sci 1994; 35:907–915.
31. Ray K, Baldwin VJ, Acland GM, Blanton SH, Aguirre GD. Cosegregation of codon 807 mutation of the canine rod cGMP phosphodiesterase beta gene and rcd1. Invest Ophthalmol Vis Sci 1994; 35:4291–4299.
32. Schmidt SY, Aguirre GD. Reductions in taurine secondary to photoreceptor loss in Irish setters with rod-cone dysplasia. Invest Ophthalmol Vis Sci 1985; 26:679–683.
33. Thanos CG, Bell WJ, O'Rourke P, et al. Sustained secretion of ciliary neuro factor to the vitreous, using the encapsulated cell therapy-based NT-501 intraocular devices. Tissue Eng 2004; 10(11/12):1617–1622.

9
Photodynamic Therapy

Ivana K. Kim and Joan W. Miller
Department of Ophthalmology, Harvard Medical School, Massachusetts Eye and Ear Infirmary, Boston, Massachusetts, U.S.A.

INTRODUCTION

In 1900, Raab's mentor von Tappeiner made the initial observations that led to the evolution of photodynamic therapy (PDT) (1). He serendipitously discovered that the combination of acridine red and light resulted in lethal toxicity to the paramecium Infusoria. The first clinical application of this kind of photochemical reaction was reported by Raab's mentor von Tappeiner who treated skin tumors with the combination of topical eosin and sunlight (2). Since this early work, the administration of photosensitizers in conjunction with the delivery of visible light has developed into a viable treatment option for various conditions in fields including dermatology, oncology, and ophthalmology.

The principle of PDT relies on the selective accumulation of a photosensitizing drug in diseased tissue with subsequent activation of the drug by illumination of the target tissue with a specific wavelength of light. Absorption of light converts the photosensitizer into a photoactive triplet state, which generates reactive oxygen species via two types of reactions. Free radicals and superoxide ions result from the transfer of hydrogen or electrons in Type I reactions. Energy transfer between the triplet state of the photosensitizer and molecular oxygen in Type II reactions generates singlet oxygen, which is thought to be the primary mediator of tissue damage in PDT (3). As the reactive distance of singlet oxygen is 0.01–0.02 µm, cytotoxic effects are precisely targeted (4).

The exact determinants of photosensitizer localization to both tumors and choroidal neovascular membranes are still under investigation. However, one factor is the "enhanced permeability and retention effect" resulting from certain anatomic characteristics of these target tissues (3). The compromised endothelial cell barrier in the neovasculature of tumors and in the subretinal space enables macromolecules such as photosensitizing compounds to reach the perivascular space by simple diffusion. These molecules then accumulate in such spaces because of the poorly developed lymphatic drainage in these target tissues. Other considerations include the interaction of various photosensitizer formulations with serum proteins involved in their transport. Liposomes may transfer drug to low-density lipoproteins (LDLs) in serum which, in turn, can serve as selective carriers (5). Proliferating endothelial cells and various tumor cells express increased numbers of LDL receptors related to the demand for cholesterol, which is required for cell membrane synthesis (6–8). Thus, photosensitizer associated with LDL may be preferentially delivered to choroidal neovascularization (CNV) or tumors.

Early investigations with photosensitizers were based on the observation that porphyrins localized in tumors and focused on tumor detection using hematoporphyrin (9). This work then led to the purification of a hematoporphyrin derivative with enhanced affinity for tumor tissue. Subsequent drug development has resulted in a class of molecules, including newer porphyrin derivatives, phthalocyanins, chlorins, and texaphyrins with high selectivity for diseased tissues, absorption peaks at longer wavelengths enabling higher light penetration of tissues, and faster clearance resulting in less cutaneous photosensitization. However, the hydrophobicity of most of these agents poses a challenge for intravenous administration and requires special formulations such as liposomes, oil-dispersions, or nanoparticles for effective delivery. The texaphyrins are water-soluble and do not necessitate such delivery systems.

Although several photosensitizers have been investigated for ocular applications, the only one approved for clinical use in ocular disease is verteporfin, a liposomal preparation of a second-generation porphyrin, benzoporphyrin derivative monoacid (BPD). BPD is lipophilic, and must be dissolved in an organic solvent such as dimethyl sulfoxide (DMSO) or prepared in a liposomal formulation for intravenous use. It has an absorption maximum near 690 nm and is cleared rapidly from the body. Allison and co-workers demonstrated its effectiveness as a photosensitizing agent in vitro and in vivo (10–14). The first ocular PDT experiments using BPD were carried out in a rabbit model of choroidal melanoma by Schmidt-Erfuth et al. (15) working under the direction of Gragoudas. They were able to induce complete necrosis of tumors, using 2 mg/kg of BPD dissolved in DMSO and complexed to LDL when irradiation was performed three hours after BPD injection. They applied 692 nm light with an irradiance of $150 \, mW/cm^2$ and a fluence of $100 \, J/cm^2$. Complications included transient vitritis and a self-limited exudative retinal detachment in 50% of treated eyes, which largely resolved within 48 hours.

PRECLINICAL STUDIES OF VERTEPORFIN FOR EXPERIMENTAL CNV

Based on the finding that vascular damage was a prominent feature after PDT with verteporfin, a series of experiments was designed to evaluate the effect of PDT on normal choroidal vessels in the rabbit eye (16). Closure of choriocapillaris and some large choroidal vessels was achieved using BPD in DMSO at a dose of 2 mg/kg, with light doses of 10, 50 or $100 \, J/cm^2$, irradiance of $100 \, mW/cm^2$, and irradiation within 30 minutes or at three hours. Damage to the retinal pigment epithelium (RPE) and photoreceptors was observed in all cases, but less toxicity to these structures was observed with the lowest light dose combined with the longer delay prior to irradiation.

After demonstrating that it was possible to close choriocapillaris with PDT using BPD, Miller and Gragoudas next investigated the ability of PDT to achieve selective occlusion of CNV without damaging retinal vessels, large choroidal vessels, or neurosensory retina. Using the well-characterized laser-injury model of CNV in the monkey, they reported both angiographic and histologic closure of CNV after PDT. The first CNV studies were performed with BPD dissolved in DMSO and mixed with LDL as liposomal BPD (verteporfin) was not available for preclinical studies. Initial treatment parameters included a drug dose of 1 and 2 mg/kg, fluence of 50, 75, 100 and 150 J/cm^2 and irradiance of 150 mW/cm^2 (17). By 24 hours after PDT, angiography demonstrated early hypofluorescence in the area of treatment, indicating occlusion of the CNV, with late staining beginning at the edge of the treatment spot. Staining of retinal vessels was sometimes observed, particularly when irradiation was performed early (within five minutes of BPD injection). Histology at 24 hours after PDT showed necrotic or absent endothelium in the CNV, with accumulation of fibrin, platelets, neutrophils and red blood cells. The choriocapillaris was similarly injured, with necrotic or absent endothelium and visible thrombi. Findings also included necrotic RPE and some pyknosis of the outer nuclear layer (ONL). The lower drug dose of 1 mg/kg caused less injury to surrounding structures than 2 mg/kg, and irradiation >5 minutes after drug injection appeared to avoid damage to retinal and larger choroidal vessels. Decreasing the drug dose from 2 to 1 mg/kg resulted in an increase in the fluence necessary to close CNV from 50 to 100 J/cm^2.

The irradiance of 150 mW/cm^2 was chosen for the initial studies because PDT had typically been performed at irradiances of ≤ 200 mW/cm^2, to avoid thermal effect and potential pain during treatment (18,19). However, irradiances at these levels led to treatment times over 10–15 minutes, which seemed impractical for ophthalmic clinical practice (17). Therefore, PDT using BPD was applied to normal primate eyes with irradiances of 150–1800 mW/cm^2, looking for denaturation of collagen fibrils by electron microscopy as an indicator of thermal injury (20). These studies demonstrated no evidence of denaturation at irradiances as high as 1800 mW/cm^2, although some damage to choroidal vessels was noted at 1200 mW/cm^2. PDT of CNV using irradiances of 300 and 600 mW/cm^2 was effective, and PDT of normal eyes at 600 mW/cm^2 confirmed selectivity (17,21). Subsequently, investigations of PDT for age-related macular degeneration using other photosensitizers have adopted 600 mW/cm^2 as the standard irradiance parameter (22). It is possible that the mild hyperthermia that may be produced with higher irradiances could act synergistically with PDT to potentiate cell killing (23).

Additional studies later refined the dosimetry using verteporfin, the liposomal formulation of BPD that would be used in clinical practice (21). A series of experiments was designed to use verteporfin PDT in experimental CNV, modifying drug dose and timing of irradiation with fluence and irradiance fixed at 150 J/cm^2 and 600 mW/cm^2, respectively. Verteporfin was tested at doses of 0.25, 0.375, 0.5, and 1 mg/kg with irradiation performed between 5 and 120 minutes after drug administration. Effective CNV closure was achieved at all tested drug doses, based on angiographic assessment one day after PDT (Fig. 1). As the drug dose was reduced, the effective time window for irradiation was shortened. At these light doses, a minimum threshold was reached at a drug dose of 0.25 mg/kg, with effective closure occurring only when irradiation was performed within 20 minutes of injection.

As the thermal laser injury used to create experimental CNV necessarily showed damage to the outer retina, the selectivity of PDT was investigated in normal eyes (21). The level of acceptable damage was defined as pyknosis of the ONL of $\leq 50\%$, mild disruption of the outer segments, choriocapillaris occlusion, and RPE necrosis. The closure of the choriocapillaris in normal choroid followed a similar pattern as the closure of CNV, in terms of drug and light dose and timing of irradiation (Fig. 2). As with closure of CNV, 0.25 mg/kg seemed to be the threshold dose for closure of choriocapillaris, achieved with almost no effect on the overlying retina. When PDT was performed using higher drug doses of 0.5 and 0.375 mg/kg, retinal structure remained well preserved. However, the RPE was affected at all doses, and there was also mild damage to photoreceptor inner and outer segments, ranging

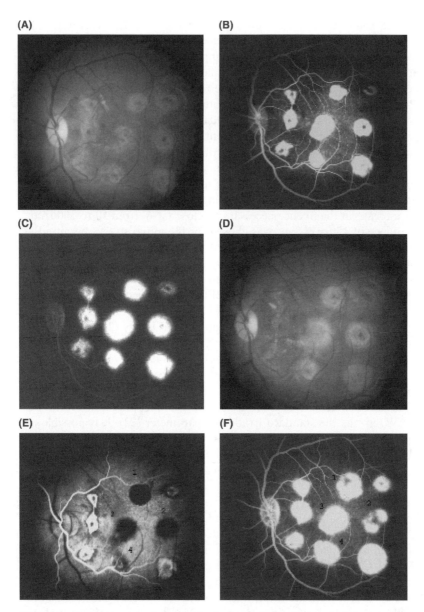

Figure 1 Photodynamic therapy (PDT) of experimental choroidal neovascularization (CNV). (A) Fundus photograph of CNV prior to PDT. (B, C) Fluorescein angiogram (*shown here in gray scale*) of CNV prior to PDT showing (B) early hyperfluorescence and (C) late leakage from areas of CNV. (D) Fundus photograph 24 hours after PDT showing mild retinal whitening in the treated areas. (E, F) Fluorescein angiogram (*shown here in gray scale*) 24 hours after PDT. Lesions were irradiated after administration of 0.5 mg/kg verteporfin using 150 J/cm^2 and 600 mW/cm^2. Lesion 1 was irradiated 10 minutes following dye injection; lesion 2 at 20 minutes, lesion 3 at 30 minutes, and lesion 4 at 50 minutes. Lesions 1, 2, and 3 show early hypofluorescence in the treated area while lesion 4 (E) demonstrates only a rim of hypofluorescence. (F) All treated lesions show staining in the later frame, which characteristically developed from the edge of the lesions. *Source*: From Ref. 21.

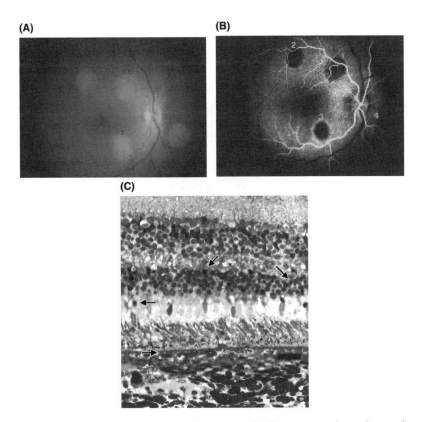

Figure 2 Effect of photodynamic therapy (PDT) on normal monkey retina and choroid. Lesions were irradiated after intravenous administration of 0.375 mg/kg verteporfin using 150 J/cm² and 600 mW/cm². Lesion 1 was irradiated 10 minutes after dye injection; lesion 2 at 20 minutes, lesion 3 at 30 minutes, and lesion 4 at 40 minutes. (**A**) Fundus photograph 24 hours after PDT of normal retina and choroid demonstrating mild deep retinal whitening in irradiated areas. (**B**) Fluorescein angiogram (*shown here in gray scale*) revealing early hypofluorescence in irradiated areas. (**C**) Light micrograph of retina and choroid 24 hours following PDT. The lesion shown was irradiated 20 minutes following dye injection. There is complete closure of the choriocapillaris and damage to the RPE (*Bruch's membrane = small arrow*). The outer retinal shows swelling with some pyknosis of outer nuclear layer nuclei (*arrow heads*). There is mild swelling and minimal pyknosis in the inner retina. Bar = 25 μm. *Abbreviation*: RPE, retinal pigment epithelium. *Source*: From Ref. 21.

from minimal swelling to more pronounced vacuolization and disarray. The lower drug dosages resulted in more selective closure of the choriocapillaris with minimal damage to the adjacent tissues. Irradiation within 10 minutes of verteporfin administration at a variety of doses caused damage of retinal vessels and was deemed unacceptable. Using a verteporfin dose of 0.375 mg/kg, irradiation at 20 minutes or longer after drug administration led to consistent closure of choriocapillaris with RPE necrosis, pyknosis in the ONL of ≤50%, and no damage to retinal or large choroidal vessels. Based on the data for closure of experimental CNV combined with the effects on normal retina and choroid, optimal treatment parameters appeared to be verteporfin 0.375 mg/kg (approximately 6 mg/m²), 150 J/cm², 600 mW/cm², and irradiation between 20 and 50 minutes after administration of verteporfin. Longer-term studies demonstrated persistent closure of CNV (up to four weeks) in eyes that had been treated with these parameters, and histopathology demonstrated a fibrous scar covered by proliferating RPE, with few open capillaries (Fig. 3). The damage noted acutely to the RPE and choriocapillaris appeared to recover by four to seven weeks after treatment (Fig. 4) (24).

While all ocular preclinical studies of verteporfin PDT were conducted using bolus injections of verteporfin, the clinical trials of verteporfin in dermatology were performed using a 30-minute intravenous infusion. Although no systemic effects had been noted in the animals studied, experiments were performed to evaluate the efficacy of an intravenous infusion given concerns regarding rapid administration of a liposomal preparation (25). An infusion of liposomal verteporfin was administered over 10 minutes (fast infusion) or 32 minutes (slow infusion). The light doses were kept constant at $150 \, \text{J/cm}^2$ and $600 \, \text{mW/cm}^2$. All CNV were occluded angiographically after PDT when irradiation was performed within 45 minutes of the start of drug infusion, falling off to 50% closure when irradiation was performed between 55 and 75 minutes, after which PDT became ineffective in CNV closure. The infusion studies suggested that effective CNV closure could be achieved if irradiation was performed within 55 minutes of the start of drug administration, but the earliest safe time for irradiation was not identified in these studies.

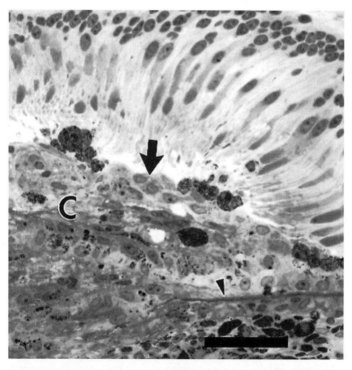

Figure 3 Light micrograph of experimental CNV four weeks after PDT. A layer of proliferated RPE (*arrow*) surrounds the CNV (C) with pigment-laden cells overlying the RPE. The outer nuclear layer, photoreceptor inner segments, and a few rudimentary outer segments remain in the area of CNV. The CNV extends through Bruch's membrane (*arrow head*) and contains few capillaries. The spaces seen are acinar structures in RPE cells. Bar = 50 μm. *Abbreviations*: PDT, photodynamic therapy; RPE, retinal pigment epithelium; CNV, choroidal neovascularization. *Source*: From Ref. 24.

(A)   (B)

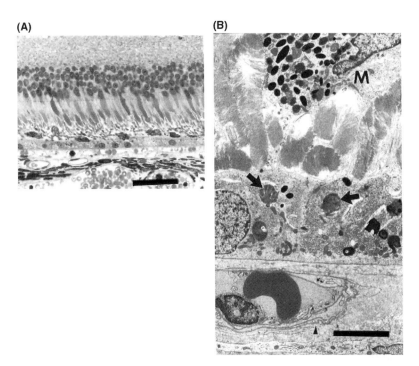

Figure 4 (**A**) Light micrograph of normal monkey retina and choroid six weeks after PDT. The inner segments appear normal, but the outer segments are shortened and distorted. The RPE appears single-layered and hypopigmented. There are pigment-layered cells overlying the RPE. The choriocapillaris and choroid are patent. Bar = 50 μm. (**B**) Transmission electron micrograph of normal retina and choroid four weeks after PDT. A pigment-laden cell (M) lies among disorganized outer segments. The RPE is lightly pigmented, has elongated microvilli and rudimentary basal infolding, and contains several lysosomes with outer segment material (*arrows*). The choriocapillaris appears reperfused with reduplication of basement membrane (*arrow head*). Bar = 5 μm. *Abbreviations*: PDT, photodynamic therapy; RPE, retinal pigment epithelium. *Source*: From Ref. 24.

Data from early clinical trials of verteporfin PDT for CNV indicated that CNV reperfused and retreatments would be required. Therefore, the recovery of normal ocular structures following multiple PDT treatments of monkey eyes was evaluated (26). Three sequential PDT treatments were performed in the same area of normal retina and choroid at two-week intervals. Three doses of liposomal BPD were studied (6, 12, and 18 mg/m^2) with the light doses kept constant (689 nm, 600 mW/cm^2, 100 J/cm^2). Histopathologic examination was performed at two and six weeks after the third treatment. Minimal damage was seen in the group treated at 6 mg/m^2, with recovery comparable to the single treatment at 0.375 mg/kg (approximately 6 mg/m^2). However, the higher drug doses induced significant cumulative damage to normal retina, choroid and optic nerve, emphasizing that selectivity of PDT with verteporfin is only relative.

In summary, preclinical studies demonstrated that the treatment of CNV with verteporfin PDT was relatively selective, resulting in some reversible damage to choriocapillaris and RPE but minimal effect on photoreceptors. Inner retina as well as retinal and choroidal vessels could be preserved. Intravenous injection of the verteporfin performed rapidly in bolus fashion or in slower infusions proved equally effective. Higher irradiances were tested and found to be safe and effective, making the treatment more practical. Additionally, it was noted that treatment effects were more sensitive to small changes in drug dose than changes in light doses and that decreasing drug doses shortened the effective time window for light application. Drug, light, and timing parameters were identified that appeared optimal for subsequent clinical study. Finally, longer-term studies and analysis of repeated treatments indicated that the effect on normal structures recovered over time within a range of drug and light doses.

OTHER PHOTOSENSITIZERS

While verteporfin has been the most extensively characterized photosensitizer for use in ocular applications, other drugs have also been evaluated in animal models and to some extent in clinical trials. Tin ethyl purpurin (SnET2) is a hydrophobic compound prepared in a lipid emulsion and is transported by low-density lipoprotein (LDL) and high density lipoprotein (HDL) in plasma (27). Occlusion of choriocapillaris was achieved with PDT using SnET2 in eyes of pigmented rabbits at light doses as low as $5 \, J/cm^2$ and a drug dose of $0.5 \, mg/kg$ (28). Light was applied 15–45 minutes following intravenous injection of SnET2, using 664 nm at a fluence of $300 \, mW/cm^2$. Retinal pigment epithelial and outer photorecepter damage was observed. Similar studies with SnET2 in a monkey model of CNV demonstrated closure of neovascularization using a drug dose of $1 \, mg/kg$ and 664 nm light at an irradiance of $600 \, mW/cm^2$ and fluences between 35 and $70 \, J/cm^2$ (22). Selectivity of SnET2 PDT in this model was not analyzed.

Lutetium texaphyrin (Lu-tex) is a water-soluble, porphyrin-related molecule which has been studied as a photosensitizer in experimental models of atherosclerosis and various tumors (29). In vitro work in our laboratory revealed preferential uptake of Lu-Tex by bovine capillary endothelial cells when compared with retinal pigment epithelial cells, suggesting the possibility of increased selectivity (R. Z. Renno and J. W. Miller, unpublished data). When Lu-Tex PDT was applied to the monkey model of experimental CNV, closure of the neovascular membranes was achieved at doses of $1–2 \, mg/kg$ with irradiation 10–40 minutes after injection using 732 nm light at $50–100 \, J/cm^2$ (30). Limited toxicity to retinal and choroidal structures, similar to that seen with verteporfin was observed.

ATX-S10 is a water-soluble chlorin that has been studied in experimental models of CNV in both rats and primates. Initial studies in rats demonstrated effective and relatively selective occlusion of laser-induced CNV using $16 \, mg/kg$ ATX-S10 followed by 670 nm irradiation immediately after dye injection at a fluence of $7.4 \, J/cm^2$ or two to four hours later at $22.0 \, J/cm^2$ (31). Subsequent angiography using $16 \, mg/kg$ ATX-S10 in rats demonstrated maximum localization of the photosensitizer to CNV at 1.5 hours after injection and irradiation with $22.0 \, J/cm^2$ at this time resulted in effective closure of CNV (32). Histologic evaluation of normal eyes treated with the same parameters showed some damage to photoreceptor outer segments and pigment-laden cells overlying the RPE, but good overall preservation of retinal and choroidal structures. Angiography studies in monkeys have revealed preferential accumulation of ATX-S10 in CNV 30 minutes after injection and several effective treatment regimens have been identified: (1) $30–74 \, J/cm^2$ applied at 30–74 minutes after $8 \, mg/kg$ injection of ATX-S10, (2) $1–29 \, J/cm^2$ at 30–74 minutes after $12 \, mg/kg$, or (3) $30–74 \, J/cm^2$ at 75–150 minutes after $12 \, mg/kg$ (33).

Preclinical studies using another chlorin, mono-L-aspartyl chlorin e6 (NPe6) have been conducted in rabbits and monkeys (34). PDT using Npe6 was performed in the monkey model of laser-induced CNV with dye doses ranging from 0.5 to 10 mg/kg and fluences from 7.5 to 225.0 J/cm^2 (35). While successful occlusion was obtained at all dye doses, optimal parameters were judged to involve treatment 5–30 minutes following dye injection with dye doses of either 0.5 or 1.0 mg/kg. Higher fluences were required with the lower photosensitizer dose and with increasing time between injection and irradiance. Histologic analysis seven days after treatment revealed numerous vacuoles in the cytoplasm of RPE cells, but the neurosensory retina remained intact.

The phthalocynanin, AlPcS4 or CASPc, has also been shown to be an effective agent in PDT of experimental CNV in monkeys (36). Complete closure of CNV was achieved using a dose of 3 mg/kg and irradiation with 675 nm light at 34 J/cm^2 and 283 mW/cm^2 applied 30 minutes after drug administration. This photosensitizer has also been used in studies of laser-targeted photoocclusion in rats (see Chapter 10) (37). The drug encapsulated in heat-sensitive liposomes was locally released using an argon laser at 5.7 W/cm^2 after intravenous administration. The released photosensitizer was then activated with 675 nm light at 270 mW/cm^2. Choriocapillaris remained occluded for a 30-day follow-up period while larger choroidal vessels and retinal vessels were unaffected. Preliminary histologic evaluation revealed no damage to RPE.

FUTURE DIRECTIONS

The successful treatment of experimental CNV with verteporfin PDT led to a series of randomized clinical trials which demonstrated a visual benefit of verteporfin PDT for patients with subfoveal CNV. However, this benefit is achieved with multiple retreatments and the rate of vision loss is still substantial. A recent report from the Verteporfin in Photodynamic Therapy group showed that 29% of patients who received PDT for occult subfoveal CNV lost six or more lines of vision after two years and 55% of these patients lost three or more lines of vision (38). Similarly, the TAP Extension Study found that 37.5% of patients with predominantly classic CNV who were treated with verteporfin PDT lost three or more lines of vision after two years (39).

Strategies for improving the treatment of ocular neovascular disorders include optimizing PDT, developing anti-angiogenic agents, and combining PDT with anti-angiogenic therapy. We have demonstrated a synergistic effect on bovine capillary endothelial cells with the combination of angiostatin and PDT using both Lu-tex and verteporfin (40). This effect was also confirmed in vivo in a rat model of CNV (41). We have also investigated the effect of inhibition of vascular endothelial growth factor (VEGF) in conjunction with PDT in experimental CNV.

Ranibizumab (1) rhufab V2 is a recombinant humanized monoclonal antibody fragment against VEGF developed by Genentech. Intravitreal administration of ranibizumab has been shown to inhibit experimental CNV in the monkey (42). Further work from our laboratory in the same model has shown that the combination of verteporfin PDT and intravitreal ranibizumab is safe and may result in a greater reduction in angiographic leakage than PDT alone (43). Clinical trials using ranibizumab for neovascular AMD are ongoing. Another anti-VEGF agent, pegaptanib, an anti-VEGF aptamer produced by Eyetech Pharmaceuticals, is also currently in clinical trials for neovascular AMD, both alone and in combination with PDT, and triamcinoolone acetonide has also been recently combined with PDT in clinical trials (see Chapter 16).

Improved treatment outcomes may also be achieved by enhancing the selectivity of PDT. Increasing the specificity of drug delivery is a key component to advancing

PDT as a therapeutic modality. Antibody-based targeting is one method currently under investigation. In vitro studies using various tumor cell lines have shown that photosensitizers conjugated to monoclonal antibodies can achieve a higher photo-toxic effect at lower doses than with drug or antibody alone (44,45). In vivo work in a mouse rhabomyosarcoma model yielded similar results (46).

While direct attachment of photosensitizing drugs to monoclonal antibodies is possible, the number of molecules that can be bound to one antibody is limited due to loss of or alterations in antigenic specificity. The use of spacers such as dextran, polyglutamic acid, or polyvinyl alcohol (PVA) has been proposed to address these issues (47). This method of conjugation allows high molar ratios of drug to antibody while conferring water-solubility to the final compound. Jiang et al. (44) linked BPD to 5E8, a monoclonal antibody against a cell-surface glycoprotein, using PVA and demonstrated 15-fold higher phototoxicity with the conjugate than with BPD alone.

For current ocular applications, the intended target for photosensitizer delivery is the neovascular endothelium. One strategy is to bind the photosensitizer to a molecule directed at binding sites on the CNV endothelium, such as VEGF receptors or integrins. Work in our laboratory focused on a peptide ATWLPPR, which has been shown to bind specifically to the VEGFR2 receptor also known as KDR or FLK-1. This peptide completely inhibits VEGF binding to VEGFR2 (48). We produced a targeted photosensitizer by binding verteporfin to a PVA linker and then to the homing peptide ATWLPPR (49). For controls we used verteporfin–PVA, which is a large but untargeted molecule, and also commercially available verteporfin. In vivo experiments were carried out in the laser-injury model of CNV in the rat for which dosimetry for verteporfin PDT has been optimized (50).

We found that PDT using both targeted verteporfin and verteporfin–PVA were effective in CNV closure. One day following treatment with targeted verteporfin, fluorescein angiography demonstrated no perfusion or leakage from CNV. Both large molecules were more efficient than unbound verteporfin in achieving CNV closure. PDT was also performed to normal retina and choroid to assess selectivity. No angiographic changes were seen 1 day after PDT using VEGFR2-targeted PDT. Histologically, the eye treated with VEGFR2-targeted verteporfin showed preserved retina and very minimal changes to RPE. In contrast, treatment of normal retina and choroid using the verteporfin–PVA control showed hyperfluorescence on angiography and retinal damage on light microscopy.

In addition to tissue-specific targeting, increasing knowledge regarding the importance of the subcellular localization of photosensitizers has raised the potential for intracellular drug targeting. There is evidence that PDT using drugs such as BPD which localize in mitochondria results in a rapid release of cytochrome c into the cytosol which initiates the apoptotic cascade (51). Photosensitizers, such as NPe6, which localize to lysosomes can induce apoptosis or necrosis, and those which accumulate in the plasma membrane can activate pathways that either lead to cell rescue or cell death (51,52). Some have suggested that targeting drug to the cell nucleus, which is particularly sensitive to damage from reactive oxygen species, could increase the efficiency of PDT (53). A better understanding of the cellular mechanisms involved in the response to PDT will allow for identification of specific intracellular targets for photosensitizer delivery as well as combination therapies directed toward modulation of signaling pathways such as those leading to apoptosis. Such advances in the delivery and design of drugs used in PDT hold the promise of better visual outcomes for a greater number of patients.

REFERENCES

1. Raab O. Uber die Wirkung fluoreszierender Stoffe auf Infusorien. Z Biol 1900; 39:524–546.
2. von Tappeiner H, Jesionek A. Therapeutische Versuche mit fluoreszierenden Stoffen. Muench Med Wochenschr 1903; 47:2042–2044.
3. Konan YN, Gurny R, Allemann E. State of the art in the delivery of photosensitizers for photodynamic therapy. J Photochem Photobiol B 2002; 66:89–106.
4. Moan J, Berg K. The photodegradation of porphyrins in cells can be used to estimate the lifetime of singlet oxygen. Photochem Photobiol 1991; 53:549–553.
5. van Leengoed HL, Cuomo V, Versteeg AA, van der Veen N, Jori G, Star WM. In vivo fluorescence and photodynamic activity of zinc phthalocyanine administered in liposomes. Br J Cancer 1994; 69:840–845.
6. Maziere JC, Santus R, Morliere P, et al. Cellular uptake and photosensitizing properties of anticancer porphyrins in cell membranes and low and high density lipoproteins. J Photochem Photobiol B 1990; 6:61–68.
7. Rudling MJ, Angelin B, Peterson CO, Collins VP. Low density lipoprotein receptor activity in human intracranial tumors and its relation to the cholesterol requirement. Cancer Res 1990; 50:483–487.
8. Denekamp J. Vascular endothelium as the vulnerable element in tumours. Acta Radiol Oncol 1984; 23:217–225.
9. Ackroyd R, Kelty C, Brown N, Reed M. The history of photodetection and photodynamic therapy. Photochem Photobiol 2001; 74:656–669.
10. Allison BA, Pritchard PH, Richter AM, Levy JG. The plasma distribution of benzoporphyrin derivative and the effects of plasma lipoproteins on its biodistribution. Photochem Photobiol 1990; 52:501–507.
11. Richter AM, Waterfield E, Jain AK, Sternberg ED, Dolphin D, Levy JG. In vitro evaluation of phototoxic properties of four structurally related benzoporphyrin derivatives. Photochem Photobiol 1990; 52:495–500.
12. Richter AM, Cerruti-Sola S, Sternberg ED, Dolphin D, Levy JG. Biodistribution of tritiated benzoporphyrin derivative (3H-BPD-MA), a new potent photosensitizer, in normal and tumor-bearing mice. J Photochem Photobiol B 1990; 5:231–244.
13. Richter AM, Waterfield E, Jain AK, et al. Photosensitising potency of structural analogues of benzoporphyrin derivative (BPD) in a mouse tumour model. Br J Cancer 1991; 63:87–93.
14. Richter AM, Yip S, Waterfield E, Logan PM, Slonecker CE, Levy JG. Mouse skin photosensitization with benzoporphyrin derivatives and Photofrin: macroscopic and microscopic evaluation. Photochem Photobiol 1991; 53:281–286.
15. Schmidt-Erfurth U, Bauman W, Gragoudas E, et al. Photodynamic therapy of experimental choroidal melanoma using lipoprotein-delivered benzoporphyrin. Ophthalmology 1994; 101:89–99.
16. Schmidt-Erfurth U, Hasan T, Gragoudas E, Michaud N, Flotte TJ, Birngruber R. Vascular targeting in photodynamic occlusion of subretinal vessels. Ophthalmology 1994; 101:1953–1961.
17. Miller JW, Walsh AW, Kramer M, et al. Photodynamic therapy of experimental choroidal neovascularization using lipoprotein-delivered benzoporphyrin. Arch Ophthalmol 1995; 113:810–818.
18. Gomer CJ, Doiron DR, Jester JV, Szirth BC, Murphree AL. Hematoporphyrin derivative photoradiation therapy for the treatment of intraocular tumors: examination of acute normal ocular tissue toxicity. Cancer Res 1983; 43:721–727.
19. Gomer CJ, Doiron DR, White L, et al. Hematoporphyrin derivative photoradiation induced damage to normal and tumor tissue of the pigmented rabbit eye. Curr Eye Res 1984; 3:229–237.
20. Moulton RS, Walsh AW, Miller JW. Response of retinal and choroidal vessels to photodynamic therapy using benzoporphyrin derivative monoacid. Invest Ophthalmol Vis Sci 1993; 34:S1169.

21. Kramer M, Miller JW, Michaud N, et al. Liposomal benzoporphyrin derivative verteporfin photodynamic therapy. Selective treatment of choroidal neovascularization in monkeys. Ophthalmology 1996; 103:427–438.

22. Baumal C, Puliafito CA, Pieroth L. Photodynamic therapy (PDT) of experimental choroidal neovascularization with tin ethyl etiopurprin. Invest Ophthalmol Vis Sci 1996; 37:S122.

23. Waldow SM, Dougherty TJ. Interaction of hyperthermia and photoradiation therapy. Radiat Res 1984; 97:380–385.

24. Husain D, Kramer M, Kenny AG, et al. Effects of photodynamic therapy using verteporfin onexperimental choroidal neovascularization and normal retina and choroid up to 7 weeks after treatment. Invest Ophthalmol Vis Sci 1999; 40:2322–2331.

25. Husain D, Miller JW, Michaud N, Connolly E, Flotte TJ, Gragoudas ES. Intravenous infusion of liposomal benzoporphyrin derivative for photodynamic therapy of experimental choroidal neovascularization. Arch Ophthalmol 1996; 114:978–985.

26. Reinke MH, Canakis C, Husain D, et al. Verteporfin photodynamic therapy retreatment of normal retina and choroid in the cynomolgus monkey. Ophthalmology 1999; 106:1915–1923.

27. Polo L, Reddi E, Garbo GM, Morgan AR, Jori G. The distribution of the tumour photosensitizers Zn(II)-phthalocyanine and Sn(IV)-etiopurpurin among rabbit plasma proteins. Cancer Lett 1992; 66:217–223.

28. Peyman GA, Moshfeghi DM, Moshfeghi A, et al. Photodynamic therapy for choriocapillaris using tin ethyl etiopurpurin (SnET2). Ophthalmic Surg Lasers 1997; 28:409–417.

29. Woodburn KW, Engelman CJ, Blumenkranz MS. Photodynamic therapy for choroidal neovascularization: a review. Retina 2002; 22:391–405.

30. Arbour JD, Connolly E, Graham K, Gragoudas E, Miller JW. Photodynamic therapy of experimental choroidal neovascularization in a monkey model using intravenous infusion of lutetium texaphyrin. Invest Ophthalmol Vis Sci 1999; 40(suppl):401.

31. Obana A, Gohto Y, Kaneda K, Nakajima S, Takemura T, Miki T. Selective occlusion of choroidal neovascularization by photodynamic therapy with a water-soluble photosensitizer, ATX-S10. Lasers Surg Med 1999; 24:209–222.

32. Hikichi T, Mori R, Nakajima S, et al. Dynamic observation of selective accumulation of a photosensitizer and its photodynamic effects in rat experimental choroidal neovascularization. Retina 2001; 21:126–131.

33. Obana A, Gohto Y, Kanai M, Nakajima S, Kaneda K, Miki T. Selective photodynamic effects of the new photosensitizer ATX-S10(Na) on choroidal neovascularization in monkeys. Arch Ophthalmol 2000; 118:650–658.

34. Mori K, Yoneya S, Ohta M, et al. Angiographic and histologic effects of fundus photodynamic therapy with a hydrophilic sensitizer (mono-L-aspartyl chlorin e6). Ophthalmology 1999; 106:1384–1391.

35. Mori K, Yoneya S, Anzail K, et al. Photodynamic therapy of experimental choroidal neovascularization with a hydrophilic photosensitizer: mono-L-aspartyl chlorin e6. Retina 2001; 21:499–508.

36. Kliman GH, Puliafito CA, Stern D, Borirakchanyavat S, Gregory WA. Phthalocyanine photodynamic therapy: new strategy for closure of choroidal neovascularization. Lasers Surg Med 1994; 15:2–10.

37. Asrani S, Zou S, D'Anna S, et al. Feasibility of laser-targeted photoocclusion of the choriocapillary layer in rats. Invest Ophthalmol Vis Sci 1997; 38:2702–2710.

38. Verteporfin in Photodynamic Therapy Study Group. Verteporfin therapy of subfoveal choroidal neovascularization in age-related macular degeneration: two-year results of a randomized clinical trial including lesions with occult with no classic choroidal neovascularization—verteporfin in photodynamic therapy study group. Am J Ophthalmol 2001; 131:541–560.

39. Blumenkranz MS, Bressler NM, Bressler SB, et al. Verteporfin therapy for subfoveal choroidal neovascularization in age-related macular degeneration: three-year results

of an open-label extension of 2 randomized clinical trials—TAP report no. 5. Arch Ophthalmol 2002; 120:1307–1314.

40. Renno RZ, Delori FC, Holzer RA, Gragoudas ES, Miller JW. Photodynamic therapy using Lu-Tex induces apoptosis in vitro, and its effect is potentiated by angiostatin in retinal capillary endothelial cells. Invest Ophthalmol Vis Sci 2000; 41:3963–3971.

41. Terada Y, Michaud NA, Connolly EJ, et al. Enhanced photodynamic therapy using angiostatin with verteporfin PDT in a laser-injury rat model. Invest Ophthalmol Vis Sci 2003; 44:1749 (E-Abstract).

42. Krzystolik MG, Afshari MA, Adamis AP, et al. Prevention of experimental choroidal neovascularization with intravitreal anti-vascular endothelial growth factor antibody fragment. Arch Ophthalmol 2002; 120:338–346.

43. Husain D, Kim I, Gauthier D, et al. Safety and efficacy of intravitreal ranibizumab in combination with verteporfin PDT on experimental choroidal neovascularization in the monkey. Arch Ophthalmol 2005; 123:506–516.

44. Jiang FN, Allison B, Liu D, Levy JG. Enhanced photodynamic killing of target cells by either monoclonal antibody or low density lipoprotein mediated delivery systems. J Control Release 1992; 19:41–58.

45. Mew D, Lum V, Wat CK, et al. Ability of specific monoclonal antibodies and conventional antisera conjugated to hematoporphyrin to label and kill selected cell lines subsequent to light activation. Cancer Res 1985; 45:4380–4386.

46. Mew D, Wat CK, Towers GH, Levy JG. Photoimmunotherapy: treatment of animal tumors with tumor-specific monoclonal antibody–hematoporphyrin conjugates. J Immunol 1983; 130:1473–1477.

47. Jiang FN, Jiang S, Liu D, Richter A, Levy JG. Development of technology for linking photosensitizers to a model monoclonal antibody. J Immunol Methods 1990; 134:139–149.

48. Binetruy-Tournaire R, Demangel C, Malavaud B, et al. Identification of a peptide blocking vascular endothelial growth factor (VEGF)-mediated angiogenesis. EMBO J 2000; 19:1525–1533.

49. Renno RZ, Terada Y, Haddadin MJ, et al. Selective photodynamic therapy by targeted verteporfin delivery to experimental choroidal neovascularization mediated by a homing peptide to vascular endothelial growth factor receptor-2. Arch Ophthalmol 2004; 122:1002–1011.

50. Zacks DN, Ezra E, Terada Y, et al. Verteporfin photodynamic therapy in the rat model of choroidal neovascularization: angiographic and histologic characterization. Invest Ophthalmol Vis Sci 2002; 43:2384–2391.

51. Moor AC. Signaling pathways in cell death and survival after photodynamic therapy. J Photochem Photobiol B 2000; 57:1–13.

52. Roberts WG, Liaw LH, Berns MW. In vitro photosensitization II. An electron microscopy study of cellular destruction with mono-L-aspartyl chlorin e6 and photofrin II. Lasers Surg Med 1989; 9:102–108.

53. Rosenkranz AA, Jans DA, Sobolev AS. Targeted intracellular delivery of photosensitizers to enhance photodynamic efficiency. Immunol Cell Biol 2000; 78:452–464.

10

Thermal-Sensitive Liposomes

Sanjay Asrani
Duke University Eye Center, Durham, North Carolina, U.S.A.

Morton F. Goldberg and Ran Zeimer
Wilmer Ophthalmological Institute, Johns Hopkins University,
Baltimore, Maryland, U.S.A.

INTRODUCTION

Liposomes are microscopic lipid bubbles designed to entrap drugs. They have been used locally as well as systemically for targeting of drugs to specific organs or for prolonging drug effect. The encapsulation of drugs in liposomes has been shown to reduce the toxicity, provide solubility in plasma, and enhance permeability through tissue barriers. Some applications related to cancer and infectious diseases have reached clinical use, while others are currently in Phase I–III human clinical trials.

A method has been developed to target drugs locally in the eye via a light-based mechanism. The method, called laser-targeted delivery (LTD) (1–3), consists of encapsulating a drug in heat-sensitive liposomes, injecting them intravenously, and releasing their content at the site of choice by noninvasively warming up the targeted tissue with a laser pulse directed through the pupil of the eye. The specific temperature needed for the phase transition is 41°C (105.8°F), which causes the liposomes to release their contents in the blood in <0.1 second. LTD can be conceptualized as a noninvasive "catheterization" of a specific microvasculature. Similar to cardiac catheterization, LTD provides the means for local delivery of an agent.

Laser-targeted delivery benefits from the basic advantages of liposomal delivery. By virtue of being encapsulated, the drug is confined to the liposomes, thereby reducing exposure of nontargeted organs. In addition, agents with a short half-life in plasma (anti-angiogenic factors, neuroprotective agents, anti-inflammatory compounds, etc.) are shielded from the blood components and can reach their target in their original form. LTD also possesses certain unique advantages, such as a well-defined thermal mechanism and a predetermined temperature to release the liposomal contents. This is in contrast to the targeting approaches which depend on complex cell surface interactions that may be altered in human diseases.

The current methods of drug administration to the retina and choroid are based on topical, periorbital, intravitreal, and systemic administrations. The first two methods are hampered by relatively low penetration, the second and the third by their invasive nature, and the last by exposure of the whole body to the drug. The difficulty of drug targeting is one of the major reasons for the paucity of pharmacological therapies available for the management of retinal and choroidal diseases. This review concentrates on the potential applications of LTD in therapy and diagnosis of ocular diseases.

METHODOLOGY OF LASER-TARGETED DRUG DELIVERY

Principle of Laser-Targeted Drug Delivery

The principle of LTD is illustrated in Figure 1 in its application for the diagnosis and therapy of choroidal neovascularization (CNV) in age-related macular degeneration (AMD). Following an intravenous injection, liposomes circulate in the blood stream. During LTD, an infrared laser beam irradiates the CNV and its surrounding tissues and is absorbed by blood in the CNV and the choriocapillaris, as well as by pigment in the retinal pigment epithelium (RPE) and choroid. The liposomes are consequently

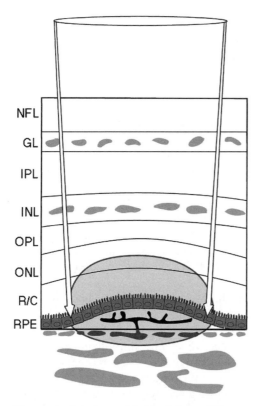

Figure 1 Principle of LTD. Schematic representation of heat and dye distribution during laser-targeted drug delivery following a laser pulse in an eye with CNV. The energy deposited in the tissues causes heating, as illustrated by the oval. The bolus of dye released in the CNV vessels is retained longer than that in the choriocapillaris because of slower flow within the CNV. The CNV and the tissues in its immediate vicinity reach the releasing temperature of the liposomes, but the retinal vessels do not. *Abbreviations*: LTD, laser-targeted delivery; CNV, choroidal neovascularization. *Source*: From Ref. 4, Figure 1.

warmed up most efficiently in these anatomic locations. These tissues are thus the first to reach the temperature necessary to cause phase transition in the circulating liposomes (41°C), resulting in release of their content. After 200 msec, the laser beam used for release is turned off, and the tissues rapidly cool secondary to rapid flow of blood, which stops further release of active agent from the liposomes. For a heated area 800 μm in diameter, the retinal temperature rise is only 2.8°C at a distance of 150 μm from the RPE, where the outermost retinal capillaries are located. Thus, liposomes circulating in the retinal vessels do not get warmed up by the required 4°C, and thus do not release their content within the retina. A few milliseconds after the release of the agents from the liposomes, the active substances are cleared from normal blood vessels, but they persist within the CNV due to its slow circulation (described in detail in a later section). If the active agent is a photosensitizer, its activation a few milliseconds later by a sensitizing wavelength causes closure only primarily in the CNV.

Liposome Preparation

The lipids typically are dipalmitoylphosphatidylphosphocholine (DPPC) and dipalmitoylphosphatidylglycerol (DPPG). In more recent experiments, distearoyl-phosphoethanolamine methoxypolyethyleneglycol 2000 (DSPE-MPEG) has been added to increase the circulation time by reducing the removal of liposomes through the reticuloendothelial system (5). The preparation followed the method of Hope et al. (6). Briefly, it consists of drying the lipids previously dissolved in methylene chloride and ethanol to a film by rotary evaporation under vacuum. A solution of the agent to be encapsulated is added, and the preparation is subjected to five cycles of rapid freezing at −20°C and thawing at 55°C. This is followed by repeated forced filtration through a stack of two 0.2-μm polycarbonate filters placed in a thermobarrel extruder. This yields large, unilamellar vesicles with relatively homogenous size of 120 nm. To remove the unencapsulated dye, the preparation is dialyzed against Ringer's lactate through a molecular porous membrane or filtered through a gel column.

Temperature Profile

Liposomes containing fluorescent dye (6-carboxyfluorescein, CF) were prepared. The temperature profile was studied by measuring the concentration of the free (unencapsulated) CF as the liposomes were incubated at various temperatures for 10 minutes in Ringer's lactate solution plus 1% human serum. Due to self-quenching at high concentrations, CF encapsulated in the liposomes does not contribute to the fluorescence of the sample. This permitted the assessment of the free CF concentration with a fluorophotometer without having to separate the supernatant from the liposomes. Complete release was defined as the fluorescence intensity after the dissolution of the liposomes with a detergent (TritonX 100). The free dose fraction was found to be 2% at room temperature, 5% at body temperature (37°C), and 83% at 41°C. This indicates that a sharp transition can be achieved in vitro at the intended temperature, yielding a 17-fold increase of free dye after release from the liposomes.

Pharmacokinetics

The pharmacokinetic behavior of the liposomes was studied in vivo. Five rats were injected with dye encapsulated in liposomes, and blood samples were collected every

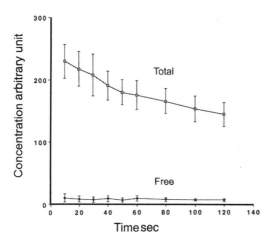

Figure 2 Pharmacokinetics of dye-encapsulated liposomes. The pharmacokinetics is illustrated by liposomes containing a fluorescent dye. The concentration in the blood is represented by the fluorescence of the sample. The upper curve represents the total concentration (encapsulated and free) in the blood, while the lower one is for the free, unencapsulated dye. Note the slow decay of the liposomes and the relatively small fraction of free dye in the blood. *Source*: From Ref. 7, Figure 2.

10 minutes. The concentration of free dye was assessed after filtration, and the total concentration was measured after lysis with detergent and warming. The results, shown in Figure 2, indicate that the intact liposomes are cleared slowly from the blood, and 75% of the dose remained encapsulated at 60 minutes. Figure 2 also indicates that the encapsulated dose remained 20-fold higher than the dose of the free dye. Thus, during LTD, a 20-fold increase in dye concentration occurs at the targeted location.

Instrumentation for Laser Delivery and Visualization

A fundus camera was modified to provide video angiograms and to deliver one laser beam used to release the content of the liposomes and another laser beam to activate the photosensitizer. The first laser was also used to illuminate the fundus. The output of the charge-coupled device (CCD) camera was fed into a video image enhancer and recorded on magnetic tapes with a high-frequency video recorder. Later, the tape was played back, and video sequences were digitized with a frame grabber for subsequent analysis.

An argon laser (filtered to deliver only at 488 nm) was used to release the liposomes' content for purposes of angiography. The power of the laser was increased gradually until a bright fluorescent bolus was observed. Typically, this was achieved with a power of 16 mW applied on a 600-μm spot for a duration of 200 msec. In the case of photo-occlusion (described later), a diode laser emitting at 675 nm was used to activate the photosensitizer. The delivery of both laser beams was controlled by shutters activated by a computer.

The instrument has been developed further to be applicable in Phase I and II clinical trials. The optics have been specifically designed to yield a compact optical head. The illumination for angiography is provided by light-emitting diodes, and the lasers for release and activation consist of diode lasers incorporated into the optical head. The imaging sensor is a digital CCD with enhanced resolution. The operation of the instrument, the acquisition and recording of images, and the processing are all computerized. These improvements make the operation similar to that of current digital cameras and thus well suited for clinical studies.

POTENTIAL THERAPEUTIC APPLICATIONS OF LTD

Laser-Targeted Photo-Occlusion

CNV accounts for the majority of legally blind eyes with AMD (8). The efficacy of thermal photocoagulation is very limited, as it is applicable only to a minority of cases; the recurrence rate is high; and it causes permanent damage to the adjacent neurosensory retina, RPE and normal choriocapillaris. Photodynamic therapy (PDT) has been recently introduced to treat CNV. It consists of injecting intravenously a light-sensitive agent (a photosensitizer) that damages and occludes CNV when exposed to light at an appropriate wavelength. The Treatment of Age-related Macular Degeneration with Photodynamic Therapy (TAP) study group has demonstrated the benefit of this therapy for a well-defined category of CNV (9). Unfortunately, this category is present in a minority of cases. PDT therapy is currently limited by the need for repeated treatments and by the collateral damage to normal tissues, such as choroidal vessels and the RPE, that are essential to preservation of vision. It has also been postulated that the transient choroidal hypoperfusion and consequent ischemia may represent an angiogenic stimulus for the recurrence and progression of CNV following PDT (10–13). A well-conducted histological study demonstrated that thrombosis following PDT was incomplete in about half of the treated eyes (12). Additionally, regrowth of occluded vessels began as soon as one week after PDT (10).

The limitations of PDT are possibly the result of needing to keep the dose of the photosensitizer and the irradiating light low enough to avoid collateral damage. The low dose is most likely sufficient to cause thrombus formation but insufficient to achieve the desired effect of vascular wall damage. Such an occlusion may thus be temporary, due to dissolution of the clot and vessel re-canalization.

Additionally, the inherent nature of the disease, which allows leakage of the photosensitizer into adjacent tissues, results in their damage as well. Laser-targeted photo-occlusion (LTO) delivers a photosensitizer specifically to CNV by releasing a bolus of the photosensitizer only in the vasculature and in the vicinity surrounding the CNV, limiting its presence to the lumen. Studies have demonstrated that CNVs are perfused with slower flow than in the normal choriocapillaris (4,14). The activation is therefore delayed until the photosensitizer has cleared from the normal vasculature. These unique features aid in specific targeting of the treatment to the CNV.

Extensive experiments in nonhuman primates, rabbits, rats, and dogs have demonstrated that LTO possesses the following additional features:

1. By irradiating with the activating infra-red beam immediately following the release, the damage can be limited to the vessels that contain the photosensitizer, thus avoiding accumulation of the photosensitizing agent in the interstitial tissues and their subsequent damage upon irradiation (15).
2. The average washout time of dye from the normal choriocapillaris was 0.9 second (average of 93 locations of eight rats) (14). Thus, during LTO, the diode laser (photosensitizing laser wavelength) was activated 1 second after the argon laser pulse and bolus release, thereby ensuring clearance of most of the photosensitizer from normal choriocapillaris. This permits

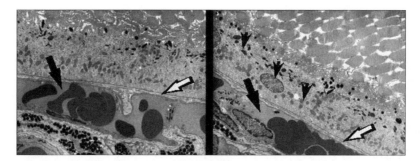

Figure 3 Histopathology of a region treated with laser-targeted photo-occlusion. Electron microscopy of a region immediately adjacent to an area treated with laser-targeted photo-occlusion resulting in occluded CNV. These sections are <50 μm from the center of the occluded CNV but within the area treated by LTO. Note the integrity of the RPE and its nucleus (*arrowheads*) and Bruch's membrane (*white arrows*) and the well perfused choriocapillaris (*black arrows*). *Abbreviations*: CNV, choroidal neovascularization; LTO, laser targeted photo-occlusion; RPE, retinal pigment epithelium. *Source*: From Ref. 15, Figure 7.

delivery of a high amount of energy to activate the photosensitizer in the CNV, while sparing the normal choriocapillaris.

3. LTO has been shown to be relatively free of damage to the normal vessels and to the RPE (16,17). CNV lesions were created in a rat model, some of which were treated with LTO and some left untreated. Light and electron microscopy showed that three of four untreated CNVs had more than one lumen open and no occlusion, and that one CNV had spontaneously occluded vessels (18). Microscopic examination of eight LTO-treated CNVs showed six with no CNV, one with partial occlusion and one without occlusion. LTO-treated areas next to the CNVs showed normal photoreceptors, RPE, Bruch's membrane, and choriocapillaris (Fig. 3) (15). The preservation of the RPE may be particularly important in AMD, because it is already abnormal and may be more sensitive to additional injury. The intact nature of Bruch's membrane is also important in AMD because breaks are believed to be associated with further proliferation of the CNV into the subretinal space.

4. LTO promises to offer effective treatment to both kinds of CNV ("classic" as well as "occult"), as action is based on the presence of photosensitizer in the CNV at the time of irradiation. Large CNVs would also be amenable to treatment for the same reason (4).

5. LTO shares the basic advantages of any other liposomal delivery system: it protects most organs (they are not exposed to the agent, thereby reducing systemic toxicity).

LTO could also be applied to retinal neovascularization which occurs in diseases such as diabetes and sickle cell disease. Most of these new vessels which proliferate into the preretinal space and vitreous can be made to regress by pan-retinal thermal photocoagulation. However, in cases of persistent neovascularization, which results in recurrent vitreous hemorrhage, LTO may potentially be used.

Laser-Targeted Drug Delivery to Retinal Tissues

Bacterial and fungal endophthalmitis, viral retinitis, toxoplasmosis, uveitis and other inflammatory disorders are among the posterior segment diseases amenable to drug treatment. Low amounts of drugs reach the retina and choroid after topical applications, because drug penetration through the outer eye wall is relatively poor. Intravitreal injections and implants can be used to raise drug concentration in the retina and choroid, allowing for prolonged drug therapy; however, the dose of some drugs is limited, because the entire retina is exposed. Conventional methods of systemic drug administration are also restrictive, as side effects may arise because of exposure of the whole body.

It has been previously shown that a bolus of dye can be released inside retinal vessels using LTD. The amount of drug delivered to the surrounding retina is minimal, because the bolus is cleared rapidly by the blood stream, and the blood–retinal barrier prevents drug penetration. However, in many retinal diseases amenable to drug therapy, the blood–retinal barrier is disrupted, and thus, targeted delivery to the parenchymal retina may be possible. To test this hypothesis, moderate argon laser pulses were applied to retinal vessels of Dutch belted rabbits to induce breakdown of the blood–retinal barrier (19). Carboxyfluorescein encapsulated in liposomes was released upstream of the damaged vascular segment, and angiograms were recorded. The penetration of the marker into the parenchymal (perivascular) retinal tissue was evaluated by comparing the intensity of the fluorescence in the area around the damaged vessel to that of an adjacent control area. The results showed that: the dye penetration increased with a greater breakdown of the blood–retinal barrier (the penetration being restricted to those specific areas), and the dye gradually diffused far from the site of release. The possibility of targeting drugs in the retina around vessels with a disrupted barrier is exciting, as it may open a unique way of enhancing the therapeutic effect within diseased portions of the retina, while minimizing side effects systemically and in remote, normal retinal (and other intraocular) locations from potentially toxic agents.

Other Applications

Significant progress can be anticipated in the development of genetic material that could be used in the treatment of retinal diseases. LTD could be of great value in targeting genetic material to a given site. For example, management of neovascularization based on blocking angiogenic agents (e.g., anti-VEGF antibodies) could be more targeted and possibly more effective than the current intraviteral or systemic approaches. Progress is being made toward the identification of growth factors specific to CNV. Once the growth factors and their receptors have been identified, new treatments could be devised, based on competition or blockage of these factors and/ or their receptors. LTD could be a preferential delivery method in the eye because of local targeting, shielding of other organs from the active agent, and prevention of degradation or inactivation of the active agent by blood components.

DIAGNOSTIC APPLICATIONS

Angiography

Conventional fluorescein angiography has been a very useful clinical tool to assess the ocular vasculature. However, it has a number of limitations. First, the dye rapidly fills both the retinal and choroidal vessels; thus, the visualization of small vascular beds, such as CNV, is often obscured by the lack of contrast caused by the bright fluorescence emanating from the large volume of dye present in the underlying normal choroidal vessels. Second, visualization and detection of abnormalities such as CNV are based on leakage of the dye or staining of vascular walls or both. This method of detection may not be reliable, because, at certain stages of the disease, the vessels may neither leak nor stain. Third, the excitation and fluorescence may be diminished by subretinal blood, turbid fluid, pigment, or fibrous tissue, thereby reducing the intensity of the angiographic image of the CNV (20,21).

Indocyanine green (ICG) is beneficial in some cases, because the excitation and emission wavelengths of this dye are longer than those from fluorescein, and the light penetrates turbid media better (22). However, the enhanced penetration of light in ICG angiography permits large underlying choroidal vessels to be visualized more effectively, masking details of adjacent smaller vascular structures. ICG angiography also shares with fluorescein angiography the disadvantage of relying on leakage and staining for diagnostic interpretation. The poor understanding of the staining and pooling mechanisms of this dye has hampered the interpretation of ICG angiograms.

Laser-targeted angiography (LTA), which consists of laser-targeted drug delivery using carboxyfluorescein liposomes, has the potential of overcoming some of the problems of conventional angiography because of the following advantages:

1. The local release of a fluorescent bolus permits the visualization of selected vascular beds without interference from overlying or underlying beds. LTA permits delivery of substances to the subretinal vasculature without causing retinal exposure to the substance (23,24). Conversely, the retinal vasculature, if diseased, can also be specifically targeted (3,25). Visualization of vessels in conditions somewhat similar to those present in AMD has been demonstrated, as shown in Figure 4.
2. As the visualization is independent of staining and leakage, but rather relies solely on the presence of a transient, brief bolus of flurorescein in the lumen, the CNV can be visualized rather easily, as long as it is patent.
3. The short release of a bolus of dye, accompanied by rapid washout, ensures that the dye will not accumulate outside the vessels and mask the CNV.
4. The bolus angiograms can be repeated for at least 45 minutes, that is, as long as the liposomes are circulating in the blood. This provides opportunities to correct errors in alignment and to perform angiography of both eyes.
5. The hemodynamics of the CNV, delineated by the progress of the bolus, may allow identification of the vessels feeding the CNV. The dynamic nature of LTA has been successfully exploited to measure hemodynamic parameters of the macro- and microcirculation of the retina and of the choroid (3,23). Identification of all the feeding vessels could allow the clinician to limit thermal photocoagulation exclusively to these vessels and, if occlusion can be achieved, large areas of normal retina could be spared, thus limiting collateral damage and potentially preserving visual function.

LTA to Visualize the Retinal and Choroidal Vasculature

Our experiments in cynomolgus monkeys and baboons, and those of others have indicated that LTA holds promise of becoming clinically useful to visualize capillary abnormalities not seen otherwise and to identify local dye leakage (3,25,26). Differentiation between retinal and subretinal leakage would be achieved, because the dye can be released only in the retinal vasculature, and retinal leakage would be all that is visualized.

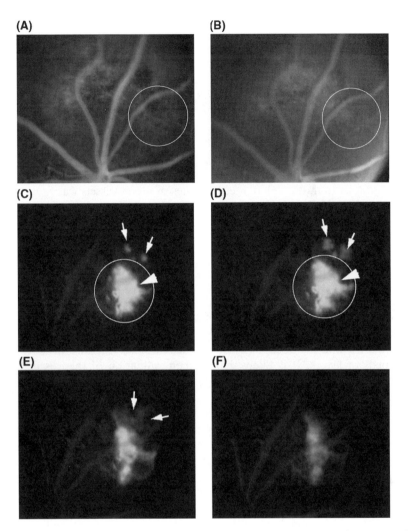

Figure 4 Fluorescein angiography and LTA of an occult CNV in a rat model. The conventional fluorescein angiograms, obtained at (**A**) 29 sec and (**B**) 3.5 min after injection, reveal the presence of a patchy fluorescent area that does not evolve over time and that provides no indication of CNV. In contrast, LTA (after release in the area marked by the circle) reveals a CNV with its exact location. (**C, D, E**) A brightly fluorescent abnormal pattern of vessels (CNV) (*arrowhead*) and fluorescent patches (*arrows*) (obtained 50, 110 and 430 msec, respectively, after the end of dye release). These patches evolve rapidly into a lobular pattern characteristic of choriocapillaris. (**F**) The fluorescent bolus clears from the normal choriocapillaris while remaining in the CNV (image obtained after 1.2 sec). *Abbreviations*: LTA, laser-targeted angiography; CNV, choroidal neovascularization. *Source*: From Ref. 16, Figure 3.

The choriocapillaris is a planar network of choroidal capillaries, presumably providing cooling and nutrition to the external portions of the retina. A number of pathologies are associated with abnormalities in the choriocapillaris, but its visualization is impeded by the presence of overlying highly pigmented RPE cells and by the background fluorescence of large underlying choroidal vessels. One potential clinical application of LTA and LTO is the visualization and management of CNV mentioned above.

Blood Flow

The measurement of retinal blood flow is important, as it provides insight into retinal physiology and leads to better understanding of the onset and progression of retinal vascular diseases that are common causes of vision loss.

The existing methods to evaluate retinal blood flow are limited to large vessels or are subjective. The application of LTA to the measurement of blood flow was demonstrated in rabbits and nonhuman primates (2,3,27). A number of parameters relevant to hemodynamics were evaluated from the progression of the dye front in the arteries, through the capillary bed and into the veins. Blood vessel diameters were measured, a map of the blood flow in the macula was drawn, and the relationship between flow and diameter in mother–daughter branches of blood vessels was found to be consistent with Murray's law (which predicts the optimum branching pattern for a vascular bed) (3,28).

Local response of the primate retinal microcirculation to increased metabolic demand was also studied (29). Light flicker was found to increase arterial blood flow and to induce local changes in the hemodynamics of the microcirculation. The findings suggested that the changes were related to the degree of neuronal activity and indicated the presence of a regulatory response that involves redirection of blood flow in the microcirculation. Similar shunting of flow has been observed during hemodilution in cerebral and coronary tissue (30). A similar mechanism may take place in diabetes; that is, shunting of capillary flow aimed at preserving flow in some selected capillary beds or layers. It is likely that further evaluation of these issues would contribute to the understanding of early functional changes preceding diabetic retinopathy.

In addition to providing information on blood flow in large retinal vessels, LTA also permitted assessment of the microcirculation. This was based on the measurement of the capillary transit time. As expected, the capillary transit time changed as a function of blood pressure and, interestingly, showed a twofold variation within the cardiac cycle (3).

A number of diseases are associated with choriocapillaris abnormalities, but their visualization is impeded in conventional angiography. Using LTA, for the first time, detailed visualization of blood flow patterns in the choroid and choriocapillaris under physiologic conditions was possible (23). In the macular area, using LTA, the choriocapillaris showed individual, in the rat, lobules, which were polygonal in shape and 200–300 µm in diameter. Each lobule was fed from its center by an arteriole, perfused radially by capillaries, and drained by a peripheral venular annulus. Each of the numerous arterioles perfused a well-defined cluster of lobules. Adjacent arterioles typically supplied separate clusters, which fit together like a jigsaw puzzle: the significance of such an arrangement is not known. The fovea was supplied by one or more branch arteriole, which were always nasal to it. At the optic nerve head, well-defined clusters of lobules created a doughnut around the optic disc.

SAFETY OF LIGHT-TARGETED DRUG DELIVERY

The information available so far indicates that LTD is safe. Intravenously administered liposomes are in use today in humans for cancer chemotherapy, as vehicles for delivery of immunomodulators, and for gene therapy (31). The liposomes used in our preparation are composed of phospholipids such as DPPC, DPPG, and polyethylene glycol. These lipids are amongst the safest used for the preparation of liposomes and have been used in clinical trials and clinical applications involving intravenous injections of liposomes (31–35). Large unilamellar liposomes, which have a high encapsulation efficiency (which also limits the amount of lipid required), have been successfully administered systemically in patients (31,36).

All the tests in humans have indicated so far that liposomal delivery does not introduce side effects other than those linked to the specific drug, which is encapsulated. We have performed a pilot toxicology test of our preparation in rats with a fluorescent dye and have not found any morbidity or mortality, or any change in biochemical parameters or histology of the liver, spleen, and kidneys (37). A formal toxicology study in the dog, using 10 times the dose intended for humans, showed no systemic side effect of clinical significance (unpublished data).

The laser power density and exposure time used to cause liposomal release of drug in the choroid are within national standards for the safe use of lights (38). Consequently, histopathologic and angiographic examinations of eyes following multiple dye releases in the choroid have indicated a lack of observable damage (23). However, the light intensity used for liposomal release of drugs into retinal vessels is higher. Damage to the RPE has been observed in this circumstance, but it is localized to a small area away from the fovea that is possibly small enough not to be noticeable by the patient (similar to extra-macular focal photocoagulation). Thus, if LTD is used for therapeutic purposes in the retina, the observed damage may be considered clinically insignificant.

LIMITATIONS

LTD is dependant upon clear ocular media. Significant media opacities, such as cataract and vitreous hemorrhage, will hamper LTD. Additionally, LTD is advantageous for transient bolus drug delivery and not for chronic therapy. LTD may have limited applications in diseases such as CMV retinitis and uveitis which require chronic therapy. Though multiple doses of the drug can be released at an ocular site during a single session of LTD, the number of sessions may be limited by the total quantity of lipid injected and possible liver toxicity. These aspects need to be further evaluated by larger toxicology studies.

There is a need for tight control of manufacturing parameters, particularly liposome size, uniformity of liposomes, stability of bioactive drugs during the encapsulation process, sterility and endotoxin control. Hydrophilic drugs are easily encapsulated, but lipophyllic drugs may need to be modified to render them encapsulable.

CONCLUSION

LTD is a promising method to deliver therapeutic and diagnostic agents to the retina and choroid. The first applications are likely to be the diagnosis and treatment of AMD. LTD is an acute drug delivery method and has potential for those drugs that need to be delivered infrequently. Further application of LTD will depend on the availability of agents that can be delivered as a bolus but have lasting effects. Agents, such as genetic and biologic material that modify cell behavior, are under development, and could be candidates for LTD.

ACKNOWLEDGMENTS

Supported by Research Grants EY 07768, EY 10017 and Core Grant EY 1765 (JHU) from the National Institutes of Health, Bethesda, Maryland, the Lew R. Wasserman Merit Award, an Alcon Research Institute Award, a Biomedical Research Award from the Whitaker Foundation, an unrestricted research grant from Research to Prevent Blindness, Inc., New York, NY and gifts from the McGraw family and the McDuffie family.

The material presented here is the product of efforts by the many collaborators who appear as authors in the cited references.

DISCLOSURE OF FINANCIAL INTEREST

Dr. Zeimer is entitled to sales royalties from PhotoVision Pharmaceuticals, Inc., Jenkintown, PA, which is developing products related to the research described in this paper. In addition, he serves as a consultant to the company. The terms of this arrangement have been reviewed and approved by the Johns Hopkins University in accordance with its conflict of interest policies.

REFERENCES

1. Zeimer R, Khoobehi B, Niesman MR, Magin RL. A potential method for local drug and dye delivery in the ocular vasculature. Invest Ophthalmol Vis Sci 1988; 29:1179–1183.
2. Zeimer R, Khoobehi B, Peyman G, Niesman MR, Magin RL. Feasibility of blood flow measurement by externally controlled dye delivery. Invest Ophthalmol Vis Sci 1989; 30:660–667.
3. Guran T, Zeimer R, Shahidi M, Mori M. Quantitative analysis of retinal hemodynamics using targeted dye delivery. Invest Ophthalmol Vis Sci 1990; 31:2300–2306.
4. Asrani S, Zou S, D'Anna S, Phelan A, Goldberg MF, Zeimer R. Selective visualization of choroidal neovascular membranes. Invest Ophthalmol Visual Sci 1996; 37:1642–1650.
5. Allen TM, Hansen CB, Martin F, Redemann C, Yau-Young A. Liposomes containing synthetic lipid derivatives of poly (ethylene glycol) show prolonged circulation half-lives in vivo. Biochim Biophys Acta 2000; 1066:29–36.
6. Hope MJ, Bally MB, Webb G, Cullis PR. Production of large unilamellar vesicles by a rapid extrusion procedure. Characterization of size distribution, trapped volume and ability to maintain a membrane potential. Biochim Biophys Acta 1985; 812:55–65.
7. Zeimer R, Goldberg MF. Novel ophthalmic therapeutic modalities based on noninvasive light-targeted drug delivery to the posterior pole of the eye. Adv Drug Deliv Rev 2001; 52:49–61.
8. Ferris FL, Fine SL, Hyman LA. Age-related macular degeneration and blindness due to neovascular maculopathy. Arch Ophthalmol 1984; 102:1640–1642.
9. Treatment of Age-related Macular Degeneration with Photodynamic Therapy (TAP) Study Group. Photodynamic therapy of subfoveal choroidal neovascularization in age-related macular degeneration with verteporfin: one-year results of 2 randomized clinical trials—TAP report 1. Arch Ophthal 2000; 117:1329–1345.
10. Michels S, Schmidt-Erfurth U. Sequence of early vascular events after photodynamic therapy. Invest Ophthalmol Vis Sci 2003; 44:2147.
11. Schmidt-Erfurth U, Teschner S, Noack J, Birngruber R. Three-dimensional topographic angiography in chorioretinal vascular disease. Invest Ophthalmol Vis Sci 2001; 42: 2386–2394.

12. Schmidt-Erfurth U, Laqua H, Schlotzer-Schrehard U, Viestenz A, Naumann GO. Histopathological changes following photodynamic therapy in human eyes. Arch Ophthalmol 2002; 120:835–844.

13. Schmidt-Erfurth U, Michels S, Barbazetto I, Laqua H. Photodynamic effects on choroidal neovascularization and physiological choroid. Invest Ophthalmol Vis Sci 2002; 43:830–841.

14. D'Anna S, Nishiwaki H, Grebe R, Zeimer R. Comparison between blood flow in experimental choroidal neovascularization and in normal choriocapillaris. Invest Ophthalmol Vis Sci (suppl) 1998; 39(4):S998 (abstract).

15. Nishiwaki H, Zeimer R, Goldberg MF, D'Anna SA, Vinores SA, Grebe R. Laser targeted photo-occlusion of rat choroidal neovascularization without collateral damage. Photochem Photobiol 2002; 75:149–158.

16. Nishiwaki H, D'Anna S, Grebe R, Zeimer R. Laser targeted photodynamic therapy occludes experimental choroidal neovascularization without visible damage to RPE and choriocapillaris. Invest Ophthalmol Vis Sci (suppl) 1998; 39(4):S276 [abstract].

17. Grebe R, Asrani S, Zou S, et al. Effects on the RPE following choriocapillaris occlusion induced by laser targeted photo-occlusion and conventional photodynamic therapy. Invest Ophthalmol Vis Sci/ARVO 1996; 37(3):S976 (abstract).

18. Zeimer R, Nishiwaki H, Grebe R, D'Anna S. Electron microscopic assessment of the effect of laser targeted photodynamic therapy on choroidal neovascularization and adjacent RPE and choriocapillaris. Invest Ophthalmol Vis Sci (suppl) 1999; 40(4):401 [abstract].

19. Ogura Y, Guran T, Shahidi M, Mori M, Zeimer R. Feasibility of targeted drug delivery to selective areas of the retina. Invest Ophthalmol Vis Sci 1991; 32:2351–2356.

20. Bressler NM, Bressler SB, Fine SL. Age-related macular degeneration. Surv Ophthalmol 1988; 32:375–413.

21. Bressler NM, Alexander J, Bressler SB, Maguire MG, Hawkins BS, Fine SL. Subfoveal neovascular lesions in age-related macular degeneration: guidelines for evaluation and treatment in the macular photocoagulation study. Arch Ophthalmol 1991; 109:1242–1257.

22. Destro M, Puliafito CA. Indocyanine green videoangiography of choroidal neovascularization. Ophthalmology 1989; 96:846–853.

23. Kiryu J, Shahidi M, Mori M, Ogura Y, Asrani S, Zeimer R. Noninvasive visualization of the choriocapillaris and its dynamic filling. Invest Ophthalmol Vis Sci 1994; 35:3724–3732.

24. Asrani S, Zou S, D'Anna S, Goldberg MF, Zeimer R. Noninvasive visualization of the blood flow in the choriocapillaris of the normal rat. Invest Ophthalmol Vis Sci 1996; 37:312–317.

25. Zeimer R, Guran T, Shahidi M, Mori MT. Visualization of the retinal microvasculature by targeted dye delivery. Invest Ophthalmol Vis Sci 1990; 31:1459–1465.

26. Khoobehi B, Peyman GA, Vo K. Laser-triggered repetitive fluorescein angiography. Ophthalmology 1992; 99:72–79.

27. Khoobehi B, Aly OM, Schuele KM, Stradtmann MO, Peyman G. Determination of retinal blood velocity with respect to the cardiac cycle using laser-triggered release of liposome-encapsulated dye. Lasers Surg Med 2000; 10:469–475.

28. Fung YC. Biodynamics Circulation. 86–91. New York: Springer-Verlag, , 1984:233–237.

29. Kiryu J, Asrani S, Shahidi M, Mori M, Zeimer R. Local response of the primate retinal microcirulation to increased metabolic demand induced by flicker. Invest Ophthalmol Vis Sci 1995; 36:1240–1246.

30. Hickam JB, Frayser R. Studies of the retinal circulation in man: observations on vessel diameter, arteriovenous oxygen difference, and mean circulation time. Circulation 1966; 33:302–316.

31. Allen TM. Liposomes. Opportunities in drug delivery. Drugs 1997; 54(suppl 4):8–14.

32. Mayhew E, Ito M, Lazo R. Toxicity of non-drug-containing liposomes for cultured human cells. Exp Cell Res 1987; 171:195–202.

33. Gabizon A, Peretz T, Sulkes A, et al. Systemic administration of doxorubicin-containing liposomes in cancer patients: a Phase I study. Eur J Cancer Clin Oncol 1989; 25:1795–1803.
34. Perez-Soler R, Lopez-Berestein G, Lautersztain J, et al. Phase I clinical and pharmacological study of liposome-entrapped *cis*-bis-neodecanoato-*trans-R,R*-1,2-diaminocyclohexane platinum (II). Cancer Res 1990; 50:4254–4259.
35. Vosika GJ, Cornelius DA, Gilbert CW, et al. Phase I trial of the ImmTher, a new liposome-incorporated lipophilic disaccharide tripeptide. J Immunother 1991; 10:256–266.
36. Rahman A, Treat J, Roh J, et al. A Phase I clinical trail and pharmacokinetic evaluation of liposome-encapsulated doxorubicin. J Clin Oncol 1990; 8:1093–1100.
37. Asrani S, D'Anna S, Alkan-Onyuksel H, Wang W, Goodman D, Zeimer R. Systemic toxicology and laser safety of laser targeted angiography with heat sensitive liposomes. J Ocul Pharmacol Ther 1995; 11:575–584.
38. The Laser Institute of America. American National Standard for the safe use of lasers, ANSI Z-136.1-1993. Orlando: The Laser Institute of America, 1993.

11

Gene Therapy for Retinal Disease

Albert M. Maguire and Jean Bennett
F.M. Kirby Center for Molecular Ophthalmology, Scheie Eye Institute, University of Pennsylvania, Philadelphia, Pennsylvania, U.S.A.

DESCRIPTION OF DRUG DELIVERY SYSTEM

Gene therapy can be considered a method of drug delivery. Gene therapy takes advantage of the host organism's gene transcription/translation machinery to locally produce bioactive substances. In gene therapy, a nucleic acid compound (DNA or RNA) is delivered to the target tissue via a vector system. The nucleic acid can directly affect gene expression by binding to homologous nucleic acid sequences within the cell. Such a phenomenon is used in "antisense" strategies. Alternatively, the target cells will transcribe the delivered DNA transgene (or the reverse-transcribed RNA template) to produce RNA, which may be bioactive itself, or may be translated to a protein with bioactive properties. The greatest technical challenge is to achieve efficient and stable expression of the introduced cDNA. This is largely a function of the vector system used to introduce the exogenous nucleic acid sequence (Table 1). In some instances, injection of naked DNA oligonucleotides alone into host tissue will result in a biological effect. Robinson et al. injected DNA oligonucleotides that were antisense to a target vascular endothelial growth factor (VEGF) molecule and obtained evidence for therapeutic effect in a mouse model of retinopathy of prematurity (1). Cellular uptake and stability of such molecules is inefficient, however. Therefore, a variety of physiochemical and biological means have been developed with which to enhance the efficiency of nucleic acid delivery into various target cells. Examples include encapsulation in liposomes, addition of lipid/cationic compounds to the nucleic acids, electroporation, use of immunoliposomes, high-pressure injection, and bombardment of tissue with gold particles coated with DNA using a "gene gun" (Table 1). The most commonly employed vector system in gene therapy, however, is that of genetically modified viruses (Table 1). Viruses are highly efficient at transfer of exogenous DNA into host tissue.

In some instances, expression of exogenous nucleic acids can be stable over time, far exceeding the performance of the nonviral systems. Adeno-associated virus (AAV), for example, has been used to deliver a fluorescent marker protein, green fluorescent protein (GFP), which is produced in the target retinal cells over the course of months and years (Fig. 1). In contrast, gene expression after delivery using nonviral methods generally persists for days or weeks. Viruses are engineered to

Table 1 Characteristics of Gene Transfer Vehicles Under Consideration for Ocular Gene Therapy

Vector	Nucleic acid	Cargo size limit	Target cells	Stability	Immune response	Cellular receptor	Intake protein
Oligos (1)	DNA	<25 bp	NR	±	NR	NR	NR
Lipofection/liposomes (2)	DNA/RNA/liposomes/(Sendai virus)	No limit		±	NR	NR	NR
Immunoliposomes (3)	DNA + pegylated liposomes targeted to the transferrin receptor	No limit	Ciliary body, iris, corneal epithelium	≥48 hr	NR	Transferrin receptor	NR
Electroporation	DNA/RNA	No limit	Ganglion cells (4,5)	≥1 wk	NR	NA	NA
Gene gun	DNA	No limit	Corneal epithelium (6)	±	NR		
Adenovirus (Ad5-based vectors)	DNA	7.5 kb; >35 kb = "gutted"	RPE, Muller, corneal endothelial, trabecular meshwork, iris	−(peak = 3–4 days)	++ Cellular	Coxsackie virus and adenovirus receptor (7)	α_v integrin (7)
Adenovirus (Ad3-based vectors) (8)	DNA	7.5 kb	Ciliary body; iris, inner retina	≥7 days	++ Cellular	NR	
Adenovirus (Ad37-based vectors) (8)	DNA	7.5 kb	Photoreceptors, ciliary body; RPE, Muller cells	≥7 days	++ Cellular	50 kDa molecule expressed on Chang C cells; sialic acid (9)	α_v integrin (8)

Vector	Genome	Cargo capacity	Cells transduced	Duration	Immune response	Receptor	
AAV; serotype 2	DNA (single-stranded)	4.8 kb; cargo can be expanded through "trans-splicing" by using 2 AAV vectors (10)	Photoreceptors, RPE cells, ganglion cells, Muller cells (11–15)	+++ (>3 yr)	Humoral	Heparin sulfate; $\alpha_v\beta5$ integrin (16)	bFGF receptor (17)
AAV; serotype 5	DNA (single-stranded)	4.8 kb	RPE; photoreceptors (11,14,18)	+++ (>3 yr)	Humoral	2,3-Sialic acid (18,19)	NR
AAV; serotype 1	DNA (single-stranded)	4.8 kb	RPE cells, cone photoreceptors (11,14,18)	+++ (>3 yr)	Humoral	NR	NR
AAV; serotype 4 (20)	DNA (single-stranded)	4.8 kb	RPE cells	≥6 mos	NR	NR	NR
Lentivirus (21)	RNA	7.5 kb	Depends on envelope; primarily RPE (18)	+++	Humoral	NR	NR
Herpes	DNA	40 kb	Blood vessels (22); RPE (23)	±	Cellular	NR	NR
Baculovirus	DNA	NR	RPE, cornea, lens, retina (24)	±	NR	NR	NR
Retrovirus	RNA	NR	Fibroblasts (25)	+++	NR	NR	NR

Note: NR = not reported; NA = not applicable. Additional references are provided in the text.
Abbreviations: AAV, adeno-associated virus; RPE, retinal pigment epithelium; bFGF, basic fibroblast growth factor.

(A)

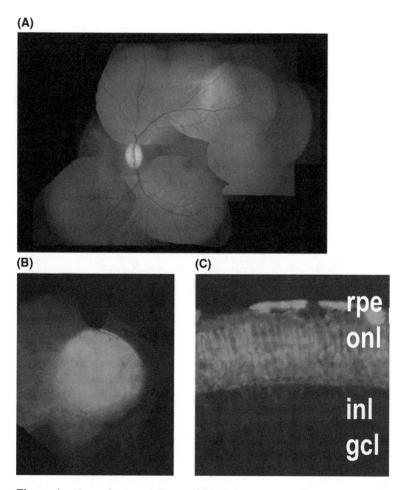

(B)   **(C)**

Figure 1   (*See color insert*) Expression of the gene encoding EGFP is stable after subretinal injection in the monkey. (**A**) Montage of fundus views of EGFP four months after injection of AAV2/2. CMV.EGFP; (**B**) Green fluorescence is visible in the blue light–illuminated retinal whole mount from the eye shown in (**A**). (**C**) Green fluorescence is present in RPE cells and rod photoreceptors of the tissue shown in (**A,B**). *Abbreviations*: AAV, adeno-associated virus; CMV, cytomegalovirus; EGFP, enhanced green fluorescent protein; RPE, retinal pigment epithelium; onl, outer nuclear layer; inl, inner nuclear layer; gcl, ganglion cell layer.

minimize any pathogenic effect on the target tissues, as well as other features to optimize their use in gene therapy applications. For example, hybrid viral systems have been developed in an attempt to alter the tropism of the vectors (11,14,18).

SPECTRUM OF DISEASES FOR WHICH THIS DELIVERY SYSTEM MIGHT BE APPROPRIATE

In principle, gene therapy is not limited to any particular spectrum of diseases. Gene therapy is limited by the type of compound that can be delivered, i.e., nucleic acids or their protein/peptide transgene products. The particular vector that carries this compound determines the target cell type in which it is expressed, as well as the onset, intensity, and duration of expression. The vector system can be tailored to the pathophysiology of the disease process. For example, a process such as choroidal neovascularization (CNV) may require an intense, short-lived application of an anti-angiogenic compound such as pigment epithelium–derived factor (PEDF). Recombinant adenoviral serotype 5 (Ad5) vectors provide just such an expression profile and, in addition, have high efficiency of transduction in retinal pigment epithelium (RPE) and Muller cells (26,27). A clinical trial employing adenoviral vector delivery of the PEDF-encoding cDNA is underway to test the treatment of CNV, secondary to age-related macular degeneration (AMD) (28).

Chronic diseases requiring long-term correction of protein/enzyme abnormality may require a vector system with stable expression of transgene product at more physiologic levels. AAV vectors provide long-term expression in photoreceptors, RPE cells (Fig. 1), and other cell types at levels that are quantitatively less than that of recombinant adenoviral vectors (12,13,15,18,29). AAV vector systems have shown promise for gene replacement therapy in autosomal recessive retinal degeneration [i.e., Leber congenital amaurosis caused by RPE65 mutations (30), retinitis pigmentosa due to peripherin/retinal degeneration slow (rds), and rhodopsin mutations (31,32)]. Thus, the spectrum of diseases for which gene therapy may be appropriate is potentially applicable to any subacute process for which a therapeutic nucleic acid can be designed. Acute conditions such as bacterial infection would be poor candidates for gene therapy because there exists a delay in onset of expression of the therapeutic molecule due to either host transcription/translational machinery or to vector-related biology.

Numerous in vivo studies have been performed with both gene expression markers and with therapeutic compounds. These have demonstrated the proof-of-principle of gene therapy for retinal disease. Studies evaluating onset, intensity, and stability of gene expression and efficiency of gene transduction have been performed using marker systems with bacterial LacZ–encoding β-galactosidase and with the jellyfish-derived gene-encoding GFP. These marker studies have proved extremely valuable in characterizing different vector systems with respect to potential clinical application. The physicochemical systems such as the addition of certain lipid vehicles to the recombinant DNA and ballistic delivery of DNA-impregnated gold particles (gene gun) have generally been found to be limited by poor transduction efficiency and transient gene expression (6). Recombinant viruses demonstrate much more favorable profiles with respect to intensity and duration of expression and therefore have received the most attention vis-a-vis clinical application. Marker studies have defined fundamentally different characteristics regarding expression onset, tissue tropism, and stability of expression—see Table 1.

Studies evaluating gene therapy approaches to various ocular conditions have largely involved in vivo application in laboratory animals. Although many ocular cell types such as RPE cells and photoreceptors can be maintained in culture, the biology of these cells in vitro may change in fundamental ways that do not reflect their behavior in vivo. Cell lines in culture often show entirely different expression patterns than the analogous cells in vivo. In addition, in vitro experiments cannot be used to evaluate other variables that may determine stability of gene expression and toxicity of therapy. For example, whereas adenoviral vectors cause minimal cytopathologic effects after in vitro transfection, immune-mediated inflammatory response can cause significant toxicity with in vivo gene therapy (33–35). In vivo experiments can reveal other potential interactions distant from the target tissue. In vivo transduction of ganglion cells has demonstrated that transgene product may appear anywhere along the axonal projections in the central nervous system, e.g., lateral geniculate body and superior colliculus (36). Such information cannot be established in vitro either in isolated cells in cell culture or even in a more complex microenvironment of tissue or organ culture. Ex vivo gene therapy has been explored for corneal diseases and as a way to achieve long-term delivery of exogenous growth factors. In the first instance, corneal buttons are treated in ex vivo conditions to achieve transduction of target cells, e.g., corneal endothelium (37,38). The gene therapy–treated corneal "button" is then transplanted to the animal recipient. The exposed corneal cells maintain expression of the novel gene after transplantation. This form of approach may be particularly well suited for treatment of donor tissue used in various anterior segment procedures such as penetrating keratoplasty. This sort of treatment may also be useful for delivering growth factors that may sustain diseased retinal tissue. Such an approach may ultimately be useful in the transplantation of RPE cells, from which primary cultures can be made with relative ease. RPE cells that have been genetically modified ex vivo have been successfully transplanted to recipient retinas in in vivo studies (39,40).

Due to the time required for transduction events to occur, treatment of host tissue by itself may not be useful. In addition, certain structures such as neural retina lose both viability and functionality when removed from their native environment. The same barriers that prevent successful transplantation also limit the possibility of ex vivo gene therapy. Ex vivo transduction of various cell types with subsequent reintroduction into host tissues is a feasible if not cumbersome technique. Transduced donor cells have been implanted in specially designed encapsulation systems for a depo-type delivery of neurotrophic factors for the retina (see Chapter 8) (41). Encapsulation is designed to minimize immune-related injury of foreign cells. Sieving et al. have proposed to transplant encapsulated RPE cells modified to express a cDNA-encoding ciliary neurotrophic factor (CNTF) as a treatment for retinitis pigmentosa (Recombinant DNA Advisory Committee, June 18, 2003, Washington, D.C.). Treatment effect is dependent on diffusion of the transgene product (CNTF, in the Sieving study) to the target site. The cell-containing capsule can also be removed if necessary, which is an important safety consideration. It would be much more difficult to remove cells that had been exposed in vivo to a viral gene transfer agent.

ANIMAL MODELS USED TO INVESTIGATE THE APPLICABILITY OF THIS DELIVERY SYSTEM FOR THE DISEASES MENTIONED ABOVE

Various animal models have been used to test gene therapy applications. In general, animal models have been used in two ways—first, to test the expression patterns of specific vector systems and second, to test proof-of-principle of optimally designed gene therapy applications in animal models of disease. Studies investigating expression patterns of vector systems have provided critical information used in designing gene therapy compounds. The profile of in vivo gene expression for various vectors has been described, including tropism for certain cell types, onset of gene expression, efficiency of expression, intensity of expression of transgene product, duration of expression, and toxicity (Table 1). Gene expression studies employing "marker" systems such as LacZ and GFP have been replicated in various mammalian species. By and large, the pattern of gene expression with regard to cellular tropism is similar in different animal species (42). This is fortunate in that it suggests that gene transduction in animals can be used to predict that in humans. Furthermore, accurate screening of vectors can be done quickly and efficiently in mice rather than animals that are more expensive and difficult to maintain. Animal studies have revealed some important differences in gene expression across different species. Some vectors display markedly different delays in onset of gene expression. This is especially true of AAV vectors, where in murine species, gene expression occurs two to three weeks after in vivo delivery whereas in nonhuman primates, this delay can be as long as six to eight weeks (42).

One of the main concerns in gene delivery is the potential toxicity of the transgene product. Toxicity can result if transgene expression extends beyond the cell population under study. In some instances, the tropism of certain viruses or viral serotypes is used to limit gene expression to a specific cell type(s). Chimeric vectors can be created to direct expression to a specific population of cells. For example, AAV2/1, an AAV virus with a serotype 2 genome, can be packaged in a capsid from an AAV virus of serotype 1 (11,14,18). After subretinal injection, the AAV2/1 virus delivers foreign genes only to retinal pigment epithelial cells—not to photoreceptors or Muller cells, which are also adjacent to the subretinal space. Similarly, adenovirus fiber proteins can be modified so that instead of targeting RPE cells efficiently after subretinal injection, they target photoreceptors (8). Cell specificity can be further refined by inserting promoter elements that drive transgene expression in only one cell type, e.g., rhodopsin promoter for rod photoreceptors (13,43).

PHARMACOKINETIC STUDIES USING THE DELIVERY SYSTEM

All of the conventional vector systems are limited to "on" expression. Attempts have been made to develop systems where the foreign gene can be turned on and off at will. Such regulatable promoters are turned on or off by the addition of a diffusible drug (such as tetracycline). The tetracycline-regulated system depends on two separate vectors, one containing a promoter-driving expression of a transcriptional activator ("transactivator") and the other containing the regulated therapeutic transgene. After coinfection with these two vectors, the host cell contains machinery to produce the therapeutic transgene and an inducible system to turn on its expression. In its basal state, the therapeutic vector is quiescent or nearly so. The first vector produces the transactivator not normally present in the microenvironment of the host cell. The second vector carries a transgene that is not activated unless it encounters both the transactivator and tetracycline. By adding tetracycline to the system, the transactivator can bind to the promoter of the therapeutic gene, ultimately resulting in production of the therapeutic drug (Fig. 2). Of note, this inducible system can be designed in the opposite configuration, such that application of an exogenous agent can result in suppression of transgene expression. Both the tetracycline-regulated system and the rapamycin-regulated system have functioned successfully and repetitively in the retina in animal studies (44–46).

RESULTS OF ANIMAL MODEL STUDIES

Successful application of ocular gene therapy has been demonstrated in animal diseases or animal models of human disease. The most encouraging results in the field of ocular gene therapy involve treatment of genetically inherited retinal degenerations occurring in various animal species. Conventional pharmacologic approaches

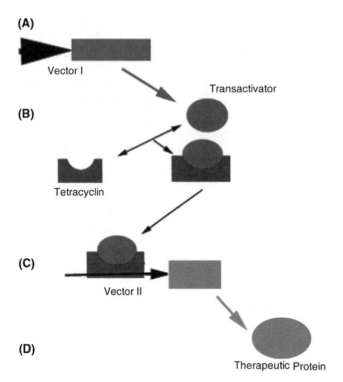

(A)

Vector I

Transactivator

(B)

Tetracyclin

(C)

Vector II

(D)

Therapeutic Protein

Figure 2 (*See color insert*) Scheme of dual vector transactivation system using tetracycline (or doxycyclin). One vector (Vector I) carries a promoter (*large arrowhead*) and a gene encoding a transactivator protein (*red circle*). The second vector (Vector II) carries a site that binds only to the transactivator when it has already bound to tetracycline. When the transactivator/ tetracycline unit has bound, a minimal promoter (*arrow*) is activated, which results in transcription of the therapeutic gene (*green rectangle*). This results in production of the therapeutic protein (*green circle*).

to treating inherited retinal diseases have been uniformly discouraging. Gene therapy offers the theoretical advantage of correcting the gene defect by introducing the normal, wild-type gene product in the gene-deficient host retina. This approach would seem especially relevant for conditions caused by recessive, lack-of-function mutations. By replacement of the absent gene product, function may be restored. Gene therapy provided the first evidence of therapeutic effect for retinal degeneration caused by a null mutation in the β-phosphodiesterase (PDEβ) gene (47). Gene replacement therapy with wild-type PDEβ results in significant delay of photoreceptor degeneration in retinal dystrophic (rd) mice, which otherwise show a near complete loss of rod photoreceptors by one month of age. This specific gene-based therapy has been successfully performed with several different vector delivery systems including adenovirus, gutted adenovirus, lentivirus, and AAV (21,47–49). Analogous results have been obtained in other rodent lines with recessive, lack-of-function mutations causing other retinal degenerations, including rds caused by mutations in the rds/peripherin (Prph2$^{Rd2/Rd2}$) gene (32), Royal College of Surgeons rats caused by mutations in the mertk1 gene (50), and RPE65 knockout mice caused by a genetically engineered knockout of the RPE65 gene [Dejneka N, Surace E, Aleman T, et al. Fetal and postnatal virus-mediated delivery of the human RPE65 gene rescues vision in a murine model of leber congenital amaurosis (submitted)].

Gene replacement therapy has also been successfully performed in naturally occurring large animal species such as cats with the lysosomal storage disease, mucopolysaccharidosis VI (51), and dogs with Leber congenital amaurosis due to a recessive mutation in the RPE65 gene (30,52,53). Gene therapy resulted in restoration in retinal function including electrophysiology and visual behavior in this RPE65 canine model. In one set of experiments, this therapeutic effect persisted for over three years of treatment (30). A second laboratory has verified the therapeutic effect delivered by a similar approach (52,53).

Gene therapy has also been applied to retinal degeneration caused by dominant mutations such as those in the rhodopsin gene. These gain-of-function mutations present more complex treatment considerations because the abnormal gene product not only needs to be replaced but its toxic effects must also be neutralized. While several pharmacologic strategies including RNA antisense, recombinant antibody fragments, and interference RNA (RNAi) may be applicable using gene therapy techniques, the most experience has been with ribozyme therapy. These specially designed RNA molecules will bind to specific regions of the mutant mRNA, resulting in its cleavage (54) (Fig. 3). Ribozyme therapy has been successfully used in mice and rats genetically engineered to have autosomal dominant retinal degeneration due to a mutant rhodopsin gene. With ribozyme therapy, the rate of photoreceptor degeneration and loss of electrophysiologic response is markedly slowed (31,55).

The vast majority of retinal degenerations do not have specifically identified gene mutations. In addition, degenerations caused by dominant genes may involve numerous unrelated mutations that cannot be addressed by any single mutation-based intervention or by single gene replacement. Nonspecific approaches have been developed for diseases with different etiologies. Exogenous application of various growth factors can produce a protective effect against retinal degenerations caused by both environmental and genetic insults (56–58). Growth factors applied by intravitreal injection have a short half-life and require repeated reinjection to maintain therapeutic levels. Using gene therapy methods, cells can be transduced to provide long-term production of these compounds. Rescue effect has been demonstrated

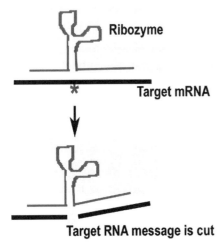

Figure 3 (*See color insert*) Scheme of ribozyme-mediated cleavage of target mRNA molecules. The ribozyme (*in red*) is designed to target mRNA sequence centered on the desired cleavage site (*asterisk*). It cleaves this mRNA sequence specifically.

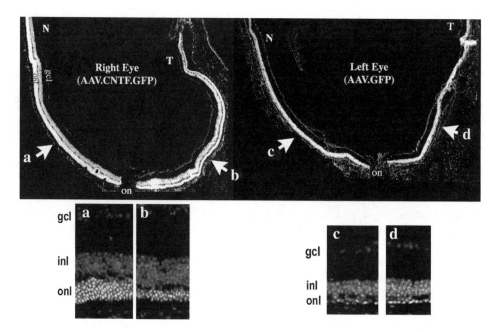

Figure 4 Gene therapy–mediated slowing of photoreceptor cell death in the rds/rds (Prph2$^{Rd2/Rd2}$) mouse. The right eye (experimental) was treated with an AAV carrying CNTF and the GFP reporter gene (control; AAV.CNTF.GFP) by intravitreal injection on the temporal (T) side at postnatal day 5. The left eye was treated by intravitreal injection on the nasal (N) side with AAV carrying GFP alone (AAV.GFP). The tissue was sectioned when animals were 8.5 months of age. Arrows indicate panels from which higher magnification views are shown below: a = nasal, experimental eye; b = temporal, experimental eye; c = nasal, control eye; d = temporal, control eye. The ONL (photoreceptor layer) in the corresponding portions of the experimental eye (a, b) is thicker than in either portion of the control eye (c, d). *Abbreviations*: AAV, adeno-associated virus; CNTF, ciliary neurotrophic factor; GFP; green fluorescent protein; gcl, ganglion cell layer; inl, inner nuclear layer; onl, outer nuclear layer; on, optic nerve. *Source*: From Ref. 62.

in several retinal degeneration models with CNTF (59–61) (for example, refer Fig. 4). Injection of vectors containing the CNTF gene results in local sustained production of the CNTF molecule. As shown in Figure 4, preservation of the outer nuclear layer (ONL) is apparent throughout the retina in an eye that had received treatment with the experimental virus (AAV.CNTF.GFP) at postnatal day 5 and, which was then evaluated for rescue effect at 8.5 months of age. In contrast, the contralateral eye, which had received an injection of an AAV carrying a control gene (GFP), has only a single row of nuclei remaining in the ONL.

While the high levels of CNTF provided by viral vector–mediated gene delivery confer a protective effect on photoreceptor cells by histologic measures, paradoxically, these levels can have a negative effect with respect to photoreceptor physiology (62–64). Other growth factors such as basic fibroblast growth factor (bFGF; FGF-2), FGF-5, and glial-derived neurotrophic factor (GDNF) have shown promise in animal models of retinal degeneration when applied using a gene therapy approach (61,65–67).

Great interest has been shown in gene-based treatments of ocular neovasculari zation. Many human retinal diseases have either retinal neovascularization or CNV as a final common pathway. Neovascularization results in retinal injury and blindness. Conditions such as diabetic retinopathy and AMD are the leading causes of blindness throughout the U.S. population and development of effective therapy is a great public health unmet need. The etiology of neovascular disease is rarely related to a single gene defect. However, the final common pathway for ocular neovascularization may be controlled by key factors that can be altered/influenced by gene therapy interventions. VEGF is a potent angiogenic protein that is associated with ocular neovascularization (68–70). Gene therapy methods allowing prolonged inhibition of VEGF production/activity are successful in suppressing neovascularization in animal models. PEDF is an endogenous protein with potent antiangiogenic activity and is postulated to have an important role in preventing CNV. Its ability to suppress neovascularization has been demonstrated in animal models (71–75). A phase 1 (safety/toxicity) human clinical trial to treat exudative AMD is currently underway, in which gene therapy delivery of PEDF is being used. A recombinant adenovirus carrying the PEDF transgene is injected into the vitreous in hopes that Muller and other cells will be transduced and produce PEDF that will suppress CNV (76). A number of other gene therapy paradigms have been shown to be successful in animal models of retinal neovascularization and many of these will likely one day be tested in human clinical trials. Examples are gene therapy using AAV to deliver PEDF, angiostatin, endostatin, tissue inhibitor of metalloproteinases 3, antisense molecules to VEGF, or sFlt-1 and delivery of RNAi specific for blocking production of VEGF (71,77–82).

TECHNIQUES FOR IMPLANTING OR PLACING THE IMPLANT IN HUMANS (IF DONE)

One significant advantage of gene therapy delivery of pharmacologic agents is that therapeutic levels can be maintained without the need for reinjection or implantation devices. Most vectors do need to be placed in direct contact with their target tissues because the gene therapy compounds, either naked nucleic acids or nucleic acids in combination with viral/nonviral vectors, are unable to diffuse across tissue barriers. Thus, delivery of gene therapy agents for photoreceptor diseases requires surgical delivery to the subretinal space (83). Several studies have shown that subretinal injection results in minimal tissue damage and the technique is routinely employed in human vitreoretinal surgery.

Treatment of the inner retina may be effected by standard techniques used for intravitreal injection. Such approaches, together with treatment of the ocular outflow tract, will be necessary for the treatment of glaucoma. Delivery of neurotrophic factors to ganglion cells via recombinant viruses has been shown to ameliorate disease in animal models of glaucoma (84). Delivery of growth factor receptors has also been evaluated in diseases in which uncontrolled proliferation leads to visual deficit (25,85). Of note, after subretinal injection, transduction of inner retinal cells is sometimes seen. This is presumably due to the leak of vector material through the open retinotomy used to gain access to the subretinal space. In contrast, intravitreal injection does not result in transduction of any other retinal cells because the vector cannot diffuse across the intact neural retina. Some vector injected into the vitreous will diffuse into the anterior segment, thus exposing cells in the ciliary region, iris, lens epithelium, corneal endothelium, and trabecular meshwork (86). Transduction of these cell types depends on the tropism of the vectors system, its concentration, and its length of contact with the target tissue.

The mode of delivery is important when considering the degree of immune response and inflammatory reaction elicited by any given vector. In general, delivery limited to the subretinal space results in the least inflammatory and immune-mediated response. In contrast, intravitreal or anterior chamber delivery will result in a much greater response for the same amount of material, e.g., plaque-forming units of the viral vector (34,87). Recombinant adenovirus can result in a significant cell-mediated response. This difference in response is felt to be due to two factors. First, the subretinal space is an immune-privileged environment where suppression of immune-mediated responses can occur—the so-called "immune deviation" (88–90). Second, the highly reactive uveal tract is exposed to a much greater degree in the anterior portion of the eye. The retinal blood vessels and the RPE possess diffusion barriers not present in structures such as the iris and ciliary processes. Other (non-adenoviral) vectors do not cause a cell-mediated response but can cause a humoral immune response (36,42) [Karakousis P, Anand V, Wakefield J, Kappes J, Maguire A, Bennett J. Favorable immune response following subretinal administration of lentivirus (submitted)]. AAV and lentivirus, however, result in a "suppressive" immune response [Karakousis P, Anand V, Wakefield J, Kappes J, Maguire A, Bennett J. Favorable immune response following subretinal administration of lentivirus (submitted)] (91). These anatomic, vector-related, and immunologic differences should be borne in mind when considering the potential toxicity of gene therapy delivery. In certain cases, it may be desirable to harness immunologic responses for therapeutic effect. Recombinant adenovirus carrying the cDNA-encoding thymidine kinase is presently being delivered to the vitreous in a phase 1 human clinical trial for retinoblastoma (92). In this instance, the immune response to both the virus and the cancerous cells targeted by the virus may be an adjunct to the therapeutic effect of killing the thymidine kinase–expressing cells.

FUTURE HORIZONS

Development of gene therapy as a viable, clinically successful methodology will require demonstration of proof-of-principle in a clinical trial. Currently, while gene therapy has demonstrated outstanding and, in some cases, unprecedented success in animal models of human disease, there are only two ongoing trials in the field of ocular therapeutics—as described above, one for CNV and the other for retinoblastoma (76,92). To a large degree, this reflects a concern about potential morbidity of gene therapy–based treatments. In widely publicized stories, there were two unexpected events. One was a death of a mildly affected patient in a gene therapy trial for a metabolic disease [ornithine transcarbamylase (OTC) deficiency] (93). In the second, two children who had been treated successfully with gene therapy for severe combined immune deficiency (SCID)—a disease that is lethal in early childhood—later developed leukemia (94).

Since the OTC trial, there has been rapid advancement in vector development with marked improvements in the safety (and efficacy) of new viral delivery systems. There is also a better understanding of the immunological responses to gene therapy vectors. The adverse events in the SCID trial are more recent. They appear to be a likely result of the choice of vector (retrovirus) and the ex vivo selection strategy used to modify the affected stem cells of the SCID patients. The risk:benefit ratio for gene therapy treatment of these children is still deemed favorable in light of the fact that without treatment, they would not be alive.

Initiation of a human clinical trial involving ocular gene therapy should be done cautiously—especially because ocular disease is rarely lethal. Nevertheless, ocular therapy has the unique advantage that the volume of tissue that requires treatment is a fraction of that for other clinical applications. For example, log units less of vector are used for retinal delivery versus intrahepatic injection. Vector administered to the eye does not escape to systemic sites. Therefore, the degree of systemic exposure to potentially immunogenic vector and transgene product is substantially diminished with ocular application. For these and other reasons, gene therapy seems ideally suited for the treatment of eye disease. If this technology is to become available for clinical use, a new industry for vector production must be developed and prove economically viable. Hopefully, alliances of academic investigators and biotechnology/pharmaceutical companies will lead to the identification of new sight-saving gene-based products for diseases that presently cannot be treated satisfactorily.

REFERENCES

1. Robinson G, Pierce E, Rook S, Foley E, Webll R, Smith L. Oligodeoxynucleotides inhibit retinal neovascularization in a murine model of proliferative retinopathy. Proc Natl Acad Sci USA 1996; 93:4851–4856.
2. Hangai M, Kaneda Y, Tanihara H, Honda Y. In vivo gene transfer into the retina mediated by a novel liposome system. Invest Ophthalmol Vis Sci 1996; 37:2678–2685.
3. Zhu C, Zhang Y, Pardridge W. Widespread expression of an exogenous gene in the eye after intravenous administration. Invest Ophthalmol Vis Sci 2002; 43:3075–3080.
4. Mo X, Yokoyama A, Oshitari T, et al. Rescue of axotomized retinal ganglion cells by *BDNF* gene electroporation in adult rats. Invest Ophthalmol Vis Sci 2002; 43:2401–2405.
5. Yokoyama A, Oshitari T, Negishi H, Dezawa M, Mizota A, Adachi-Usami E. Protection of retinal ganglion cells from ischemia-reperfusion injury by electrically applied Hsp27. Invest Ophthalmol Vis Sci 2001; 42:3283–3286.
6. Tanelian D, Barry M, Johnston S, Le T, Smith G. Controlled gene gun delivery and expression of DNA within the cornea. Biotechniques 1997; 23:484–488.
7. Nemerow G. Cell receptors involved in adenovirus cell entry. Virology 2000; 274.
8. Von Seggern DJ, Aguilar E, Kinder K, et al. In vivo transduction of photoreceptors or ciliary body by intravitreal injection of pseudotyped adenoviral vectors. Mol Ther 2003; 7:27–34.
9. Arnberg G, Mei YF, Wadell G. Fiber genes of adenoviruses with tropism for the eye and the genital tract. Virology 1997; 227:239–244.
10. Reich S, Kuroki A, Fosnot J, et al. VEGF directed siRNA inhibits experimental choroidal neovascularization. Mol Vision 2003; 9: 210–216.
11. Surace E, Auricchio A, Reich S, et al. Delivery of adeno-associated viral vectors to the fetal retina: impact of viral capsid proteins on retinal neuronal progenitor transduction. J Virol 2003; 77:7957–7963.
12. Bennett J, Duan D, Engelhardt JF, Maguire AM. Real-time, noninvasive in vivo assessment of adeno-associated virus-mediated retinal transduction. Invest Ophthalmol Vis Sci 1997; 38:2857–2863.
13. Flannery J, Zolotukhin S, Vaquero MI, LaVail MM, Muzyczka N, Hauswirth WW. Efficient photoreceptor-targeted gene expression in vivo by recombinant adeno-associated virus. Proc Natl Acad Sci USA 1997; 94:6916–6921.
14. Rabinowitz J, Rolling F, Li C, et al. Cross-packaging of a single adeno-associated virus (AAV) type 2 vector genome into multiple AAV serotypes enables transduction with broad specificity. J Virol 2002; 76:791–801.

15. Ali RR, Reichel MB, Thrasher AJ, et al. Gene transfer into the mouse retina mediated by an adeno-associated viral vector. Hum Mol Genet 1996; 5:591–594.

16. Summerford C, Bartlett JS, Samulski RJ. αVβ5 integrin: a co-receptor for adeno-associated virus type 2 infection. Nature Med 1999; 5:78–82.

17. Ferguson I, Schweitzer J, Johnson EMJ. Basic fibroblast growth factor: receptor-mediated internalization, metabolism, and anterograde axonal transport in retinal ganglion cells. J Neurosci 1990; 10:2176–2189.

18. Auricchio A, Kobinger G, Anand V, et al. Exchange of surface proteins impacts on viral vector cellular specificity and transduction characteristics: the retina as a model. Hum Mol Genet 2001; 10:3075–3081.

19. Walters R, Yi S, Keshavjee S, et al. Binding of adeno-associated virus type 5 to 2,3-linked sialic acid is required for gene transfer. J Biol Chem 2001; 276:20,610–20,616.

20. Weber M, Rabinowitz J, Provost N, et al. Recombinant adeno-associated virus serotype 4 mediates unique and exclusive long-term transduction of retinal-pigmented epithelium in rat, dog, and nonhuman primate after subretinal delivery. Mol Ther 2003; 7:774–781.

21. Takahashi M, Miyoshi H, Verma IM, Gage FH. Rescue from photoreceptor degeneration in the rd mouse by human immunodeficiency virus vector-mediated gene transfer. J Virol 1999; 73:7812–7816.

22. Chowers I, Banin E, Hemo Y, et al. Gene transfer by viral vectors into blood vessels in a rat model of retinopathy of prematurity. Invest Ophthalmol Vis Sci 2001; 85:991–995.

23. Liu X, Brandt C, Gabelt BA, Bryar P, Smith M, Kaufman P. Herpes simplex virus mediated gene transfer to primate ocular tissues. Exp Eye Res 1999; 69:385–395.

24. Haeseleer F, Imanashi Y, Saperstein D, Palczewski K. Gene transfer mediated by recombinant baculovirus into mouse eye. Invest Ophthalmol Vis Sci 2001; 42:3294–3300.

25. Ikuno Y, Leong FL, Kazlauskas A. Attenuation of experimental proliferative vitreoretinopathy by inhibiting the platelet-derived growth factor receptor. Invest Ophthalmol Vis Sci 2000; 41:3107–3116.

26. Bennett J, Wilson J, Sun D, Forbes B, Maguire A. Adenovirus vector-mediated in vivo gene transfer into adult murine retina. Invest Ophthalmol Vis Sci 1994; 35:2535–2542.

27. Li T, Adamian M, Roof DJ, et al. In vivo transfer of a reporter gene to the retina mediated by an adenoviral vector. Invest Ophthalmol Vis Sci 1994; 35:2543–2549.

28. Campochiaro P, Gehlbach P, Haller J, Handa J, Nguyen Q, Sung J. An open-label, phase I, single administration, dose-escalation study of ADGVPEDF.11D (ADPEDF) in neovascular age-related macular degeneration (AMD). Hum Gene Ther 2001:2029–2032.

29. Grant C, Ponnazhagan S, Wang XS, Srivastava A, Li T. Evaluation of recombinant adeno-associated virus as a gene transfer vector for the retina. Curr Eye Res 1997; 16:949–956.

30. Acland GM, Aguirre GD, Ray J, et al. Gene therapy restores vision in a canine model of childhood blindness. Nat Genet 2001; 28:92–95.

31. Lewin AS, Drenser KA, Hauswirth WW, et al. Ribozyme rescue of photoreceptor cells in a transgenic rat model of autosomal dominant retinitis pigmentosa. Nature Med 1998; 4:967–971.

32. Ali R, Sarra GM, Stephens C, et al. Restoration of photoreceptor ultrastructure and function in retinal degeneration slow mice by gene therapy. Nature Genet 2000; 25:306–310.

33. Di Polo A, Aigner LJ, Dunn RJ, Bray GM, Aguayo AJ. Prolonged delivery of brain-derived neurotrophic factor by adenovirus-infected Muller cells temporarily rescues injured retinal ganglion cells. Proc Natl Acad Sci USA 1998; 95:3978–3983.

34. Hoffman LM, Maguire AM, Bennett J. Cell-mediated immune response and stability of intraocular transgene expression after adenovirus-mediated delivery. Invest Ophthalmol Vis Sci 1997; 38:2224–2233.

35. Reichel M, Ali R, Thrasher A, Hunt D, Bhattacharya S, Baker D. Immune responses limit adenovirally mediated gene expression in the adult mouse eye. Gene Ther 1998; 5:1038–1046.

36. Dudus L, Anand V, Acland GM, et al. Persistent transgene product in retina, optic nerve and brain after intraocular injection of rAAV. Vision Res 1999; 39:2545–2553.

37. Larkin DF, Oral HB, Ring CJ, Lemoine R, George AJ. Adenovirus-mediated gene delivery to the corneal endothelium. Transplantation 1996; 61:363–370.

38. Fehervari Z, Rayner SA, Oral HB, George AJ, Larkin DF. Gene transfer to ex vivo stored corneas. Cornea 1997; 16:459–464.

39. Hansen K, Sugino I, Yagi F, et al. Adeno-associated virus encoding green fluorescent protein as a label for retinal pigment epithelium. Invest Ophthalmol Vis Sci 2003; 44: 772–780.

40. Lai C, Gouras P, Doi K, et al. Tracking RPE transplants labeled by retroviral gene transfer with green fluorescent protein. Invest Ophthalmol Vis Sci 1999; 40:2141–2146.

41. Tao W, Wen R, Goddard M, et al. Encapsulated cell-based delivery of CNTF reduces photoreceptor degeneration in animal models of retinitis pigmentosa. Invest Ophthalmol Vis Sci 2002; 43:3292–3298.

42. Bennett J, Anand V, Acland GM, Maguire AM. Cross-species comparison of in vivo reporter gene expression after rAAV-mediated retinal transduction. Meth Enzymol 2000; 316:777–789.

43. Bennett J, Zeng Y, Bajwa R, Klatt L, Li Y, Maguire AM. Adenovirus-mediated delivery of rhodopsin-promoted *bcl*-2 results in a delay in photoreceptor cell death in the *rd/rd* mouse. Gene Ther 1998; 5:1156–1164.

44. Dejneka NS, Auricchio A, Maguire AM, et al. Pharmacologically regulated gene expression in the retina following transduction with viral vectors. Gene Ther 2001; 8:442–446.

45. Auricchio A, Rivera V, Clackson T, et al. Pharmacological regulation of protein expression from adeno-associated viral vectors in the eye. Mol Ther 2002; 6:238–242.

46. McGee LH, Rendahl KG, Quiroz D, et al. AAV-mediated delivery of a tet-inducible reporter gene to the rat retina. Invest Ophthalmol Vis Sci 2000; 41:S396.

47. Bennett J, Tanabe T, Sun D, et al. Photoreceptor cell rescue in retinal degeneration (*rd*) mice by in vivo gene therapy. Nature Med 1996; 2:649–654.

48. Kumar-Singh R, Farber D. Encapsidated adenovirus mini-chromosome-mediated delivery of genes to the retina: application to the rescue of photoreceptor degeneration. Hum Mol Genet 1998; 7:1893–1900.

49. Jomary C, Vincent K, Grist J, Neal M, Jones S. Rescue of photoreceptor function by AAV-mediated gene transfer in a mouse model of inherited retinal degeneration. Gene Ther 1997; 4:683–690.

50. Vollrath D, Feng W, Duncan J, et al. Correction of the retinal dystrophy phenotype of the RCS rat by viral gene transfer of Mertk. Proc Natl Acad Sci USA 2001; 98:12,584–12,589.

51. Ho T, Maguire A, Aguirre G, Salvetti A, Haskins M, Bennett J. Phenotypic rescue after adeno-associated virus (AAV)-mediated delivery of arylsulfatase B (ASB) to the retinal pigment epithelium (RPE) of feline mucopolysaccharidosis VI (MPS VI). J Gene Med 2002; 4:613–621.

52. Narfstrom K, Katz ML, Bragadottir R, et al. Functional and structural recovery of the retina after gene therapy in the RPE65-/-null mutation dog. Invest Ophtahlmol Vis Sci 2003; 44:1663–1672.

53. Narfstrom K, Katz ML, Ford M, Redmond TM, Rakoczy E, Bragadottir R. In vivo gene therapy in young and adult RPE65 dogs produces long-term visual improvement. J Hered 2003; 94:31–37.

54. Haseloff J. Simple RNA enzymes. Nature 1988; 334:585–591.

55. LaVail MM, Yasumura D, Matthes MT, et al. Ribozyme rescue of photoreceptor cells in P23H transgenic rats: long-term survival and late stage therapy. Proc Natl Acad Sci USA 2000; 97:11,488–11,493.

56. Faktorovich E, Steinberg R, Yaumura D, Matthes M, LaVail M. Basic fibroblast growth factor and local injury protect photoreceptors from light damage in the rat. J Neurosci 1990; 12:3554–3567.

57. LaVail MM, Unoki K, Yasamura D, Matthes MT, Yancopoulos GD, Steinberg RH. Multiple growth factors, cytokines, and neurotrophins rescue photoreceptors from the damaging effects of constant light. Proc Natl Acad Sci USA 1992; 89:11,249–11,253.
58. Steinberg R, Matthes M, Yasumura D, et al. Slowing by survival factors of inherited retinal degenerations in transgenic rats with mutant opsin genes. Invest Ophthalmol Vis Sci 1997; 38:S226.
59. Cayouette M, Gravel C. Adenovirus-mediated gene transfer of ciliary neurotrophic factor can prevent photoreceptor degeneration in the retinal degeneration (rd) mouse. Hum Gene Ther 1997; 8:423–430.
60. Liang FQ, Dejneka NS, Cohen DR, et al. AAV-mediated delivery of ciliary neurotrophic factor (CNTF) prolongs photoreceptor survival in the rhodopsin knockout mouse. Mol Ther 2000; 3:241–248.
61. Peterson WM, Flannery JG, Hauswirth WW, et al. Enhanced survival of photoreceptors in P23H mutant rhodopsin transgenic rats by adeno-associated virus (AAV)-mediated delivery of neurotrophic genes. Invest Ophthalmol Vis Sci 1998; 39:S1117.
62. Liang FQ, Aleman TS, Dejneka NS, et al. Long-term protection of retinal structure but not function using rAAV. CNTF in animal models of retinitis pigmentosa. Mol Ther 2001; 4:461–472.
63. Bok D, Yasumura D, Matthes M, et al. Effects of adeno-associated virus-vectored ciliary neurotrophic factor on retinal structure and function in mice with a P216L rds/peripherin mutation. Exp Eye Res 2002; 74:719–735.
64. Schlichtenbrede F, MacNeil A, Bainbridge J, et al. Intraocular gene delivery of ciliary neurotrophic factor results in significant loss of retinal function in normal mice and in the Prph2$^{Rd2/Rd2}$ model of retinal degeneration. Gene Ther 2003; 10:523–527.
65. McGee SL, Abel H, Hauswirth W, Flannery J. Glial cell line derived neurotrophic factor delays photoreceptor degeneration in a transgenic rat model of retinitis pigmentosa. Mol Ther 2001; 4:622–629.
66. Akimoto M, Miyatake SI, Kogishi JI, et al. Adenovirally expressed basic fibroblast growth factor rescues photoreceptor cells in RCS rats. Invest Ophthalmol Vis Sci 1999; 40:273–279.
67. McGee LH, Lau D, Zhou S, et al. Rescue of photoreceptor degeneration in S334ter(4) mutant rhodopsin transgenic rats by adeno-associated virus (AAV)-mediated delivery of basic fibroblast growth factor. Invest Ophthalmol Vis Sci 1999; 40:S936.
68. Aiello L, Avery R, Arrigg P, et al. Vascular endothelial growth factor in ocular fluid of patients with diabetic retinopathy and other retinal disorders. N Eng J Med 1994; 331:1480–1487.
69. Neely K, Gardner T. Ocular neovascularization—clarifying complex interactions. Am J Pathol 1998; 153:665–670.
70. Kwak N, Okamoto N, Wood JM, Campochiaro PA. VEGF is major stimulator in model of choroidal neovascularization. Invest Ophthalmol Vis Sci 2000; 41:3158–3164.
71. Auricchio A, Behling K, O'Connor E, et al. Inhibition of retinal neovascularization by intraocular viral-mediated delivery of anti-angiogenic agents. Mol Ther 2002; 6:490–494.
72. Duh EJ, Yang HS, Suzuma I, et al. Pigment epithelium-derived factor suppresses ischemia-induced retinal neovascularization and VEGF-induced migration and growth. Invest Ophthalmol Vis Sci 2002; 43:821–829.
73. Mori K, Duh E, Gehlbach P, et al. Pigment epithelium-derived factor inhibits retinal and choroidal neovascularization. J Cell Physiol 2001; 188:253–263.
74. Mori K, Gehlbach P, Yamamoto S, et al. AAV-mediated gene transfer of pigment epithelium-derived factor inhibits choroidal neovascularization. Invest Ophthalmol Vis Sci 2002; 43:1994–2000.
75. Raisler BJ, Berns KI, Grant MB, Beliaev D, Hauswirth WW. Adeno-associated virus type-2 expression of pigmented epithelium- derived factor or Kringles 1–3 of angiostatin reduce retinal neovascularization. Proc Natl Acad Sci USA 2002; 99:8909–8914.

76. Rasmussen H, Chu K, Campochiaro P, et al. Clinical protocol. An open-label, phase I, single administration, dose-escalation study of ADGVPEDF. 11D (ADPEDF) in neovascular age-related macular degeneration (AMD). Hum Gene Ther 2001; 12:2029–2032.

77. Lai YK, Shen WY, Brankov M, Lai CM, Constable IJ, Rakoczy PE. Potential long-term inhibition of ocular neovascularisation by recombinant adeno-associated virus-mediated secretion gene therapy. Gene Ther 2002; 9:804–813.

78. Lai CM, Spilsbury K, Brankov M, Zaknich T, Rakoczy PE. Inhibition of corneal neovascularization by recombinant adenovirus mediated antisense VEGF RNA. Exp Eye Res 2002; 75:625–634.

79. Lai CC, Wu WC, Chen SL, et al. Suppression of choroidal neovascularization by adeno-associated virus vector expressing angiostatin. Invest Ophthalmol Vis Sci 2001; 42:2401–2407.

80. Bainbridge JW, Mistry A, De Alwis M, et al. Inhibition of retinal neovascularisation by gene transfer of soluble VEGF receptor sFlt-1. Gene Ther 2002; 9:320–326.

81. Reich S, Fosnot J, Kuroki A, et al. Small interfering RNA targeting VEGF effectively inhibits ocular neovascularization in a mouse model. Mol Vision 2003; 9:210–216.

82. Reich S, Bennett J. Gene therapy for ocular neovascularization: a cure in sight. Curr Opin Genet Dev 2003; 13:317–322.

83. Liang FQ, Anand V, Maguire AM, Bennett J. Intraocular delivery of recombinant virus. In: Rakoczy PE, ed. Methods in Molecular Medicine: Ocular Molecular Biology Protocols. 47. Totowa, New Jersey: Humana Press Inc, 2000:125–139.

84. Vorwerk CK, Dejneka NS, Schuettauf F, Bennett J, Naskar R, Dreyer EB. In vivo gene therapy with AAV-bFGF protects against excitotoxicity. IOVS 2000; 41:S16.

85. Mori K, Gehlbach P, Ando A, et al. Intraocular adenoviral vector-mediated gene transfer in proliferative retinopathies. Invest Ophthalmol Vis Sci 2002; 43:1610–1615.

86. Budenz D, Bennett J, Alonso L, Maguire A. In vivo gene transfer into murine trabecular meshwork and corneal endothelial cells. Invest Ophthalmol Vis Sci 1995; 36:2211–2215.

87. Borras T, Tamm ER, Zigler JS Jr. Ocular adenovirus gene transfer varies in efficiency and inflammatory response. Invest Ophthalmol Vis Sci 1996; 37:1282–1293.

88. Jiang LG, Jorquera M, Streilein JW. Subretinal space and vitreous cavity as immunologically privileged sites for retinal allografts. Invest Ophthalmol Vis Sci 1993; 34:3347–3354.

89. Anand V, Duffy B, Yang Z, Dejneka NS, Maguire AM, Bennett J. A deviant immune response to viral proteins and transgene product is generated on subretinal administration of adenovirus and adeno-associated virus. Mol Ther 2002; 5:125–132.

90. Anand V, Chirmule N, Fersh M, Maguire AM, Bennett J. Additional transduction events after subretinal readministration of recombinant adeno-associated virus. Hum Gene Ther 2000; 11:449–457.

91. Bennett J. Immune response following intraocular delivery of recombinant viral vectors. Gene Ther 2003; 10:977–982.

92. Hurwitz R, Brenner M, Poplack D, Horowitz M. Retinoblastoma treatment. Science 1999; 284:2066.

93. Wade N. Patient dies while undergoing gene therapy. New York Times. New York, New York, 1999.

94. Stolberg SG. Trials are halted on a gene therapy. New York Times. New York, New York, 2002:1.

12
Biodegradable Systems

Hideya Kimura
Nagata Eye Clinic, Nara, Japan

Yuichiro Ogura
Ophthalmology and Visual Science, Nagoya City University Graduate School of Medical Science, Nagoya, Aichi, Japan

FUNDAMENTALS OF BIODEGRADABLE POLYMERIC DEVICES

Drug delivery systems using biodegradable polymers can provide a significant advantage over nonbiodegradable systems because the entire device is eventually absorbed by the body, eliminating the need for subsequent removal. Polylactic acid (PLA) and copolymers with glycolic acid (PLGA) have been the most promising as biodegradable materials and have been used successfully in absorbable sutures for many years (1). The degradation products (lactic and glycolic acids) are metabolized via the Krebs' cycle to carbon dioxide and water. PLA- and PLGA-based systems are usually matrices in which drug is dispersed within the polymers and is released both by diffusion through the polymer and as the polymers degrade. The drug release from biodegradable polymeric devices depends on the molecular weight of the polymers, the composition of the monomer copolymer, and drug loading (2). In this chapter, we provide an overview of biodegradable systems, and describe our experience with drug delivery systems in the form of scleral plugs and intrascleral implants.

Control of Drug Release

We have reported the drug release profile from scleral plugs in detail (3). Scleral plugs were made of PLA or PLGA and contained various amounts of ganciclovir (GCV) (Figs. 1 and 2). In vitro release studies demonstrated a triphasic release pattern: an initial burst, a second stage that is derived from diffusional release before erosion and swelling of the polymer begins, and a sudden burst due to swelling and disintegration of the polymeric matrix. The initial burst may have resulted from the rapid release of drugs deposited on the surface and in the water channels in the matrix. Higher drug loading may have induced a higher initial burst through numerous water channels. During the second stage, the drug was released slowly, possibly controlled by both diffusion of drug through the polymer matrix and degradation of the polymer. Typically the release rate increases as the molecular weight and PLGA

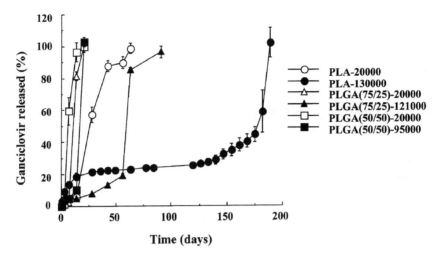

Figure 1   Effects of the molecular weight and LA/GA ratio of PLA (PLGA) on in vitro release from scleral plugs at a constant ganciclovir loading of 10%. The values shown are mean ± SD of $n = 5$. *Abbreviations*: PLA, polylactic acid; PLGA, polyglycolic acid. *Source*: From Ref. 4.

lactide content decreases. The third phase, characterized by rapid drug release coincides with bulk erosion of the polymer and disintegration of the device.

In general, it is difficult to achieve constant release (so called zero-order) from biodegradable polymeric devices because of the polymer biodegradation. It is also difficult to both prolong the duration and increase the drug release rate in the diffusional phase. Recently however, we reported that scleral plugs prepared by blending PLAs with different molecular weight ranges had a prolonged duration and better release rate in the diffusional phase and less final burst than implants made from a single molecular weight range (Fig. 3) (4). The scleral plugs were prepared by blending PLA-70,000 (molecular weight: 70,000) and PLA-5000 (molecular

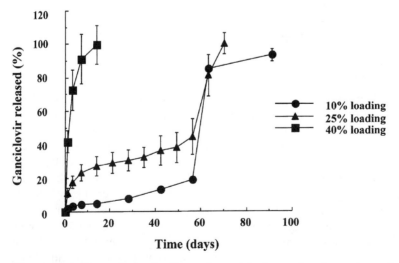

Figure 2   Effects of the loading of ganciclovir on in vitro release from the PLGA (75/25)-121,000 scleral plug. The values shown are mean ± SD of $n = 5$. *Abbreviation*: PLGA, polyglycolic acid. *Source*: From Ref. 4.

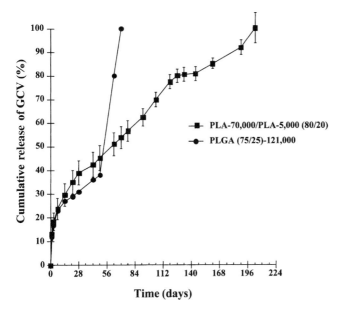

Figure 3 Cumulative release GCV from the scleral plugs of PLA-70,000 and PLA-5000 (whose content ratio was 80:20) containing 25% of GCV. The values shown are mean ± SD. The duration of GCV release was prolonged further compared with the plug made of PLGA (75/25)-121,000. *Abbreviations*: GCV, ganciclovir; PLA, polylactic acid; PLGA, polyglycolic acid. *Source*: From Ref. 4.

weight: 5000) at weight ratios of 80/20. These plugs included 25% GCV. The high-molecular-weight PLA may play a substantial role in the framework of the device and restrict the degradation rate of the low-molecular-weight PLA. Also the low-molecular-weight PLA may regulate drug release by slowing pore formation during the diffusional phase. The degradation rate of other biodegradable devices such as microspheres and intrascleral implants is also controlled in a similar manner.

Drug Delivery Systems

General Overview

Several different intraocular drug delivery systems using biodegradable polymers such as microspheres (5–9), intraocular implants (10–13), scleral plugs (3,4,14–22), and intrascleral implants (23) have been developed.

Moritera et al. (5) first reported an intravitreal drug delivery system using bio-degradable polymer microspheres. PLA microspheres containing doxorubicin hydro-chloride (6) and PLGA microspheres containing retinoic acid (7) have been reported for the treatment of proliferative vitreoretinopathy (PVR). GCV-loaded PLGA microspheres have been developed using a new oil-in-oil emulsion technique with fluorosilicone (8). Interestingly, after intravitreal injection of PLA nanoparticles with the mean size of 310 nm, nanoparticles transversed the retina and reached the retinal pigment epithelium (9). Targeted drug delivery to the retina and retinal pigment epithelium could be feasible using PLA nanoparticles.

Surodex® (Oculex Pharmaceuticals, Inc.) is a PLGA rod containing dexa-methasone, which is implanted at cataract surgery for treatment of postsurgical inflammation (10). In a multicenter, randomized, double-masked, parallel group

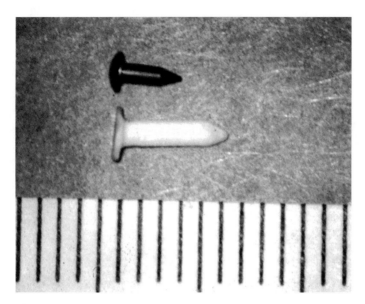

Figure 4 Scleral plug made of biodegradable polymers. The plug weighs 8.5 mg and is 5.0 mm long. *Source*: From Ref. 14.

study, Surodex® safely and effectively suppressed postoperative inflammation after uncomplicated cataract surgery (11). Posurdex® (Allergan, Inc.), which has a similar design, is implanted in the vitreous cavity to deliver dexamethasone to the posterior segment of the eye. Clinical trials for Posurdex® are ongoing for the treatment of macular edema associated with diabetes and other conditions (see Chapter 19). For the treatment of PVR, two intravitreal implants have been reported; a PLGA rod containing 5-fluorouracil (12) and a multiple drug delivery implant consisted of three cylindrical segments, each of which contained one of the following drugs: 5-fluorouridine, triamcinolone, or human recombinant tissue plasminogen activator (13).

The scleral plug is a device that is implanted through a sclerotomy at the pars plana; it releases the drug intravitreally (Fig. 4). Its shape is similar to that of a metallic scleral plug, which is used temporarily during pars plana vitrectomy. Controlled release of doxorubicin hydrochloride [adriamycin (ADR)] (15,16), GCV (3,4,17,19,21), fluconazole (18), 5-fluorouracil (20), and tacrolimus (FK506) (22) have been reported.

The intrascleral implant is a device that is implanted in the sclera; it delivers the drug through the sclera to the intraocular tissues (Fig. 5). Transscleral delivery may be an effective method of achieving therapeutic concentrations of drugs in the posterior segment (24–27). The intrascleral implant that incorporated betamethasone phosphate (BP) successfully delivered the drug to the retina/choroid and vitreous (28). The concentration of BP was maintained at a level that should suppress inflammation in the retina–choroid for more than eight weeks, and did not produce any ocular toxicity.

Relative Advantages and Disadvantages of Different Biodegradable Systems

Microspheres can be administered into the vitreous cavity by injection as a suspension. Although this is an advantage of this system, it can be disadvantageous as a large quantity of microspheres cannot be given by intravitreal injection and microspheres may cause a temporary disturbance in vitreous transparency. In

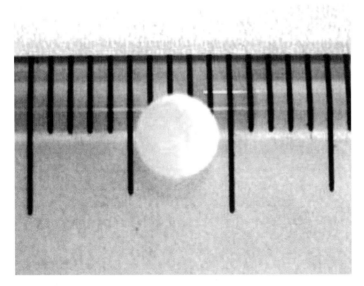

Figure 5 Intrascleral implant made of biodegradable polymers. The implant weighs 7 mg and is 0.5 mm thick and 4 mm in diameter. *Source*: From Ref. 23.

contrast, relatively large amounts of the drug can be loaded into scleral plugs, intrascleral implants, and intravitreal devices without decreasing vitreous transparency. Furthermore, scleral plugs can be applied at the sclerotomy sites at the end of pars plana vitrectomy as an adjunctive therapy. Intrascleral implants are less invasive than microspheres and scleral plugs, as complications such as endophthalimitis, vitreous hemorrhages, retinal detachment, and potential risks of intraocular systems, are virtually eliminated. In intravitreal drug delivery systems such as microspheres and scleral plugs, the drug is released intravitreally, reaches the surface of the retina, and diffuses into the retina–choroid. Transvitreal permeation into the retina is limited for relatively large molecules, such as tissue plasminogen activator (70 kDa), because the inner limiting membrane is a barrier to penetration (28). In contrast, large molecules such as immunoglobulin (150 kDa) have been reported to penetrate the retina through a transscleral route (26). Accordingly, we speculate that intrascleral implants may be more useful for site-specific treatment in the retina–choroid and for intraocular delivery of large molecular compounds, such as bioactive protein and antibody, than intravitreal systems.

SPECTRUM OF DISEASES FOR WHICH BIODEGRADABLE SYSTEMS MAY BE USEFUL

All intraocular disorders that require systemic administration or frequent local administration of the drug may be appropriate for these biodegradable systems. Uveitis is a chronic disorder that requires long-term medical therapy. Topical drug treatment is not effective in the treatment of posterior uveitis because of limited intraocular penetration. Systemic administration of corticosteroid or immunosuppressive agents may be effective but are associated with systemic side effects. Sustained drug delivery systems may be effective in the treatment of uveitis. In addition, specific inflammatory disorders such as cytomegalovirus retinitis or fungal endophthalmitis may be treated with sustained delivery systems of antiviral agents, antibiotics, or antifungal agents. Especially, for very chronic inflammation, repeat administration would likely be necessary even with sustained drug delivery systems.

The exudative type of age-related macular degeneration (AMD), which is associated with choroidal neovascularization, also may be a good target for biodegradable drug delivery systems. Numerous anti-angiogenic agents have been investigated in the treatment of AMD. In addition, AMD is a chronic disease and it can be expected that any pharmacological therapy will likely require long-term treatment. Therefore, sustained drug delivery may be beneficial.

Recently, macular edema associated with uveitis (29), diabetic retinopathy (30), and central retinal vein occlusion (31,32) has been treated with intravitreal injection of triamcinolone acetonide. Macular edema decreased after treatment but recurred three to six months after injection. A sustained-release steroid delivery system may be more attractive than a simple injection of triamcinolone as it could reduce or eliminate the need for multiple intravitreal injections.

PVR is a serious complication of retinal detachment surgery. Inhibition of cellular proliferation and postoperative inflammation may reduce the development of PVR. Inhibition of postoperative inflammation would eliminate one of the components of PVR and biodegradable sustained delivery systems that contain anti-inflammatory agents may be useful.

ANIMAL MODELS USED TO TEST BIODEGRADABLE DRUG DELIVERY SYSTEMS

Experimental Cytomegalovirus Retinitis

Experimental cytomegalovirus retinitis was induced by intravitreal injection of human cytomegalovirus (HCMV) solution (21). HCMV AD169 was grown on human fetal lung fibroblast monolayers. HCMV AD169 supernatant was collected and injected onto confluent monolayers of Hs68 cells. HCMV-infected cells were harvested, and their culture medium was collected. Eyes of pigmented rabbits were inoculated with 0.1 mL (5×10^6 pfu/mL) HCMV supernatant. The eyes were examined by ophthalmoscopy at one, two, three, and four weeks after HCMV inoculation. Posterior segment disease was graded on a 0+ to 4+ scale of increasing severity. The retinal and choroidal diseases were scored as follows: 0+, no abnormalities; 1+, focal white retinal infiltrates; 2+, focal-to-geographic retinal infiltrates and vascular engorgement; 3+, severe retinal infiltrates, vascular engorgement, and hemorrhage; and 4+, all the foregoing, plus retinal detachment and necrosis.

Experimental Uveitis

A relatively severe, nonspecific experimental uveitis model was created according to a modification of a previously published protocol (33,34). Pigmented rabbits were injected subcutaneously with 10 mg of *Mycobacterium tuberculosis* H37RA antigen suspended in 0.5 mL of mineral oil. One week later, a second injection of the same amount of subcutaneous antigen was given. A microparticulate suspension of *M. tuberculosis* H37RA antigen was prepared by ultrasonicating a suspension of crude extract in sterile balanced salt solution. Fifty micrograms of antigen suspended in 0.1 mL of balanced salt solution was injected into the vitreous cavity (first challenge).

To simulate chronic inflammation with exacerbations, the eyes were challenged again with the same amount of intravitreal antigen on day 14 (second challenge). On days 3, 7, 14, 17, 21, and 28 after the first challenge, slit-lamp biomicroscopy and indirect ophthalmoscopy were used to evaluate the severity of inflammation. Anterior chamber cells and flare were graded on a 0–4 scale based on the cell number and opacity observed by slit-lamp examination. The vitreous opacity was also graded on a 0–4 scale based on the examination of the posterior pole by indirect ophthalmoscopy. Aqueous protein was measured using a protein assay kit, and aqueous cell count was measured by hemocytometer.

Experimental Proliferative Vitreoretinopathy

Experimental PVR was induced by intravitreal injection of fibroblasts in the pigmented rabbit eye. The eyes received injections of 0.3 mL of sulfur hexafluoride (SF_6) gas. Seven days after the first injection, an additional 0.3 mL of SF_6 gas was injected to achieve complete compression of the vitreous. Homologous fibroblasts from Tenon's capsule were cultured. Seven days after the second gas injection, 1×10^5 cultured fibroblasts were injected over the medullary wings. The animals were then placed immediately in a supine position for one hour to allow the cells to settle on the vascularized retina. Fundus changes were observed for four weeks by indirect ophthalmoscopy. The fundus findings were graded as follows: stage 1, normal retina or wrinkling of the medullary wing; stage 2, pucker formation; and stage 3, traction retinal detachment (5).

RESULTS OF EFFICACY STUDIES

Scleral Plugs Containing GCV for Experimental Cytomegalovirus Retinitis

Scleral plugs were prepared by dissolving PLA with an average molecular weight of 70,000 and 5000 (PLA-70,000 and PLA-5000, respectively) whose content ratio was 80:20 and 25% of GCV in acetic acid.

Scleral plugs were prepared by dissolving PLA and GCV in acetic acid in the ratio 3:1. The PLA used was a blend of two molecular weight ranges, 80% had an average molecular weight of 70,000 (PLA-70,000) and 20% had an average molecular weight of 5000 (PLA-5000).

The resultant solution was lyophilized to obtain a homogeneous cake. The cake then was compressed into a scleral plug on a hot plate. In a rabbit study the scleral plug containing GCV was found to maintain GCV concentrations in the vitreous in a therapeutic range adequate to treat HCMV retinitis for more than 200 days (4). The 20 eyes of 20 pigmented rabbits that were inoculated with HCMV were divided into two groups. One week after HCMV inoculation, the control group ($n = 10$) received no treatment. In the treatment group ($n = 10$), a scleral plug containing GCV was implanted at the pars plana (21).

In the control eyes, whitish retinal exudates developed three days after HCMV inoculation and increased gradually until three weeks after inoculation. Thereafter the chorioretinitis decreased until four weeks after injection. In the treated group, scores for vitreoretinal lesions were significantly lower than those in the control group at three weeks after HCMV inoculation (Fig. 6).

Sustained release of GCV into the vitreous cavity with biodegradable scleral plugs was thus effective for the treatment of experimentally induced HCMV retinitis in rabbits.

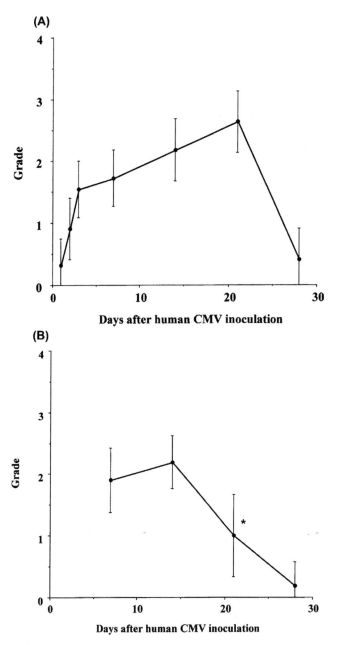

Figure 6 Clinical disease grading. (**A**) HCMV-inoculated rabbit eyes (*control*). (**B**) Treated eyes with scleral plug containing ganciclovir in HCMV-inoculated rabbit eyes. $^*P < 0.01$, unpaired *t*-test. *Abbreviation*: HCMV, human cytomegalovirus. *Source*: From Ref. 21.

Scleral Plugs Containing Tacrolimus (FK506) for Experimental Uveitis

Scleral plugs were prepared by dissolving a bioerodible polymer (99%) and FK506 (1%) in 1,4-Dioxane. We used poly(DL-lactide-co-glycolide), with a weight-averaged molecular weight of 63,000, whose copolymer ratio of DL-lactide to glycolide was 50:50 (22). In in vitro tests, the scleral plug released FK506 for more than 35 days.

Efficacy studies were conducted in the experimental rabbit uveitis model described above. After preimmunization with *M. tuberculosis* H37RA antigen, the treated eyes ($n = 8$) received scleral plugs containing FK506, and the control eyes received blank plugs. One day after implantation, 50 µg of antigen was injected into the vitreous cavity. Both the treated eyes and the control eyes were challenged again with the same amount of intravitreal antigen on day 14.

The results of anterior chamber cell, flare, and vitreous opacity clinical grading following the first challenge are shown in Figure 7. In the control eyes, several severe uveitis secondary complications including corneal neovascularization, corneal opacity, marked posterior synechia, and cataract were observed. In contrast, such complications were not seen in the treated eyes. Persistent marked vitreous opacity was observed for at least 28 days in untreated eyes, but was minimal throughout the observation period in the treated eyes. Evaluated by clinical criteria, the treated eyes had significantly less inflammation than did the control eyes. Aqueous protein concentration and aqueous cell count in the treated eyes were significantly lower than those in the control eyes.

Together, the results show that biodegradable scleral plugs containing FK506 are highly effective in suppressing the inflammation of experimental uveitis in the rabbit.

Scleral Plugs Containing ADR for Experimental PVR

Scleral plugs have been tested in a rabbit model of PVR. For these experiments, scleral plugs composed of 99% PLA (average molecular weight of 20,000) and 1% of ADR were prepared (16). The scleral plug released ADR over five weeks in vitro (Fig. 8).

Experimental PVR was induced in 22 eyes of 22 pigmented rabbits as described above. At the time of fibroblast intravitreal injection, the treatment group ($n = 11$) received scleral plugs containing ADR and the control groups ($n = 11$) received no treatment. Fundus changes were observed by indirect ophthalmoscopy.

The scleral plug decreased the incidence of traction retinal detachment from 100% to 64% at 28 days after implantation (Fig. 9). The differences in traction retinal detachment rate between control and treatment groups were significant ($P = 0.002$, two-way ANOVA).

PHARMACOKINETIC AND PHARMACODYNAMIC STUDIES

Scleral Plugs Containing GCV

For in vivo release studies, scleral plugs prepared from blends of PLA-70,000 and PLA-5000 at weight ratios of 80/20 and 25% of GCV were used. The scleral plugs containing GCV were implanted in pigmented rabbits. Animals were killed at days 1 and 3 and at weeks 1, 2, 3, 4, 6, 8, 10, 12, 14, 16, 18, 20, and 24 after implantation, and the eyes were enucleated. Five rabbits were used at each time point. The intravitreal GCV concentration was determined by high-performance liquid chromatography (HPLC).

The scleral plugs maintained a constant vitreous GCV concentration within the ED_{50} range (0.1–2.75 µg/mL) for six months without any sudden burst (Fig. 10) (4). However, further studies may be needed to evaluate effective GCV concentrations clinically, as the ED_{50}s are values determined in various conditions in vitro.

(A)

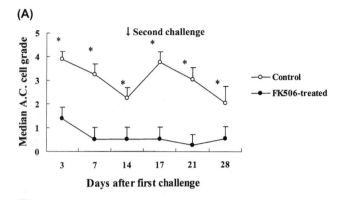

(B)

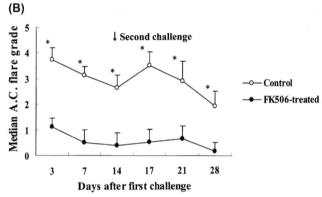

(C)

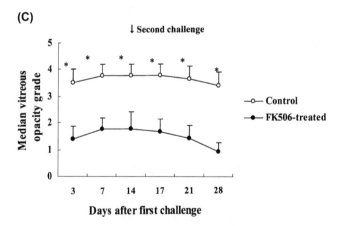

Figure 7 Clinical disease grading. **(A)** Anterior chamber cell grade. **(B)** Anterior chamber flare grade. **(C)** Fundus opacity grade (mean $\pm$ SEM, $P < 0.001$, a Mann–Whitney U nonparametric test). *Source*: From Ref. 22.

Scleral Plugs Containing ADR

To evaluate scleral plug ADR vitreous pharmacokinetics, 1% ADR-loaded PLA scleral plugs with a weight-averaged molecular weight of 20,000 were used (16). Pigmented rabbits underwent vitrectomy, and a scleral plug was implanted at the pars plana. Vitreous fluid (0.2 mL) was aspirated through the pars plana with a 30-gauge needle from the center of the vitreous cavity. Samples of vitreous humor

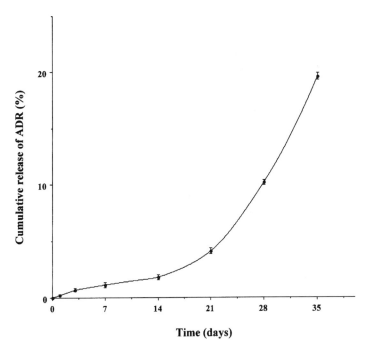

Figure 8   Profiles of in vitro release of adriamycin from the implant. The values are shown as mean ± SD. *Abbreviation*: ADR, adriamycin. *Source*: From Ref. 16.

were collected at days 1, 3, 7, 14, 21, and 28 after implantation. The concentrations of ADR in the vitreous humor were determined by HPLC.

The vitreous humor ADR concentrations are shown in Figure 11. ADR was maintained between 0.27 ± 0.06 and 0.76 ± 0.38 ng/mL between day 1 and day 7 and from a level 3.72 ± 0.57 to 8.07 ± 0.76 ng/mL between day 14 and day 21.

Intrascleral Implants Containing Betamethasone Phosphate

Intrascleral implants were prepared with 25% betamethasone phophate (BP) and 75% PLA with a weight-averaged molecular weight of 20,000 (23). Each intrascleral implant weighed approximately 7 mg and was 0.5 mm thick and 4 mm in diameter. The in vitro BP release from the implant was evaluated. The cumulative release of BP from the intrascleral implants is plotted in Figure 12. The data show biphasic release profiles, with an initial burst and a second stage. An initial burst (35%) was observed during the first day, and then BP was gradually released over 50 days.

We used 20 eyes of 20 pigmented rabbits to study in vivo release of BP from the implant and to evaluate pharmacodynamics in the ocular tissues after implantation. A scleral pocket was made at a depth of about one-half the total scleral thickness with a crescent knife 2 mm from the limbus. The scleral implant was inserted in the scleral pocket. At one, two, four, and eight weeks after implantation, animals were killed and four eyes were immediately enucleated at each time point. The concentrations of BP in the implants and samples of ocular tissues (aqueous humor, vitreous, and retina/choroid) were determined by HPLC.

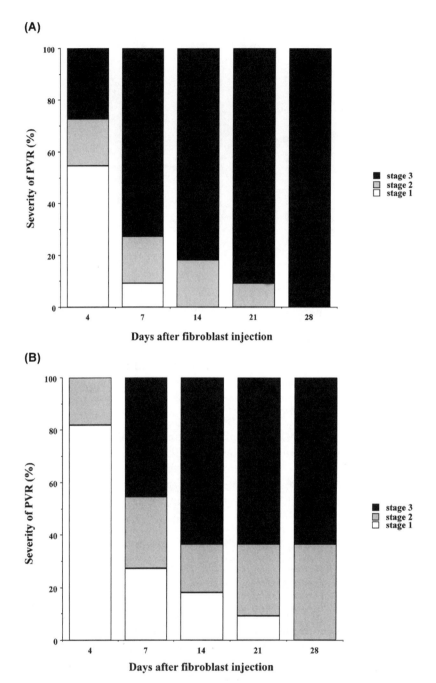

Figure 9 Effect of scleral plug containing adriamycin on experimental proliferative vitreo-retinopathy. The plugs significantly reduced the incidence of traction retinal detachment ($P = 0.002$). *Abbreviation*: PVR, prolitrative vitreoretinopathy. *Source*: From Ref. 16.

Figure 13 shows the profile of in vivo release of BP from the implant at the sclera. The profile was obtained by estimating the percentage of BP remaining versus the initial content in the implant. In contrast with the in vitro release profile, no initial burst was observed. In addition, more than 80% of BP was released at 28 days.

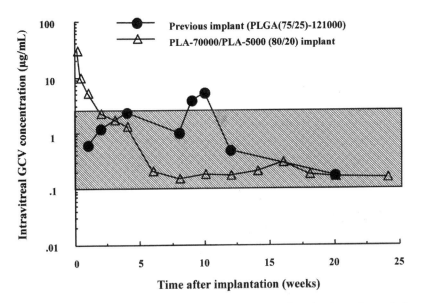

Figure 10 GCV concentrations in the vitreous after implantation of the scleral plug prepared from the blend of PLA-70,000 and PLA-5000 with a ratio of 80/20. The values are shown as mean ± SD. The shaded area indicates the ED_{50} range of GCV for CMV replication. *Abbreviations*: CMV, cytomegalo virus; GCV, ganciclovir; PLA, ploylactic acid; PLGA, polyglycolic acid. *Source*: From Ref. 4.

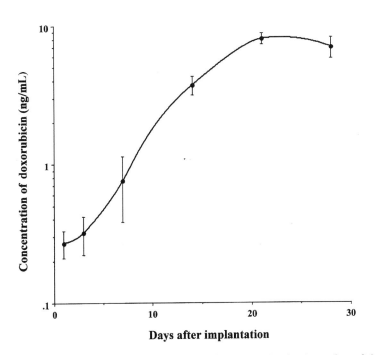

Figure 11 ADR concentrations in the vitreous after implantation of the scleral plug containing ADR. The values are shown as mean ± SD. *Abbreviation*: ADR, adriamycin. *Source*: From Ref. 16.

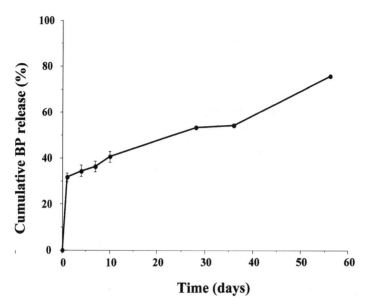

Figure 12 Profiles of in vitro release of BP from the implant. The values are shown as mean ± SD. *Abbreviation*: BP, betamethasone phosphate. *Source*: From Ref. 23.

The BP concentrations in the vitreous and the retina–choroid after implantation are shown in Figure 14. The level of BP in the retina–choroid was significantly higher than in the vitreous at all times. Both in the vitreous and in the retina–choroid, maximum concentrations were observed at two weeks after implantation. Thereafter, the levels of BP gradually decreased. The BP concentrations in the vitreous and retina–choroid remained within the concentration range capable of suppressing inflammatory responses (0.15–4.0 μg/mL) for more than eight weeks (35–40). In the aqueous humor, BP was below the detection limit during the observation period.

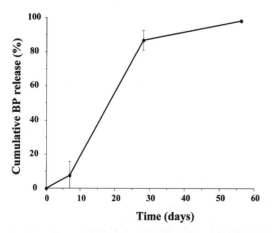

Figure 13 Profiles of in vivo release of BP from the implant. The values are shown as mean ± SD; *Abbreviation*: BP, betamethasone phosphate. *Source*: From Ref. 23.

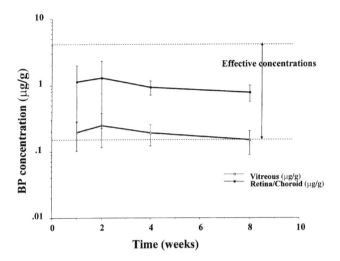

Figure 14 BP concentrations in the vitreous and the retina–choroid after intrascleral implantation of the BP-loaded implant. The values are shown as mean ± SD. *Abbreviation*: BP, betamethasone phosphate. *Source*: From Ref. 23.

SUMMARY AND FUTURE HORIZONS

In this chapter, we have described a variety of biodegradable polymeric devices, and have presented data on two systems, scleral plugs and intrascleral implants, in more detail. The scleral plugs are illustrative of intravitreal drug delivery systems in which the released drug diffuses in the vitreous and reaches the retina. The intrascleral implant is a transscleral drug delivery system in which the released drug diffuses into the retina–choroid. A drug with low molecular weight can easily diffuse into intraocular tissues after intravitreal administration. The ocular penetration of macromolecules such as antibodies is, however, generally poor. Especially in intravitreal delivery of the drug, macromolecules cannot penetrate through the internal limiting membrane of the retina (28). In contrast, macromolecules can reach the retina by transscleral delivery (25,26). Therefore, drug delivery systems may be selected according to drug characteristics or to target specific disorders.

The pathogenesis of several ocular disorders has been recently revealed by using molecular biological techniques. Experimentally, specific agents such as genes, antisense oligonucleotide therapy, antibodies, and growth factors have been reported to be effective for the treatment of ocular diseases. However, efficient delivery systems of those agents to target tissues are not available at present. Intraocular drug delivery systems using biodegradable polymers represent one promising method for that purpose.

REFERENCES

1. Kulkarni RK, Pani KC, Neuman C, Leopard F. Polylactic acid for surgical implant. Arch Surg 1966; 93:839–43.
2. Kimura H, Ogura Y. Biodegradable polymers for ocular drug delivery. Ophthalmologica 2001; 215:143–155.
3. Kunou N, Ogura Y, Hashizoe M, Honda Y, Hyon S-H, Ikada Y. Controlled intraocular delivery of ganciclovir with use of biodegradable scleral implant in rabbits. J Control Release 1995; 37:143–150.

4. Kunou N, Ogura Y, Yasukawa T, et al. Long-term sustained release of ganciclovir from biodegradable scleral implant for the treatment of cytomegalovirus retinitis. J Control Release 2000; 68:263–271.
5. Moritera T, Ogura Y, Yoshimura N, et al. Biodegradable microspheres containing adriamycin in the treatment of proliferative vitreoretinopathy. Invest Ophthalmol Vis Sci 1992; 33:3125–3130.
6. Moritera T, Ogura Y, Honda Y, Wada R, Hyon SH, Ikada Y. Microspheres of biodegradable polymers as a drug-delivery system in the vitreous. Invest Ophthalmol Vis Sci 1991; 32:1785–1790.
7. Giordano GG, Refojo MF, Arroyo MH. Sustained delivery of retinoic acid from microspheres of biodegradable polymer in PVR. Invest Ophthalmol Vis Sci 1993; 34: 2743–2751.
8. Veloso AA Jr, Zhu Q, Herrero-Vanrell R, Refojo MF. Ganciclovir-loaded polymer microspheres in rabbit eyes inoculated with human cytomegalovirus. Invest Ophthalmol Vis Sci 1997; 38:665–675.
9. Bourges JL, Gautier SE, Delie F, et al. Ocular drug delivery targeting the retina and retinal pigment epithelium using polylactide nanoparticles. Invest Ophthalmol Vis Sci 2003; 44:3562–3569.
10. Tan DT, Chee SP, Lim L, Lim AS. Randomized clinical trial of a new dexamethasone delivery system (Surodex) for treatment of post-cataract surgery inflammation. Ophthalmology 1999; 106:223–231.
11. Chang DF, Garcia IH, Hunkeler JD, Minas T. Phase II results of an intraocular steroid delivery system for cataract surgery. Ophthalmology 1999; 106:1172–1177.
12. Rubsamen PE, Davis PA, Hernandez E, O'Grady GE, Cousins SW. Prevention of experimental proliferative vitreoretinopathy with a biodegradable intravitreal implant for the sustained release of fluorouracil. Arch Ophthalmol 1994; 112:407–413.
13. Zhou T, Lewis H, Foster RE, Schwendeman SP. Development of a multiple-drug delivery implant for intraocular management of proliferative vitreoretinopathy. J Control Release 1998; 55:281–295.
14. Kimura H, Ogura Y, Hashizoe M, Nishiwaki H, Honda Y, Ikada Y. A new vitreal drug delivery system using an implantable biodegradable polymeric device. Invest Ophthalmol Vis Sci 1994; 35:2815–2819.
15. Hashizoe M, Ogura Y, Kimura H, et al. Scleral plug of biodegradable polymers for controlled drug release in the vitreous. Arch Ophthalmol 1994; 112:1380–1384.
16. Hashizoe M, Ogura Y, Takanashi T, Kunou N, Honda Y, Ikada Y. Implantable biodegradable polymeric device in the treatment of experimental proliferative vitreoretinopathy. Curr Eye Res 1995; 14:473–477.
17. Hashizoe M, Ogura Y, Takanashi T, Kunou N, Honda Y, Ikada Y. Biodegradable polymeric device for sustained intravitreal release of ganciclovir in rabbits. Curr Eye Res 1997; 16:633–639.
18. Miyamoto H, Ogura Y, Hashizoe M, Kunou N, Honda Y, Ikada Y. Biodegradable scleral implant for intravitreal controlled release of fluconazole. Curr Eye Res 1997; 16:930–935.
19. Yasukawa T, Kimura H, Kunou N, et al. Biodegradable scleral implant for intravitreal controlled release of ganciclovir. Graefes Arch Clin Exp Ophthalmol 2000; 238:186–190.
20. Yasukawa T, Kimura H, Tabata Y, Ogura Y. Biodegradable scleral plugs for vitreoretinal drug delivery. Adv Drug Deliv Rev 2001; 52:25–36.
21. Sakurai E, Matsuda Y, Ozeki H, Kunou N, Nakajima K, Ogura Y. Scleral plug of biodegradable polymers containing ganciclovir for experimental cytomegalovirus retinitis. Invest Ophthalmol Vis Sci 2001; 42:2043–2048.
22. Sakurai E, Nozaki M, Okabe K, Kunou N, Kimura H, Ogura Y. Scleral plug of biodegradable polymers containing tacrolimus (FK506) for experimental uveitis. Invest Ophthalmol Vis Sci 2003; 44:4845–4852.

23. Okabe J, Kimura H, Kunou N, Okabe K, Kato A, Ogura Y. Biodegradable intrascleral implant for sustained intraocular delivery of betamethasone phosphate. Invest Ophthalmol Vis Sci 2003; 44:740–744.

24. Ahmed I, Patton TF. Importance of the noncorneal absorption route in topical ophthalmic drug delivery. Invest Ophthalmol Vis Sci 1985; 26:584–587.

25. Ambati J, Canakis CS, Miller JW, et al. Diffusion of high molecular weight compounds through sclera. Invest Ophthalmol Vis Sci 2000; 41:1181–1185.

26. Ambati J, Gragoudas ES, Miller JW, et al. Transscleral delivery of bioactive protein to the choroid and retina. Invest Ophthalmol Vis Sci 2000; 41:1186–1191.

27. Geroski DH, Edelhauser HF. Transscleral drug delivery for posterior segment disease. Adv Drug Deliv Rev 2001; 52:37–48.

28. Kamei M, Misono K, Lewis H. A study of the ability of tissue plasminogen activator to diffuse into the subretinal space after intravitreal injection in rabbits. Am J Ophthalmol 1999; 128:739–746.

29. Antcliff RJ, Spalton DJ, Stanford MR, Graham EM, Ffytche TJ, Marshall J. Intravitreal triamcinolone for uveitic cystoid macular edema: an optical coherence tomography study. Ophthalmology 2001; 108:765–772.

30. Martidis A, Duker JS, Greenberg PB, et al. Intravitreal triamcinolone for refractory diabetic macular edema. Ophthalmology 2002; 109:920–927.

31. Greenberg PB, Martidis A, Rogers AH, Duker JS, Reichel E. Intravitreal triamcinolone acetonide for macular oedema due to central retinal vein occlusion. Br J Ophthalmol 2002; 86:247–248.

32. Ip MS, Kumar KS. Intravitreous triamcinolone acetonide as treatment for macular edema from central retinal vein occlusion. Arch Ophthalmol 2002; 120:1217–1219.

33. Cheng CK, Berger AS, Pearson PA, Ashton P, Jaffe GJ. Intravitreal sustained-release dexamethasone device in the treatment of experimental uveitis. Ophthalmology 1998; 105:46–56.

34. Jaffe GJ, Yang CS, Wang XC, Cousins SW, Gallemore RP, Ashton P. Intravitreal sustained-release cyclosporine in the treatment of experimental uveitis. Ophthalmology 1998; 105:46–56.

35. Arya SK, Wong-Staal F, Gallo RC. Dexamethasone-mediated inhibition of human T cell growth factor and gamma-interferon messenger RNA. J Immunol 1984; 133:272–276.

36. Culpepper JA, Lee F. Regulation of IL 3 expression by glucocorticoids in cloned murine T lymphocytes. J Immunol 1985; 135:3191–3197.

37. Lewis GD, Campbell WB, Johnson AR. Inhibition of prostaglandin synthesis by glucocorticoids in human endothelial cells. Endocrinology 1986; 119:62–69.

38. Grabstein K, Dower S, Gills S, Urdal D, Larsen A. Expression of interleukin 2, interferon-γ, and the IL 2 receptor by human peripheral blood lymphocytes. J Immunol 1986; 136:4503–4508.

39. Knudsen PJ, Dinarello CA, Strom TB. Glucocorticoids inhibit transcriptional and post-transcriptional expression of interleukin 1 in U937 cells. J Immunol 1987; 139:4129–4134.

40. Lee SW, Tsou AP, Chan H, et al. Glucocorticoids selectively inhibit the transcription of the interleukin 1β gene and decrease the stability of interleukin 1β mRNA. Immunology 1988; 85:1204–1208.

13

Transscleral Drug Delivery to the Retina and Choroid

Jayakrishna Ambati
*Department of Ophthalmology and Visual Sciences and Physiology,
University of Kentucky, Lexington, Kentucky, U.S.A.*

INTRODUCTION

As discussed in other chapters, drugs can be delivered locally to the retina and choroid by intravitreous injection, or intravitreous biodegradable or nonbiodegradable sustained delivery devices. However, with these methods, potential complications such as retinal detachment, posterior dislocation, endophthalmitis, vitreous hemorrhage, and cataract formation are significant. Further, polymer implants easily encase drugs with molecular weight less than about 1 kDa, but not antibodies against growth factors and cytokines, which are much larger. Even with potential advances in polymer technology that may accommodate larger molecules, a fundamental problem persists. The internal limiting membrane of the retina prevents diffusion of substances larger than about 4.5 nm in molecular radius (1,2), i.e., molecules (depending on their shape) larger than 40–70 kDa cannot diffuse into the retina from the vitreous, whereas antibodies (IgG) are about 150 kDa in size.

We and others have shown that *transscleral delivery* may be a viable modality of delivering drugs to the posterior segment (3–6). The sclera has a large and accessible surface area, and a high degree of hydration that renders it conducive to water-soluble substances. It is also hypocellular and thus has few proteolytic enzymes or protein-binding sites that can degrade or sequester drugs. The fact that scleral permeability does not appreciably decline with age (4) is serendipitous for the treatment of chronic diseases such as diabetic retinopathy and age-related macular degeneration, which affect older persons.

SCLERAL ANATOMY

The sclera is composed of collagen fibrils embedded in a glycosaminoglycan (GAG) matrix. Scleral collagen is predominantly type I (7). Collagen types III, V, and VI, VIII, and XII are also found in human sclera (8–12), while the lamina cribrosa and the vascular basement membranes in the neighboring choroid contain type IV collagen (13).

The collagen fibrils are quite heterogeneous in diameter (25–300 nm) and inter-woven into bundles of 500–600 nm diameter (14). On the external surface of the human sclera they are arranged in a reticular configuration with diameters of 80–140 nm, while the internal fibrils are arranged in an irregular rhombic pattern (15). Collagen bundles have a macroperiodicity of 35–75 nm and a microperiodicity of about 11 nm. Elastic fibers and fibroblast processes are interposed between these bundles.

There are regional differences in the orientation of scleral collagen fibrils. In the equatorial region, collagen fibrils are heterogeneous in diameter (25–300 nm) and are arranged in lamellar bundles with random orientations.

The microscopic architecture of bovine sclera is similar. The inner collagen layers demonstrate criss-crossed laminae while outer layers are more heterogeneous and larger in diameter (16). The axial periodicity of the fibrils as revealed by atomic force microscopy is 67 nm (17).

There is spatial variation of GAG composition within the sclera (18). Peripap-illary sclera is rich in dermatan sulfate. The post-equatorial region is rich in chon-droitin sulfate, while the equatorial sclera contains higher amounts of hyaluronic acid. In the thin myopic sclera, there is a reduced concentration of GAGs (19). Diffu-sion across sclera can occur through perivascular spaces, the aqueous media of the gel-like mucopolysaccharides, and across the scleral matrix itself.

There is topographic variation in the density of scleral emissaries, with the tem-poral sclera being most free of these vascular conduits (20). The landscape of scleral thickness is quite varied. The mean thickness of human sclera is 0.53 mm at the limbus, 0.39 mm at the equator, and 0.9–1.0 mm near the optic nerve (21). However, even these figures are subject to great variation, with equatorial thickness frequently below 0.1 mm. These factors would be important considerations in the placement of a transscleral drug delivery device. With a mean total surface area of 16.3 cm^2, the sclera is an inviting portal for intraocular drug delivery.

IN VITRO STUDIES OF SCLERAL PERMEABILITY

The ease with which a molecule diffuses across a tissue can be characterized by its permeability, measured in cm/sec. This value can be conceived of as the velocity of the molecule across the tissue.

Several investigators have used a two-chamber diffusion cell apparatus to characterize in vitro scleral permeability (3–5,22–25) to radioactively—or fluores-cently—labeled compounds.

The common element of most such studies is the dissection and isolation of scleral tissue, followed by placement of the sclera between two chambers represent-ing the episcleral surface and the uveal surface. One chamber contains the labeled compound, and the other chamber is sampled periodically after steady-state flux is attained. Some studies have used an apparatus where the chambers are constantly stirred. This may yield a higher apparent permeability by utilizing an unmixed depot on the tissue surface where static boundary layers may exist. However the impact of boundary layers on high molecular weight tracers is not expected to be significant, especially when temperature fluctuations are minimal (26,27).

Bovine sclera is permeable to molecules as large as albumin (69 kDa) (3). Human sclera is permeable to 70 kDa dextran (4), while rabbit sclera is permeable to 150 kDa dextran and IgG of the same molecular weight (5). More recently, organ-cultured human scleral tissues have been shown to possess permeability characteristics similar to donor human sclera (25).

We measured the permeability of rabbit sclera to a series of fluorescently labeled hydrophilic compounds with a wide range of molecular weights and radii. Sodium fluorescein, the smallest compound, had the highest permeability coefficient, whereas 150 kDa dextran, which had the largest molecular radius, had the lowest permeability coefficient. Our rabbit permeability data was quite similar to the reported permeability of human sclera. Bovine scleral permeability was less than that of the rabbit or human (Table 1).

The integrity of the fluorescent label conjugation to the parent molecule was assessed by subjecting the "uveal" chamber contents to protein precipitation. The fluorescence of the resulting supernatants was not different from that of the diffusion medium, indicating there was no significant dissociation. We confirmed that our diffusion apparatus preserved the anatomical and functional integrity of the sclera both by electron microscopy and demonstrating similar permeability characteristics of fresh sclera and stored sclera.

Scleral permeability declines exponentially with increasing molecular weight (3–5,28). However, molecular radius is a much better predictor of scleral permeability than molecular weight (5). For example, the globular protein albumin (69 kDa; 3.62 nm) has higher scleral permeability than the linear 40 kDa dextran (4.5 nm). Log-linear regression analysis demonstrated that molecular radius was a better predictor of permeability ($r^2 = 0.87$, $P = 0.001$) than molecular weight ($r^2 = 0.31$, $P = 0.16$). In an ideal aqueous medium the Stokes–Einstein equation predicts that permeability declines as a linear function of molecular radius. However, in porous diffusion through a fiber matrix such as the sclera, permeability declines roughly exponentially with molecular radius (29,30), as observed in our experiments.

We found that rabbit sclera was more permeable to bovine serum albumin and IgG than to dextrans of comparable molecular weight. This disparity may stem from differential binding of proteins and dextrans to collagen fibers in the hypocellular sclera (30). It is also interesting that proteins, despite having greater numbers of negative charges than dextrans, diffuse faster through sclera. This suggests that

Table 1 The Permeability of Sclera to Tracers of Varying Molecular Weight and Molecular Radius

Tracer	Molecular weight (Da)	Molecular radius (nm)	Permeability coefficient ($\times 10^{-6}$ cm/sec) (mean $\pm$ SD)
Sodium fluorescein	376	0.5	84.5 ± 16.1
FITC-D-4 kD	4400	1.3	25.2 ± 5.1
FITC-D-20 kD	19,600	3.2	6.79 ± 4.18
FITC-D-40 kD	38,900	4.5	2.79 ± 1.58
FITC-BSA	67,000	3.62	5.49 ± 2.12
Rhodamine D-70 kD	70,000	6.4	1.35 ± 0.77
FITC-D-70 kD	71,200	6.4	1.39 ± 0.88
FITC-IgG	150,000	5.23	4.61 ± 2.17
FITC-D-150 kD	150,000	8.25	1.34 ± 0.88

Abbreviations: FITC, fluorescein isothiocyanate; D, dextran; BSA, bovine serum albumin; IgG, immunoglobulin G.

molecular radius plays a greater role in determining scleral permeability than molecular weight or charge, similar to diffusion through extracellular tissue in the brain (31). Scleral permeability to the 70 kDa dextran is not significantly greater than to the 150 kDa dextran, suggesting that the hydrodynamic radii of these molecules within sclera is not identical to their molecular radii in aqueous solution. A similar situation exists in brain tissue, where diffusion of 40 and 70 kDa dextrans are not significantly different (32). These data are consistent with the existence of multiple diffusion passages in sclera with varying size limitations.

Surgical thinning of the sclera predictably increases its permeability; however, neither cryotherapy nor transscleral diode laser application alters permeability (4). It also increases with increasing tissue hydration (28). Scleral permeability of small molecules increases rapidly with increasing temperature (23). This low activation energy supports the existence of an aqueous pore pathway rather than a transcellular one. This paracellular pathway presumably traverses the aqueous media of the mucopolysaccharide matrix.

Various prostaglandins (33) and their ester prodrugs diffuse across human sclera (34). Hydrolysis of these ester prodrugs is less in the sclera than the cornea. Thus, the sclera is relatively deficient in protease activity compared to the cornea.

Prostaglandins have been found to enhance human scleral permeability (25), apparently via release of matrix metalloproteinases and subsequent collagen matrix remodeling (35). Increased permeability also may result from reduction of intraocular pressure. An increase of 15 mmHg pressure reduces scleral permeability to small molecules roughly by half (24). These authors demonstrated, using Peclet number analysis (36), that intraocular pressure reduces permeability not by inducing counteraction hydrostatic flow, but by compressing scleral fibers and narrowing diffusion pathways.

IN VIVO STUDIES OF SCLERAL PERMEABILITY

Anders Bill performed many of the initial studies demonstrating the movement of macromolecules across the sclera (37,38). Following injection of radio- or dye-labeled albumin, or red dextran of molecular weight 40 kDa into the suprachoroidal space of rabbits, both substances passed out through the sclera and accumulated in the extraocular tissues, suggesting that they diffused through the sclera via peri-vascular and interfibrillary spaces. In studies of uveoscleral outflow in the cynomol-gus monkey, Thorotrast particles (10 nm) were observed to enter the sclera but not latex spheres larger than 100 nm (39). Thus, the sclera has a large pore area with little restriction to flow.

The diffusion of small molecules across the sclera has been well characterized. Various sulfonamides diffuse across rabbit sclera (40,41). In patients undergoing cataract surgery, methazolamide, a hydrophilic compound, penetrated the sclera much faster than the cornea, but ethoxzolamide, which is lipophilic, had similar diffusion across the sclera and cornea (40). Other studies also have found greater scleral permeability to hydrophilic molecules than lipophilic ones (42,43).

The importance of the transscleral route in the penetration of topically administered macromolecules was demonstrated by denying corneal access by affixing a glass cylinder to the corneoscleral junction with cyanoacrylate adhesive (44). Significant intraocular levels of inulin (5 kDa) were obtained by topical administration outside the cornea in their paradigm. Neither reentry from the general circulation nor delivery by local vasculature accounted for the intraocular levels of the drug.

Contralateral tissue drug levels were <1% of those in the treated eye. Further, intra-venous administration achieved <5% of the tissue levels attained after topical dosing. These results imply that systemic absorption and intraocular reentry are not significant routes of delivery. In animals where topical dosage was applied *after* sacrifice, there was no significant difference in intraocular drug levels compared to live animals. This implied that delivery by local vasculature was not a major component.

We studied the in vivo pharmacokinetics of transscleral delivery of IgG. We used an osmotic pump, the tip of which was secured flush against bare sclera in rabbits to facilitate unidirectional movement, to deliver fluorescently labeled IgG (150 kDa) at rates on the order of μL/hr. Biologically relevant concentrations in the choroid and retina were attained for periods of up to four weeks with negligible systemic absorption (6). Levels in the vitreous and aqueous humors, and orbit were negligible. Although there was a spatial concentration gradient, the IgG concentration in the choroidal hemisphere distal to the footprint of the osmotic pump tip was half of that in the proximal hemisphere. The elimination of IgG from the choroid and retina followed first-order kinetics with half-lives of approximately two to three days.

Transscleral delivery is but a means to the end of achieving pharamacologic effect. To confirm biological activity of an agent delivered via the transscleral route, we used the following paradigm. Quite apart from its angiogenic action, VEGF induces a phenomenon known as leukostasis, the adhesion of leukocytes to the vascu-lar endothelium (45). This process is mediated by intracellular adhesion molecule-1 (ICAM-1), an endothelial cell surface receptor. Using an osmotic pump, we deliv-ered a monoclonal antibody directed against ICAM-1 onto the scleral surface of one eye of pigmented rabbits, and a control isotype antibody to the other eye. In treated eyes, VEGF-induced leukostasis, as measured by tissue myeloperoxidase (an enzyme abundant in leukocytes) activity, was inhibited in the choroid by 80% and in the retina by 70%. We also determined that there was remarkable selectivity of delivery, with an ocular concentration that was 700 times the plasma concentra-tion (6). The delivery of drug to the retina may be somewhat surprising in the face of the classical belief that the retinal pigment epithelium (RPE) forms a strict barrier to drug penetration; however, a recent study has shown that even molecules as large as 80 kDa dextrans can diffuse through this tissue (46).

Periocular (subconjunctival, peribulbar, retrobulbar) injections can deliver drugs to intraocular tissues via transcorneal or transscleral diffusion, or systemic absorption and recirculation. Small molecules such as hydrocortisone diffuse across the rabbit sclera after subconjunctival injection (47). There is one report of a large molecule (tissue plasminogen activator –70 kDa) reaching the vitreous cavity after subconjunctival injection in rabbits (48). These authors suggested that most of the penetration was via transscleral diffusion as levels in the plasma and fellow eye were much lower than in the treated eye. However, other reports in humans suggest that much of the intraocular delivery of subconjunctival or peribulbar injections is via systemic absorption (49,50).

The high ocular selectivity in our experimental design may be due to the sponge-like nature of the sclera. While bolus injections of drugs into the subconjunc-tival space may overwhelm the absorptive capacity of the sclera, leading to systemic absorption, the gradual delivery employed in our experiments may permit more complete scleral absorption as the absorptive capacity of the sclera is more closely matched to the rate of drug delivery.

Another drug delivery modality is iontophoresis, the application of an electrical current to facilitate diffusion of an ionized drug across tissue barriers. von Sallmann

(51,52) provided the first reports on transcorneal iontophoresis. This approach was used to deliver various antibiotics into the anterior chamber (53,54) and vitreous cavity (55).

Maurice was the first to report successful transscleral iontophoretic delivery of fluorescein into the rabbit vitreous humor (56). Potentially therapeutic intraocular levels of antibiotics (57–63), steroids (64), and anti-cytomegaloviral drugs (65,66), and anti-fungals (67) have been achieved via transscleral iontophoresis in animal models.

The intensity and duration of the applied current, as well as the configuration of the electrode determine the degree of intraocular penetration. Drug characteristics such as its ionization and molecular shape also affect the rate of transscleral diffusion. Hydrophilic drugs with low molecular weight that are ionized at physiological pH are best suited for iontophoretic delivery. Negatively charged molecules penetrate much better than positively charged ones (58,63).

Retinal damage appears to be an undesirable companion of transscleral iontophoresis. Retinal edema occurs within seconds of current application (56). Disruption of the normal retinal architecture, hemorrhagic necrosis, fibrosis, and RPE hyperplasia are evident beneath the iontophoretic site (68,69), and multiple applications exacerbate these injuries (70). Loss of tissue integrity, although spatially confined, may permit drug entry into the vitreous cavity or even the anterior segment tissues. Drug stability to possible electrochemical reactions is another possible limitation of this approach.

Several types of drug formulations have been reported to enhance sustained delivery of various drugs with a view to clinical use. Poly(lactic-co-glycolic) acid microspheres have been used as an effective vehicle to deliver an anti-VEGF RNA aptamer in a sustained fashion over 20 days, while preserving the drug's bioactivity (71). Fibrin sealants have been used to deliver carboplatin to the eye for up to two weeks (72). A collagen matrix gel has been used to deliver cisplatin in a controlled-release fashion that achieves higher intraocular concentrations than with the native drug (73). A biodegradable polymeric scleral plug delivering FK506 or betamethasone was found to be highly effective in suppressing inflammation in experimental uveitis for up to one month (74–76). Oligonucleotides have also been reported to traverse the sclera, with obvious implications for the burgeoning field of gene therapy (77). The clinically used anesthetic benzalkonium chloride was found to enhance transscleral penetration of molecules as large as 20 kDa dextran into the retina and choroid (78).

FUTURE DIRECTIONS

The eye presents unique challenges to the delivery of drugs. When the demand for sustained delivery to the target tissue is coupled with the desire to avoid systemic exposure, circumstances are ripe for creative approaches. Transscleral delivery is a nondestructive and minimally invasive method that achieves targeted delivery of high molecular weight compounds to the posterior segment. This drug delivery modality exhibits linear kinetics of absorption and elimination, with the potential to deliver constant doses of medication. By bridging the potential of our deepening understanding of mediators of disease to the need for focused pharmacologic intervention, transscleral delivery sits at the crossroads of molecular medicine and ophthalmology. It is robust as it is not limited to delivering anti-angiogenic drugs, but can be extended to neuroprotective agents or vectors for gene transfer, which hold promise in the treatment of glaucoma and other chorioretinal degenerations.

A long-term transscleral delivery device may be clinically feasible because the human eye is remarkably tolerant of foreign bodies overlying the sclera, such as scleral buckles used in treating retinal detachment, even for years. Moreover, human sclera is hypocellular and has a large surface area, both of which facilitate diffusion. While orbital clearance, intraocular pressure, uveoscleral outflow, choroidal blood flow, and the blood–retinal barriers pose formidable hurdles, sustained, targeted drug delivery to the choroid and retina in the clinical setting soon may be a reality.

REFERENCES

1. Kamei M, Misono K, Lewis H. A study of the ability of tissue plasminogen activator to diffuse into the subretinal space after intravitreal injection in rabbits. Am J Ophthalmol 1999; 128:739–746.
2. Mordenti J, Cuthbertson RA, Ferrara N, et al. Comparisons of the intraocular tissue distribution, pharmacokinetics, and safety of 125I-labeled full-length and Fab antibodies in rhesus monkeys following intravitreal administration. Toxicol Pathol 1999; 27:536–544.
3. Maurice DM, Polgar J. Diffusion across the sclera. Exp Eye Res 1977; 25:577–582.
4. Olsen TW, Edelhauser HF, Lim JI, Geroski DH. Human scleral permeability. Effects of age, cryotherapy, transscleral diode laser, and surgical thinning. Invest Ophthalmol Vis Sci 1995; 36:1893–1903.
5. Ambati J, Canakis CS, Miller JW, et al. Diffusion of high molecular weight compounds through sclera. Invest Ophthalmol Vis Sci 2000; 41:1181–1185.
6. Ambati J, Gragoudas ES, Miller JW, et al. Transscleral delivery of bioactive protein to the choroid and retina. Invest Ophthalmol Vis Sci 2000; 41:1186–1191.
7. Keeley FW, Morin JD, Veesely S. Characterization of collagen from normal human sclera. Exp Eye Res 1984; 39:533–542.
8. Shuttleworth CA. Type VIII collagen. Int J Biochem Cell Biol 1997; 29:1145–1148.
9. Wessel H, Anderson S, Fite D, Halvas E, Hempel J, SundarRaj N. Type XII collagen contributes to diversities in human corneal and limbal extracellular matrices. Invest Ophthalmol Vis Sci 1997; 38:2408–2422.
10. Chapman SA, Ayad S, O'Donoghue E, Bonshek RE. Glycoproteins of trabecular meshwork, cornea and sclera. Eye 1998; 12:440–448.
11. White J, Werkmeister JA, Ramshaw JA, Birk DE. Organization of fibrillar collagen in the human and bovine cornea: collagen types V and III. Connect Tissue Res 1997; 36:165–174.
12. Kimura S, Kobayashi M, Nakamura M, Hirano K, Awaya S, Hoshino T. Immunoelectron microscopic localization of decorin in aged human corneal and scleral stroma. J Electron Microsc (Tokyo) 1995; 44:445–449.
13. Marshall GE, Konstas AG, Lee WR. Collagens in the aged human macular sclera. Curr Eye Res 1993; 12:143–153.
14. Komai Y, Ushiki T. The three-dimensional organization of collagen fibrils in the human cornea and sclera. Invest Ophthalmol Vis Sci 1991; 32:2244–2258.
15. Thale A, Tillmann B, Rochels R. Scanning electron-microscopic studies of the collagen architecture of the human sclera—normal and pathological findings. Ophthalmologica 1996; 210:137–141.
16. Raspanti M, Marchini M, Della Pasqua V, Strocchi R, Ruggeri A. Ultrastructure of the extracellular matrix of bovine dura mater, optic nerve sheath and sclera. J Anat 1992; 181:181–187.
17. Fullwood NJ, Hammiche A, Pollock HM, Hourston DJ, Song M. Atomic force microscopy of the cornea and sclera. Curr Eye Res 1995; 14:529–535.
18. Trier K, Olsen EB, Kobayashi T, Ribel-Madsen SM. Biochemical and ultrastructural changes in rabbit sclera after treatment with 7-methylxanthine, theobromine, acetazolamide, or l-ornithine. Br J Ophthalmol 1999; 83:1370–1375.

19. Awetissow ES. The role of the sclera in the pathogenesis of progressive myopia (author's transl). Klin Monatsbl Augenheilkd 1980; 176:777–781.
20. Norn M. Topography of scleral emissaries and sclera-perforating blood vessels. Acta Ophthalmol (Copenh) 1985; 63:320–322.
21. Olsen TW, Aaberg SY, Geroski DH, Edelhauser HF. Human sclera: thickness and surface area. Am J Ophthalmol 1998; 125:237–241.
22. Sasaki H, Yamamura K, Tei C, Nishida K, Nakamura J. Ocular permeability of FITC-dextran with absorption promoter for ocular delivery of peptide drug. J Drug Target 1995; 3:129–135.
23. Unlu N, Robinson JR. Scleral permeability to hydrocortisone and mannitol in the albino rabbit eye. J Ocul Pharmacol Ther 1998; 14:273–281.
24. Rudnick DE, Noonan JS, Geroski DH, Prausnitz MR, Edelhauser HF. The effect of intraocular pressure on human and rabbit scleral permeability. Invest Ophthalmol Vis Sci 1999; 40:3054–3058.
25. Kim JW, Lindsey JD, Wang N, Weinreb RN. Increased human scleral permeability with prostaglandin exposure. Invest Ophthalmol Vis Sci 2001; 42:1514–1521.
26. Crank J. The Mathematics of Diffusion. Oxford: Clarendon, 1975.
27. Carslaw HS, Jaeger JC. Conduction of heat in solids. London: Oxford University Press, 1959.
28. Boubriak OA, Urban JP, Akhtar S, Meek KM, Bron AJ. The effect of hydration and matrix composition on solute diffusion in rabbit sclera. Exp Eye Res 2000; 71:503–514.
29. Cooper ER, Kasting G. Transport across epithelial membranes. J Control Release 1987; 6:23–35.
30. Edwards A, Prausnitz MR. Fiber matrix model of sclera and corneal stroma for drug delivery to the eye. AIChE J 1998; 44:214–225.
31. Tao L, Nicholson C. Diffusion of albumins in rat cortical slices and relevance to volume transmission. Neuroscience 1996; 75:839–847.
32. Nicholson C, Tao L. Hindered diffusion of high molecular weight compounds in brain extracellular microenvironment measured with integrative optical imaging. Biophys J 1993; 65:2277–2290.
33. Bito LZ, Baroody RA. The penetration of exogenous prostaglandin and arachidonic acid into, and their distribution within, the mammalian eye. Curr Eye Res 1981; 1:659–669.
34. Madhu C, Rix P, Nguyen T, Chien DS, Woodward DF, Tang-Liu DD. Penetration of natural prostaglandins and their ester prodrugs and analogs across human ocular tissues in vitro. J Ocul Pharmacol Ther 1998; 14:389–399.
35. Gaton DD, Sagara T, Lindsey JD, Gabelt BT, Kaufman PL, Weinreb RN. Increased matrix metalloproteinases 1, 2, and 3 in the monkey uveoscleral outflow pathway after topical prostaglandin F(2 alpha)-isopropyl ester treatment. Arch Ophthalmol 2001; 119:1165–1170.
36. Incropera FP, DeWitt DP. Fundamentals of Heat and Mass transfer. New York: John Wiley, 1996.
37. Bill A. The drainage of albumin from the uvea. Exp Eye Res 1964; 3:179–187.
38. Bill A. Movement of albumin and dextran through the sclera. Arch Ophthalmol 1965; 74:248–252.
39. Inomata H, Bill A. Exit sites of uveoscleral flow of aqueous humor in cynomolgus monkey eyes. Exp Eye Res 1977; 25:113–118.
40. Edelhauser HF, Maren TH. Permeability of human cornea and sclera to sulfonamide carbonic anhydrase inhibitors. Arch Ophthalmol 1988; 106:1110–1115.
41. Kao KD, Lu DW, Chiang CH, Huang HS. Corneal and scleral penetration studies of 6-hydroxyethoxy-2-benzothiazole sulfonamide: a topical carbonic anhydrase inhibitor. J Ocul Pharmacol 1990; 6:313–320.
42. Chien DS, Homsy JJ, Gluchowski C, Tang-Liu DD. Corneal and conjunctival/scleral penetration of p-aminoclonidine, AGN 190342, and clonidine in rabbit eyes. Curr Eye Res 1990; 9:1051–1059.

43. Ahmed I, Gokhale RD, Shah MV, Patton TF. Physicochemical determinants of drug diffusion across the conjunctiva, sclera, and cornea. J Pharm Sci 1987; 76:583–586.
44. Ahmed I, Patton TF. Importance of the noncorneal absorption route in topical ophthalmic drug delivery. Invest Ophthalmol Vis Sci 1985; 26:584–587.
45. Miyamoto K, Khosrof S, Bursell SE, et al. Vascular endothelial growth factor (VEGF)-induced retinal vascular permeability is mediated by intercellular adhesion molecule-1 (ICAM-1). Am J Pathol 2000; 156:1733–1739.
46. Pitkanen L, Ranta V-P, Moilanen H, Urtti A. Permeability of retinal pigment epithelium: effects of permeant molecular weight and lipophilicity. Invest Ophthalmol Vis Sci 2005; 46:641–646.
47. McCartney HJ, Drysdale IO, Gornall AG, Basu PK. An autoradiographic study of the penetration of subconjunctivally injected hydrocortisone into the normal and inflamed rabbit eye. Invest Ophthalmol 1965; 4:297–302.
48. Lim JI, Maguire AM, John G, Mohler MA, Fiscella RG. Intraocular tissue plasminogen activator concentrations after subconjunctival delivery. Ophthalmology 1993; 100:373–376.
49. Weijtens O, van der Sluijs FA, Schoemaker RC, et al. Peribulbar corticosteroid injection: vitreal and serum concentrations after dexamethasone disodium phosphate injection. Am J Ophthalmol 1997; 123:358–363.
50. Weijtens O, Feron EJ, Schoemaker RC, et al. High concentration of dexamethasone in aqueous and vitreous after subconjunctival injection. Am J Ophthalmol 1999; 128:192–197.
51. von Sallmann L. Sulfadizine iontophoresis in pyocyaneus infection of rabbit cornea. Am J Ophthalmol 1942; 25:1292–1300.
52. von Sallmann L. Iontophoretic introduction of atropine and scopolamine into the rabbit eye. Arch Ophthalmol 1943; 29:711–719.
53. Witzel SH, Fielding IZ, Ormsby HL. Ocular penetration of antibiotics by iontophoresis. Am J Ophthalmol 1956; 42(no. 2, pt. 2):89–95.
54. Hughes L, Maurice DM. A fresh look at iontophoresis. Arch Ophthalmol 1984; 102:1825–1829.
55. Fishman PH, Jay WM, Rissing JP, Hill JM, Shockley RK. Iontophoresis of gentamicin into aphakic rabbit eyes. Sustained vitreal levels. Invest Ophthalmol Vis Sci 1984; 25:343–345.
56. Maurice DM. Iontophoresis of fluorescein into the posterior segment of the rabbit eye. Ophthalmology 1986; 93:128–132.
57. Choi TB, Lee DA. Transscleral and transcorneal iontophoresis of vancomycin in rabbit eyes. J Ocul Pharmacol 1988; 4:153–164.
58. Church AL, Barza M, Baum J. An improved apparatus for transscleral iontophoresis of gentamicin. Invest Ophthalmol Vis Sci 1992; 33:3543–3545.
59. Grossman RE, Chu DF, Lee DA. Regional ocular gentamicin levels after transcorneal and transscleral iontophoresis. Invest Ophthalmol Vis Sci 1990; 31:909–916.
60. Barza M, Peckman C, Baum J. Transscleral iontophoresis of gentamicin in monkeys. Invest Ophthalmol Vis Sci 1987; 28:1033–1036.
61. Barza M, Peckman C, Baum J. Transscleral iontophoresis of cefazolin, ticarcillin, and gentamicin in the rabbit. Ophthalmology 1986; 93:133–139.
62. Burstein NL, Leopold LH, Bernacchi DB. Trans-scleral iontophoresis of gentamicin. J Ocul Pharmacol 1985; 1:363–368.
63. Yoshizumi MO, Cohen D, Verbukh I, Leinwand M, Kim J, Lee DA. Experimental transscleral iontophoresis of ciprofloxacin. J Ocul Pharmacol 1991; 7:163–167.
64. Lam TT, Edward DP, Zhu XA, Tso MO. Transscleral iontophoresis of dexamethasone. Arch Ophthalmol 1989; 107:1368–1371.
65. Lam TT, Fu J, Chu R, Stojack K, Siew E, Tso MO. Intravitreal delivery of ganciclovir in rabbits by transscleral iontophoresis. J Ocul Pharmacol 1994; 10:571–575.

66. Sarraf D, Equi RA, Holland GN, Yoshizumi MO, Lee DA. Transscleral iontophoresis of foscarnet. Am J Ophthalmol 1993; 115:748–754.
67. Grossman R, Lee DA. Transscleral and transcorneal iontophoresis of ketoconazole in the rabbit eye. Ophthalmology 1989; 96:724–729.
68. Lam TT, Fu J, Tso MO. A histopathologic study of retinal lesions inflicted by trans-scleral iontophoresis. Graefes Arch Clin Exp Ophthalmol 1991; 229:389–394.
69. Yoshizumi MO, Lee DA, Sarraf DA, Equi RA, Verdon W. Ocular toxicity of ionto-phoretic foscarnet in rabbits. J Ocul Pharmacol Ther 1995; 11:183–189.
70. Yoshizumi MO, Dessouki A, Lee DA, Lee G. Determination of ocular toxicity in multi-ple applications of foscarnet iontophoresis. J Ocul Pharmacol Ther 1997; 13:529–536.
71. Carrasquillo KG, Ricker JA, Rigas IK, Miller JW, Gragoudas ES, Adamis AP. Controlled delivery of the anti-VEGF aptamer EYE001 with poly(lactic-co-glycolic)acid microspheres. Invest Ophthalmol Vis Sci 2003; 44:290–299.
72. Simpson AE, Gilbert JA, Rudnick DE, Geroski DH, Aaberg TM Jr, Edelhauser HF. Transscleral diffusion of carboplatin: an in vitro and in vivo study. Arch Ophthalmol 2002; 120:1069–1074.
73. Gilbert JA, Simpson AE, Rudnick DE, et al. Transscleral permeability and intraocular concentrations of cisplatin from a collagen matrix. J Control Release 2003; 89:409–417.
74. Sakurai E, Nozaki M, Okabe K, Kunou N, Kimura H, Ogura Y. Scleral plug of biodegradable polymers containing tacrolimus (FK506) for experimental uveitis. Invest Ophthalmol Vis Sci 2003; 44:4845–4852.
75. Okabe K, Kimura H, Okabe J, Kato A, Kunou N, Ogura Y. Intraocular tissue distri-bution of betamethasone after intrascleral administration using a non-biodegradable sustained drug delivery device. Invest Ophthalmol Vis Sci 2003; 44:2702–2707.
76. Okabe J, Kimura H, Kunou N, Okabe K, Kato A, Ogura Y. Biodegradable intrascleral implant for sustained intraocular delivery of betamethasone phosphate. Invest Ophthal-mol Vis Sci 2003; 44:740–744.
77. Shuler RK, Dioguardi PK, Henjy C, Nickerson JM, Cruysberg LPJ, Edelhauser HF. Scleral permeability of a small, single-stranded oligonucleotide. J Ocul Pharmacol Ther 2004; 20:159–168.
78. Okabe K, Kimura H, Okabe J, et al. Effect of benzalkonium chloride on transscleral drug delivery. Invest Ophthalmol Vis Sci 2005; 46:703–708.

14

Nondegradable Intraocular Sustained-Release Drug Delivery Devices

Mark T. Cahill and Glenn J. Jaffe
Duke University Eye Center, Durham, North Carolina, U.S.A.

IMPLANTED NONDEGRADABLE SUSTAINED-RELEASE DEVICES

Description of Drug Delivery System

Nondegradable sustained-release devices can be either matrix or reservoir systems. In matrix systems, the drug is homogeneously dispersed inside the matrix material, and slow diffusion of the drug through the polymatrix material provides sustained release of the drug (Fig. 1A). Reservoir systems consist of a central core of drug surrounded by a layer of permeable or semipermeable nondegradable material and the majority of ocular nondegradable devices used clinically to date have been reservoir designs (Fig. 1B). Ocular nondegradable devices with a reservoir design typically consist of a layer of permeable nondegradable material, such as polyvinyl alcohol (PVA) typically surrounding the central drug core (1,2). The impermeable layer surrounding the drug core is usually made of ethylene vinyl acetate polymer or silicones which are used in conjunction with PVA to alter the surface area available for release (Fig. 2) (1,2).

Spectrum of Diseases for Which This Delivery System Might Be Appropriate

Their long duration of release means that nondegradable sustained-release implants are suitable for treating chronic conditions such as glaucoma, cytomegalovirus (CMV) retinitis, age-related macular degeneration, diabetes, and uveitis (1,3–8). Nondegradable sustained-release devices may also have a role in delivering proteins such as neuroprotectors in degenerative diseases such as retinitis pigmentosa as discussed in Chapter 8 (9).

Animal Models Used to Investigate the Applicability of the Delivery System

A rabbit model of uveitis has been used to test nondegradable devices containing dexamethasone, fluocinolone acetonide, cyclosporin A, and a combination of

(A)

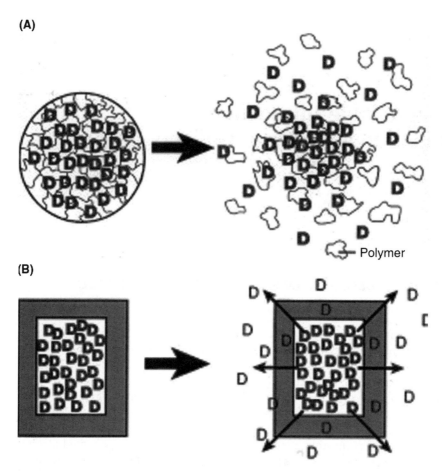

Figure 1 **(A)** Schematic drawing of a matrix design implant. **(B)** Schematic drawing of a reservoir design implant. The nondegradable polymers are used as a coating around a drug reservoir. Once the implant is in an aqueous environment, the implant starts to release the drug.

dexamethasone and cyclosporin A (6–8,10–12). Uveitis was induced by initial immunization of rabbits with H37RA-mTB antigen. The H37RA-mTB antigen is a cell wall fraction of the *Mycobacterium tuberculosis* H37RA strain that is an attenuated version of the virulent type *M. tuberculosis* H37RV strain. After the initial

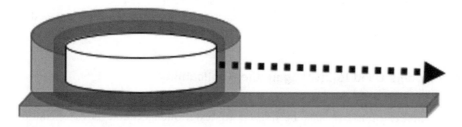

Figure 2 Diagram of an ocular nondegradable device with a reservoir design. The central drug core is surrounded by an impermeable nondegradable layer (*dark gray*) which is surrounded by an outermost semipermeable nondegradable layer (*light gray*).

immunization, eyes were challenged with intravitreal injections of the same antigen. Some eyes were rechallenged with intravitreal antigen to mimic disease exacerbations characteristic of chronic inflammation. In the models used for assessing corticosteroid implants this method resulted in a nonspecific inflammatory response that consisted of mononuclear and polymorphonuclear cellular infiltration, whereas in the model used to assess the cyclosporin A device the antigen preparation was centrifuged prior to insertion resulting in a more specific T-cell driven response (6).

Equine recurrent uveitis (ERU) is the primary cause of blindness in horses. Like uveitis in human eyes, eyes with chronic ERU have a mononuclear cell infiltration consisting predominantly of CD4+ T lymphocytes with resultant elevated levels of interleukin-2 (IL-2). An experimental model of ERU was developed to assess the effectiveness of a nondegradable device containing cyclosporin A to treat ERU. In this model, as described in the rabbit eye, peripheral immunization of a horse with H37RA-mTB antigen, followed by an intraocular challenge with the same antigen results in a predominantly monocellular inflammatory response similar to ERU (13). Eyes were also rechallenged with antigen to simulate disease exacerbations.

A rabbit model to assess the possible neuroprotective effects of central acting calcium channel blockers using a nondegradable sustained-release device has been developed (3). In this model an intraocular infusion was used to elevate the intraocular pressure (IOP). To assess the effects of the study drug, IOP was maintained at 40 mmHg for over an hour in both treatment and control eyes and was then decreased to a normal level. The procedure was repeated at 48 and 72 hours to simulate IOP spikes in human eyes (3).

While a number of animal models of diabetic macular edema exist, sustained delivery of steroids using a nondegradable device has not been tested in an animal model (14,15). Other nondegradable sustained-release drugs that have been investigated in vitro but not in animal models, include methotrexate and trimetrexate to treat intraocular lymphoma, and disease 2-methoxyestradiol to treat choroidal neovascularization (16–18).

Pharmacokinetic and Pharmacodynamic Studies Using the Delivery System

As in the majority of intraocular drug delivery implants, nondegradable devices are based on simple diffusion of the drug from the implant to the aqueous or vitreous. The vitreous is in contact with the retina, and indirectly with the choroid and sclera, the posterior chamber aqueous, and the posterior capsule of the lens. Drug molecules released from the device diffuse throughout the vitreous and then move into the surrounding boundary tissues. The vitreous concentration of any drug is dependent on four factors that include device release rate, distribution volume of the posterior segment, intravitreal metabolism, and the drug elimination rate through the boundary tissues. The distribution volume of the eye is determined by the size of the eye, the amount of protein in the eye, and the lipophilicity of the adjacent boundary tissues. Furthermore, despite its high water content, drug molecules are not uniformly distributed through the vitreous because the local concentration gradient is influenced by different elimination rates through each boundary tissue (19).

As previously indicated, intraocular sustained-release delivery systems are of two types. In matrix systems, the drug is homogeneously dispersed inside the matrix material, and slow diffusion of the drug through the polymatrix material provides sustained release of the drug. The kinetic release of drug in matrix systems is not constant and depends on the volume fraction of the drug in the matrix. This means that the greater the concentration of the drug in the matrix the greater the release from the system (20).

Reservoir systems consist of a central drug core surrounded by a layer of semipermeable nondegradable material. The release rate of the reservoir devices is based on Fick's law of diffusion and is determined by the area of release, the thickness of the semipermeable coatings, the shape of the implant, and the ease with which the drug diffuses through the semipermeable coating, which is also termed the diffusivity. The release profile follows zero-order kinetics and is characterized by a minimal initial burst of drug release followed by constant drug release over time.

Nondegradable implants have longer durations of action than biodegradable implants, and the duration of release can be extended by increasing the amount of drug in the core (2). Alternatively changing the solubility of the core drug can increase the release duration. An example is fluocinolone acetonide, a lipophilic, synthetic corticosteroid with a potency similar to dexamethasone. However, fluocinolone acetonide is 1/24th as soluble as dexamethasone, and can be released over a much longer period of time than dexamethasone with a similar sized delivery system (2). In vitro studies have demonstrated that nondegradable devices containing 2 and 15 mg of fluocinolone acetonide release the drug in a linear fashion at a mean rate of 1.9 and 2.2 μg/day, respectively. These release kinetics result in a predicted lifespan of 2.7 years for the 2 mg device and 18.6 years for the 15 mg device.

Silicone oil and polyfluorinated gases are frequently used as long-acting tamponade following surgery to repair retinal detachments (21). Vitreoretinal surgery with silicone oil or gas tamponade may be necessary during or following sustained drug delivery device implantation. For example, intravitreal ganciclovir implants have been placed during or prior to vitreoretinal surgery with silicone oil tamponade in eyes with rhegmatogenous retinal detachment associated with CMV retinitis (4,22–24). Accordingly it is important to know whether the drug release rate differs in the presence of an intraocular tamponade.

Clinical and experimental studies have demonstrated altered intravitreal pharmacodynamics when nondegradable implants containing ganciclovir are used with silicone oil. Silicone oil-filled eyes have reduced distribution volume for ganciclovir when compared to eyes without, as the drug partitions into saline but not silicone oil (P. Ashton, unpublished data). However, a rabbit study has demonstrated that intraocular concentrations of ganciclovir are similar in eyes with and without silicone oil and concluded that intravitreal drug concentration is independent of the distribution volume (25). This paradox may be explained by a reduction in the transretinal clearance rate. It has been assumed that a thin film of drug-containing aqueous exists between the silicone oil and retina interface allowing drug delivery to the whole of the retina as evidenced by adequate disease control in such eyes (23). However, the silicone oil may reduce the time that the drug spends in contact with the retina with a subsequent reduction in the transretinal drug clearance (Fig. 3) (25). A subsequent clinical study of 19 patients with CMV retinitis who had both a ganciclovir implant and silicone oil vitreous substitute demonstrated that the implant retained its effectiveness when used in conjunction with silicone oil (26).

An animal study using a rabbit model assessed the effect of intraocular gas on intravitreal drug concentrations using a nondegradable sustained-release device containing fluocinolone and 5-fluorouracil (27). Intraocular drug concentrations over a 6-week period were compared in eyes with drug devices that were injected with C_3F_8

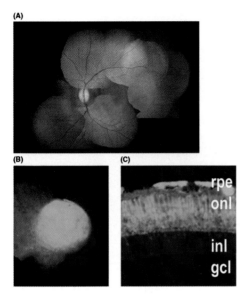

Figure 11.1 Expression of the gene encoding EGFP is stable after subretinal injection in the monkey. (*See p. 160*)

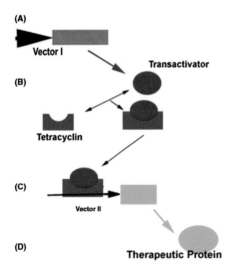

Figure 11.2 Scheme of dual vector transactivation system using tetracycline (or doxycyclin). (*See p. 164*)

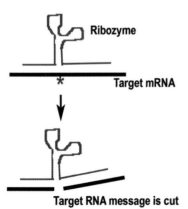

Figure 11.3 Scheme of ribozyme-mediated cleavage of target mRNA molecules. (*See p. 165*)

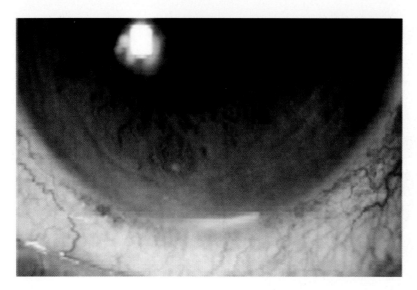

Figure 17.3 Slit-lamp photograph of an isolated incidence in which two Surodex devices are visible in the inferior angle. (*See p. 269*)

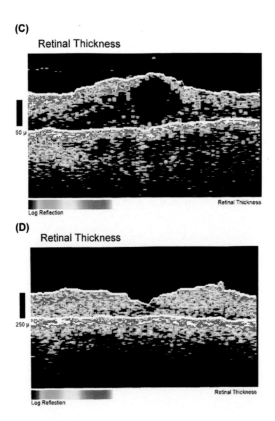

(C)
Retinal Thickness

50 μ

Log Reflection

Retinal Thickness

(D)
Retinal Thickness

250 μ

Log Reflection

Retinal Thickness

Figure 20.5 (C) OCT before injection demonstrates pronounced macular thickening. (D) OCT following injection shows macular edema resolution. (*See p. 308*)

(B)

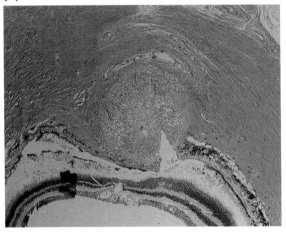

Figure 20.6 (B) Cross section of the optic nerve demonstrates sectioning of the scleral ring in a cadaver eye. (*See p. 314*)

(A)

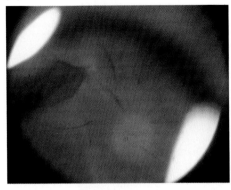

Figure 20.7 (A) In branch vein desheathing surgery, a bent MVR blade or scissors are used to incise the common adventitia at the arteriovenous crossing allowing seperation of the vessels. (*See p. 316*)

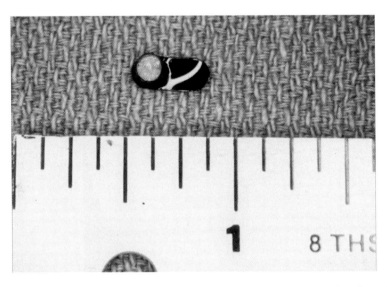

Figure 21.1 The ganciclovir implant. Note the yellow pellet of ganciclovir on the left and the strut which is secured to the sclera to the right. (*See p. 335*)

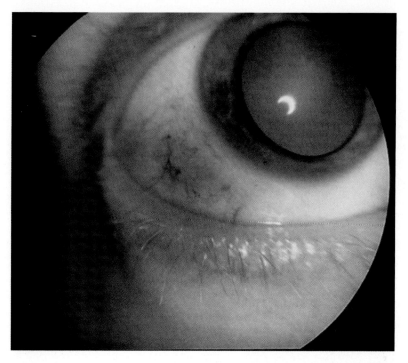

Figure 21.2 The scleral incision to place the implant into the vitreous is at the pars plana. (*See p. 336*)

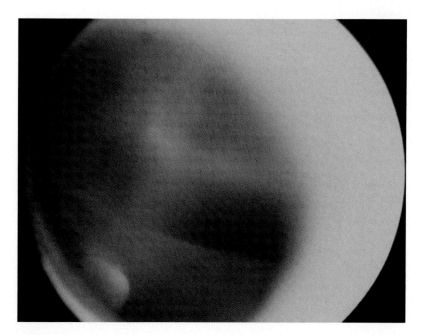

Figure 21.5 The ganciclovir implant is noted in the inferior aqueous layer in a silicone oil–filled eye. (*See p. 339*)

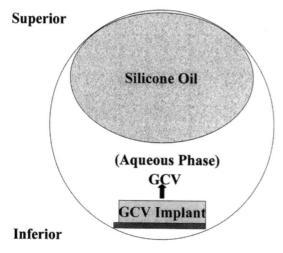

Superior

Inferior

Figure 3 Diagram of eye containing GCV implant and silicone oil demonstrating the inferior placement of the GCV device in the aqueous phase of the vitreous. There is a layer of aqueous phase between the silicone oil bubble and the retina. Note that the silicone oil resides in the nondependent portion of the vitreous cavity. *Abbreviation*: GCV, ganciclovir.

gas and control eyes with the same type of drug devices that received a sham injection. No measurable drug levels were detected in aqueous samples in either gas-filled eyes or control eyes. After two days the intravitreal drug levels were higher in gas-filled eyes than control eyes. This difference was probably the result of the smaller volume of distribution in gas-filled eyes secondary to contraction of the vitreous by the intravitreal gas bubble. After four days the intravitreal drug concentrations were similar in both gas-filled and control eyes as steady-state concentrations independent of volume of distribution were reached.

After three and six weeks, implants in gas-filled eyes had smaller amounts of drug than implants in eyes without gas. The authors hypothesized that this discrepancy was due to an increased drug clearance rate in gas-filled eyes secondary to transient disruption of the blood–retinal barrier caused by the intravitreal injection of C_3F_8 gas. Maintenance of the steady-state level would then require a higher rate of drug release from implants in the gas-filled eyes. If this hypothesis is true then nondegradable sustained-release drug devices may have a shorter lifespan in gas-filled eyes (27).

It may be difficult to reproducibly alter the implant drug release rate to respond to changes in the disease state. Ideally, a drug delivery system would release drug when the intraocular disease is active, and would release just enough drug during disease inactivity to maintain the eye in a quiescent state. In actual practice, the amount of protein in an eye will alter the release rate of drug and eyes with high levels of protein such as those seen in ocular inflammation result in higher release rates. In vitro studies have demonstrated that daily fluocinolone acetonide delivery is increased by 20% when placed in a protein-rich environment and returns to baseline release rates when placed in protein-free media (Fig. 4) (2).

If these in vitro results also hold in vivo, more fluocinolone acetonide would be released in actively inflamed eyes with great blood–ocular barrier breakdown and higher intravitreal protein concentration than in quiet eyes. It is not possible to assess the amount of drug remaining in a device by direct inspection. However, waiting for reactivation of a disease such as CMV retinitis to indicate that device replacement or

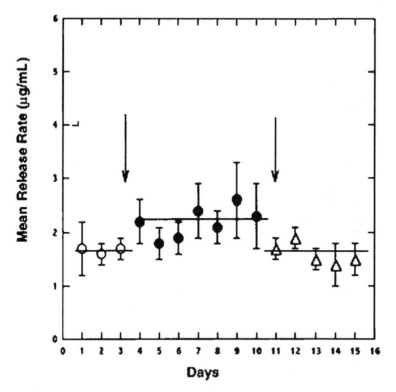

Figure 4 Fluocinolone acetonide release rates after nondegradable devices containing 2 mg drug were placed in PBS (*open circles*), then switched back to PBS with 50% plasma proteins (*closed circles*), and then switched back to PBS (*open triangles*); $n = 8$ at each time point. Data represent mean ± standard deviation. First arrow denotes time at which devices were switched from PBS to PBS + plasma protein. Second arrow refers to time at which devices switched back to PBS alone. *Abbreviation*: PBS, phosphate-buffered saline.

exchange is required can be problematic. Recurrent CMV retinitis in eyes treated with an implant alone without supplementary systemic antiviral medication can be fulminant, as there is no low-level systemic drug state to prevent disease progression (28).

Results of Animal Model Studies

Sustained intraocular dexamethasone delivery has been shown to effectively treat uveitis in a rabbit model (10). Masked observers graded intraocular inflammation and retinal function was evaluated by electroretinography (ERG). Eyes that received a sustained-release dexamethasone device had significantly reduced clinical signs of intraocular inflammation when compared with control eyes that received a sham device. Objectively, treated eyes had lower protein concentrations and leukocyte counts than control eyes. Furthermore, untreated eyes had significantly depressed ERGs and more histological evidence of tissue inflammation when compared with treated eyes (Fig. 5).

Antigen-rechallenged eyes treated with sustained-release dexamethasone also had less inflammation than control eyes and late complications including corneal neovascularization, cataract, and hypotony were less prevalent in treated eyes. A separate toxicity study also demonstrated that sustained-release dexamethasone was safe using clinical, electrophysiological, and histological parameters (12).

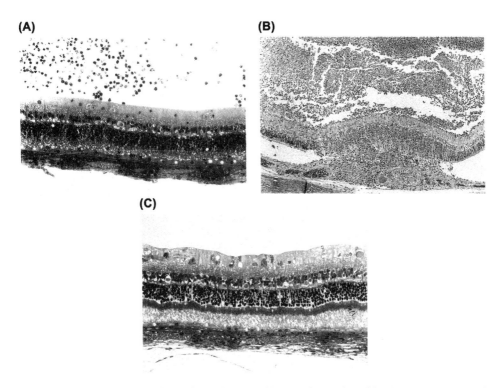

Figure 5 Histological sections of rabbit eyes with experimental uveitis. (**A**) In an untreated eye there is focal granulomatous inflammation involving vitreous, retina, and choroid. The photoreceptor layer is completely destroyed (hematoxylin and eosin ×250). (**B**) High-power view of sections of the retina in an untreated eye. There is marked inflammatory cell infiltration. Note the absence of outer segments and disorganization of the inner segments of the photoreceptor layer (toludine blue ×625). (**C**) In an eye treated with a dexamethasone sustained-release implant, inflammation is minimal and the tissue integrity is preserved (hematoxylin and eosin ×250).

Ideally a corticosteroid delivery device should provide therapeutic drug levels over the duration of a given disease. As previously mentioned fluocinolone acetonide has a low solubility in aqueous solution which is 1/24th that of dexamethasone and this low solubility potentially could allow very extended drug release without requiring an excessively bulky polymer system. To test this hypothesis an in vivo safety and pharmokinetics study in the rabbit eye was performed (7). In this study, eyes implanted with a sustained-release fluocinolone acetonide device had no clinical evidence of toxicity either from the device itself or the contained drug, when compared with fellow, control eyes. Retinal function was determined using scotopic ERG and measured as a ratio of the experimental eye B-wave amplitude to the fellow, control eye B-wave amplitude. While a slight decrease in the B-wave amplitude was seen in experimental eyes, with the exception of the 28-week time point, this difference was not statistically significantly different from baseline values and overall the retinal function remained normal (Fig. 6) (7). Furthermore, histological analysis of two eyes that had nondegradable devices implanted was similar to that of two normal fellow eyes that did not receive a sustained-release device (Fig. 7).

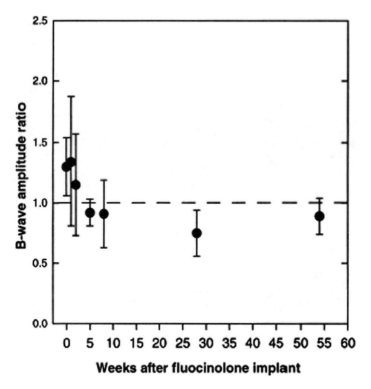

Figure 6 ERG B-wave amplitude ratios (drug device implanted eye/fellow eye) in normal rabbit eyes with a fluocinolone acetonide sustained-release device in one eye. The dark adapted B-wave amplitude ratio was initially slightly greater than one during the first three weeks of the study. Thereafter, levels remained near one for more than one year after device implantation. At 28 weeks, the B-wave amplitude ratio was lower than at other time points (approximately 0.75; $P < 0.05$); however, by 54 weeks the ratio (0.89) was once again not significantly different from 1 ($P = 0.17$). *Abbreviation*: ERG, electroretinography.

Systemic cyclosporin A is effective in treating experimentally created uveitis and chronic uveitis in humans (29–32). However, cyclosporin A has significant side effects when administered systemically and is poorly absorbed when applied topically to the eye. To avoid the systemic side effects of cyclosporin A and to circumvent its poor topical penetration a nondegradable device containing cyclosporin A was developed (11). The device can produce constant levels of cyclosporin in the vitreous and ocular tissues for as many as nine years (11,12). There was no clinical or histological evidence of toxicity after sustained release of cyclosporine in either the rabbit or primate eye (11,12). Electrophysiological tests demonstrated reversible decreased ERG B-wave amplitude in rabbit eyes but this change was not seen in the primate eye (Fig. 8). This difference may be explained by the more complete retinal vasculature in the monkey eye or by the higher susceptibility of rabbit cells to drug toxic effects (33). Anatomical similarities between the primate and human eye suggest that sustained release of cyclosporine would not be toxic in the human.

A sustained-release device containing cyclosporin A has been evaluated in a rabbit model of experimental uveitis (6). The model has been described earlier and involved initial subcutaneous immunization of the rabbits followed by an intravitreal

(A)   **(C)**

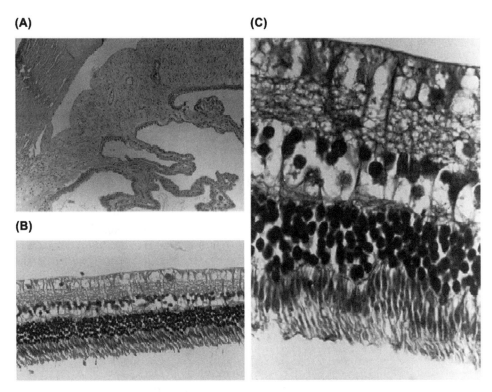

(B)

Figure 7 Histological sections of a normal rabbit eye that received a 15 mg fluocinolone acetonide sustained-release device demonstrated normal uveal anatomy. (**A**) Iris and ciliary body (hematoxylin and eosin ×12). (**B**) Retina in the region of the medullary ray. There is an artifactual retinal detachment (hematoxylin and eosin ×40). (**C**) High-power view of B (hematoxylin and eosin ×100).

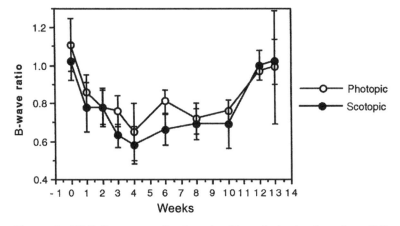

Figure 8 ERG B-wave amplitude ratios (drug device implanted eye/fellow eye) in normal rabbit eyes with a cyclosporin A/dexamethasone sustained-release device in one eye. There was a significant depression in B-wave ratios from weeks 1 to 11 ($P < 0.05$). Maximum depression occurred at four weeks when photopic ERGs showed a 35% reduction in B-wave ratio while scotopic ERGs showed a 42% reduction. Both photopic and scotopic ERGs returned to normal by week 12. The B-wave latencies under both photopic and scotopic conditions were normal throughout the course of the study. *Abbreviation*: ERG, electroretinography.

challenge of tuberculin antigen, with rechallenges in a number of animals to simulate chronic inflammation. Study eyes, which received devices containing cyclosporin A, had significantly less clinical signs of intraocular inflammation than control eyes that received a sham device. Furthermore, control eyes had significant ERG B-wave amplitude depression and had marked histological evidence of tissue inflammation and disorganization when compared with treatment eyes. Intravitreal cyclosporin A remained at therapeutic levels for six months while systemic levels of cyclosporin A were either low or nondetectable (6).

In the previously described horse model of experimental uveitis, eyes that received a cyclosporine A containing sustained-release device had less severe clinical signs of inflammation after intravitreal antigen rechallenge when compared with eyes that received a polymer-only sham device (34). Furthermore, the duration of clinical signs of inflammation was shorter in eyes that had a cyclosporin A device than those that did not have one. Aqueous and vitreous levels of protein, IL-2, and interferon gamma mRNA were significantly lower in eyes that received a cyclosporin A device when compared with those eyes with a polymer-only device. Similarly, the total number of infiltrating cells, the number of T lymphocytes, and the amount of tissue destruction was significantly less in eyes treated with cyclosporin A than controls. Interestingly, the cyclosporin A device did not completely eliminate the development of a second recurrent episode of uveitis in these animals, which may be explained by the fact that vitreous concentrations of cyclosporin A were below therapeutic levels.

Nondegradable sustained-release cyclosporin A devices have been well tolerated for up to 12 months in horses with normal ocular examinations. Scotopic ERGs were unchanged in eyes receiving drug-containing devices and those that received sham devices. Two cases of endophthalmitis were attributed to complications of the implantation surgery, while histological examination of uncomplicated eyes demonstrated only mild lymphoplasmacytic infiltration in the ciliary body and pars plana at the implantation site (34). In a subsequent study of horses with ERU, a device releasing 4 µg of cyclosporin A per day can reduce clinical evidence of intraocular inflammation, significantly reduce the number of recurrent episodes of uveitis per year and stabilize visual acuity in the majority of eyes (35).

A nondegradable device containing the centrally acting calcium channel blocker nimodipine has been tested in a previously described animal model of glaucoma. Eyes that received the device had no evidence of toxicity or inflammation when compared with control eyes that did not have an implant. Retinal function was measured using ERG at baseline and after elevation of the IOP to 40 mmHg for one hour on three occasions. In control eyes ERGs were markedly reduced after one IOP spike but ERGs remained normal in eyes with the nimodipine device even after three pressure spikes. These results suggest that long-term neuroprotection with nimopidine in eyes with glaucoma may be possible and warrants further study (3).

A recent animal study has demonstrated that it is possible to achieve therapeutic intraocular concentrations of methotrexate without clinical or electrophysiological evidence of retinal toxicity (36). Nondegradable sustained-release devices containing MTX and trimetrexate which could be used to treat intraocular lymphoma have also been developed and initial in vitro pharmacokinetic studies have been performed. However, the nondegradable devices containing these drugs have not been assessed in normal animal eyes or in animal models of disease. Similarly, a nondegradable device with a thalidomide drug core has been assessed in vitro with an aim to eventually treat choroidal neovascularization but has not been applied in an animal model of this disease.

Techniques for Implanting or Placing the Implant in Humans

Nondegradable sustained-release devices such as the ganciclovir implant and flucinolone acetonide implants are inserted through the pars plana (7,22,37). Nondegradable implants should be prepared before the eye is opened (7,22,37). Some preparation of the implants is required; the ganciclovir device suture strut must be trimmed to a length of 2.0 mm, and a hole is created 0.5 mm from the end of the strut for the anchoring suture (Fig. 9). It is not necessary to trim the fluocinolone acetonide implant suture strut. Typically a double-armed 8-0 suture passed also-full thickness sclera has been used (7,22,37). Recently there have been reports of spontaneous implant dislocation when nylon sutures have been used and less degradable suture materials such as prolene are now recommended to anchor the device to the sclera (38).

Insertion of an infusion cannula through the pars plana before placement of the implant is not necessary unless the eye has already undergone a vitrectomy. Optimally, the device should be inserted in the inferior quadrant unless there is a contraindication such as an existing implant(s) or tractional retinal detachment. As the future development of a retinal detachment is possible, particularly in patients with CMV retinitis, inferior placement of the device maximizes the likelihood that it would remain in an inferior aqueous meniscus, if vitreoretinal surgery is required. After a local peritomy is made, the ganciclovir devices are placed into the vitreous cavity via a 5–6 mm pars plana incision, 4 mm behind, and

Figure 9 Photograph of a prepared ganciclovir implant prior to implantation. A hole has been placed in the strut 1.5 mm from the base. The strut has also been trimmed so that the distance from the hole to the end of the strut is no more than 0.5 mm and there are no sharp edges.

(A)

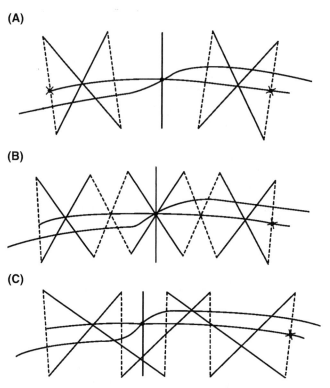

(B)

(C)

Figure 10 Diagram showing three methods of wound closure after insertion of a ganciclovir sustained-release device. (**A**) Wound closure with an X suture on either side of the anchoring suture. Note that the X suture is started within the wound so that the knot remains buried when the suture ends are trimmed. The long ends of the anchoring suture are placed under the two X sutures. (**B, C**) Wound closures with a running suture. Note that the suture is started within the wound so that the knot remains buried when the suture ends are trimmed. The long ends of the anchoring suture are placed under the running suture.

parallel to the corneoscleral limbus. It is important to inspect the wound to ensure that the pars plan has been completely incised to avoid placing the device in the suprachoroidal space. One advantage of newer nondegradable devices, such as those containing flucinolone acetonide, in comparison with the ganciclovir implant is their smaller size (7). Accordingly, these devices can be inserted through a smaller scleral incision, typically 3–3.5 mm.

Correct orientation of the two arms of the anchoring suture is important and the implants should have the drug disk facing anteriorly. The anchoring suture is closed securely and its ends are left long. The scleral wound is then closed with a running or X closure using 8-0 or 9-0 suture tied over the tails of the anchoring suture. The incision for the fluocinolone acetonide implant is typically closed with interrupted buried 9-0 prolene sutures tied tightly over the anchoring suture tails. The loose ends of the anchoring suture are positioned under the closing suture to prevent cut ends eroding through the conjunctiva (Fig. 10). The IOP should be normalized with balanced salt solution before final suture tightening to reduce postoperative astigmatism. During these steps prolapsed vitreous is removed with a vitreous cutter or a cellulose sponge and scissors, but routine complete vitrectomy may be associated with a higher rate of complications. At the end of the procedure the posi- tion of the implanted device is verified with indirect ophthalmoscopy.

Although uncommon, the insertion of nondegradable sustained-release devices can be associated with a number of complications (7,22,23,39). These include astigma- tism, wound dehiscence (1.0%), foreign body reaction, hypotony (1.9%–3.6%), ele- vated IOP (3.7%), hyphema (2.1%–8%), uveitis (1.8%–3.9%), cystoid macular edema (0.9%–5.9%), epiretinal membrane (3.7%), endophthalmitis (3.3%, 0%, 1.8%), cataract formation (3.9%, 11.7%), vitreous hemorrhage (3.3%–23%), retinal tears (3.8%), retinal detachment (5.4%–25%), and suprachoroidal placement (2.1%) (4,22–24,39–44).

The specific disease may determine whether one replaces an empty device. Patients with CMV retinitis who received a ganciclovir implant may not need another device if they are immune reconstituted (see Chapter 21). For patients with uveitis the need for further implants depends on the disease activity and can be made on a case-by-case basis. Empty devices can be either exchanged for a full one, or another device can be inserted at a site either contiguous to the initial site, or in a completely separate location (28,43,45). Typically, if the device is removed, the new implant is placed in the same site (28,45). Placing subsequent devices into the same site has the advantage of reducing the extent of the circumferential incision in the eye. However, removal of the implant can be difficult due to fibrosis of the support strut (28,45). Removal or exchange of nondegradable sustained- release implants have similar complications as those seen with first implantation (28,43,45). However, complications particularly associated with removal of nonde- gradable devices include loss of the device into the vitreous cavity, separation of the pellet from the support strut, vitreous hemorrhage, particularly with repeated same site exchanges, and thinning of the sclera resulting in wound leakage (28,45,46). An increased incidence of retinal detachment is not associated with implant exchange when compared with contiguous site or separate site implant insertion (28,43,45).

Future Horizons

The two main future developments in nondegradable sustained-release devices will be intraocular delivery of proteins and genes, which have been discussed in other chapters, and further applications of conventional medications. The potential developments using conventional medication cores are numerous. One option is to vary the drug solubility to alter drug delivery duration. Alternatively drug deliv- ery can be prolonged by increasing the amount of drug in the reservoir (which would require increased device length or width). Current drug cores have been used to treat a number of different diseases as evidenced by current studies of steroid devices to treat uveitis and diabetic macular edema (see Chapters 17 and 19). Simi- larly, other medications such as methotrexate may have a role in treating uveitis and ocular lymphoma. Different drugs could be combined to treat concomitant diseases an example of which is combining a steroid core with antiglaucoma medications to prevent steroid-associated elevations of IOP. Finally, further investigation of medications such as mycophenolate mofetil and infliximab, and tumor necrosis fac- tor antibody that are presently used to treat ocular disease systemically could result in new medication cores to treat a variety of diseases locally using sustained-release devices.

IMPLANTED MICRODIALYSIS PROBES AS SUSTAINED-RELEASE DRUG DELIVERY SYSTEMS

Description of Drug Delivery System

Microdialysis was originally developed to monitor endogenous compounds in the central nervous system and consists of a semipermeable membrane through which small molecules can diffuse without any fluid transfer (47). The microdialysis probe typically consists of an episcleral fixation plate, a guide tube and a dialysis membrane, made of a polycarbonate–polyether copolymer or polyamide (Fig. 11) (48). The probe is inserted 4–6 mm behind the limbus through a 0.9-mm opening, and sutured to the eye with the tip of the probe containing the dialysis membrane directed as much as possible into the central vitreous. In previous animal studies, connecting tubes from the probe were passed under the skin of the forehead and out between the animals' ears. Microdialysis probes can both deliver substances into the vitreous, and remove endogenous molecules to allow analysis of the intravitreal environment (48,49).

The main advantage of the microdialysis probe is that the amount of administered drug can be altered at anytime. This is in contrast to an implanted nondegradable sustained-release device which delivers the whole dose of drug at a predetermined rate unless it is removed. Furthermore, microdialysis probes deliver the drug without changing the intraocular volume and can provide information about the intravitreal environment. The main potential disadvantage of the microdialysis probe is that permanent access to the vitreous cavity increases the risk of intraocular infections. However, no cases of endophthalmitis were reported in previous animal studies. The microdialysis experiments also required that the connecting tubes remain in place for as long as the probes were in use, which would not be practical in humans.

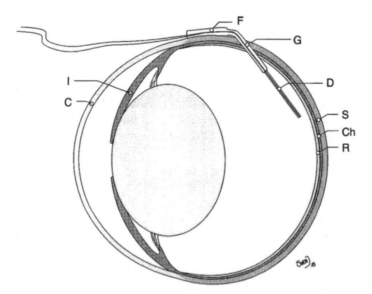

Figure 11 Semidiagrammatic representation of the microdialysis probe in the rabbit eye. The episcleral fixation plate (F) of the probe measured 5.0×3.5 mm, the guide tube (G) 4.0×1.0 mm. The dialysis membrane tube (D) measured 4.0×0.8 mm and had a wall thickness of 19 μm. *Abbreviations*: C, Cornea; I, iris; R, retina; Ch, choroid; S, sclera.

Spectrum of Diseases for Which This Delivery System Might Be Appropriate

As described below, microdialysis probes have been used to deliver therapeutic concentrations of drugs that may be useful to treat a variety of retinal diseases (50–52). Drugs that target macular edema, syphilis, CMV retinitis, proliferative vitreoretinopathy and retinal degenerative diseases have been tested. However, treatment for other conditions with this approach, particularly intraocular malignancy, could be envisioned.

Animal Models Used to Investigate the Applicability of the Delivery System

Microdialysis probes are still in the early stages of development and their use against specific diseases has not been investigated in any animal models.

Pharmacokinetic and Pharmacodynamic Studies Using the Delivery System

In vitro and in vivo experiments were carried out to determine the transport capacity of various dialysis membranes. Using an average perfusion rate of $4\,\mu L/min$ it was noted that polycarbonate membranes tended to retain some drugs in an unpredictable manner, a characteristic not seen with polyamide membranes. Furthermore, the transport capacity of the probes was not just dependent on the area of the dialysis membrane, but was also proportional to the length of the membrane, and inversely proportional to its width. This effect of probe width on the total delivery capacity is explained by the fact that flow rates are approximately 7.5 times slower in wide tubes and when compared with narrow tubes, fewer molecules are close enough to the tube wall in wide tubes to diffuse out. Practically speaking, the longer and thinner a dialysis membrane is, the more molecules it will be able to deliver. No difference in transport capacity was found in the two different types of dialysis membrane types when their dimensions were the same.

While microdialysis probes have the ability to deliver drugs to the vitreous, they can also withdraw intravitreal molecules. However, the ability to recover molecules from the vitreous is inversely proportional to the perfusion speed of the probe, which is in contrast to the higher perfusion speeds required for higher delivery capacity. This paradox is important when microdialysis probes are being considered to simultaneously deliver drugs to, and withdraw samples from, the vitreous cavity. Furthermore there is no precise method to predict the dialysis results, as many factors affect the transport of molecules across the dialysis membrane. Some of these influencing factors are known such as the length of time the probe has been in place, while others are not, and relate to the complexities of the probe geometry and its interactions with a given surrounding media, neither of which have not been fully explained mathematically (53).

Results of Animal Model Studies

Both short- and long-term studies of the effectiveness of permanently implanted probes have been undertaken in the rabbit eye, and in one study probes remained in place for up to six months (48,49,54–58). An early rabbit study of microdialysis probes determined the optimum surgical technique to insert the probes and examined histological sections of eyes that had received them. Probes were inserted through an opening in the nasal sclera (the inlet tube), passed across the eye, and exited through the temporal sclera (the outlet tube) (49). Data on the first 11 of 27 probes implanted were not reported, but in the remaining 16 eyes, three probes were lost as a result of the animal pulling the probe out. There was minimal inflammation of the eyes and no cases of endophthalmitis. Cataract was seen in earlier experiments and was attributed to accidental lens touch. Initial probes clogged within two weeks but regular perfusion of the probes in later animals prevented this complication. Histological analysis of eyes after implantation demonstrated minimal tissue reaction at wound sites and no inflammatory reaction around the probe (Fig. 12) (49).

A later experiment modified the microdialysis probe with both inlet and outlet probes mounted on a single stiff tube, which meant that only one opening was required in the eye. In total, microdialysis probes were inserted into one eye of 10 rabbits. As in the previous experiments there was minimal ocular inflammation and there were no cases of endophthalmitis. The probes were perfused each day to prevent clogging (48). The probes were left in situ for an average of 20.8 days and

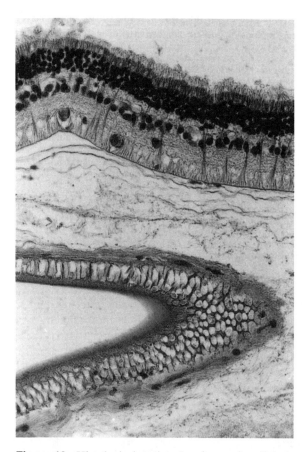

Figure 12 Histological section showing a microdialysis probe in a rabbit eye showing part of the dialysis membrane near the probe tip. The probe had been in the eye for 30 days. There is no inflammatory reaction and the structure of the overlying retina, retinal pigment epithelium, and choroid is normal. There is an artifactual retinal detachment that occurred during preparation of the section (hematoxylin and eosin ×3584).

histological examination of the eyes demonstrated minimal gliotic tissue around the wound site but no inflammation around the probe. The retina appeared normal.

A number of subsequent experiments which estimated intravitreal drug concentrations administered using the microdialysis probe were reported (50,51). Benzyl penicillin, dexamethasone, 5-fluorouracil, daunomycin, and ganciclovir were radiolabeled and 120–140 µL of each drug was infused into the vitreous. The animals were subsequently killed and the vitreous concentration of the drug was calculated using liquid scintillation spectrometry. It was possible to achieve therapeutic concentrations of each of the drugs in the vitreous without reaching the maximum transport capacity of the probes. The authors concluded that it was possible to reach clinically useful concentrations of these drugs which had known clinical applications using the microdialysis probe (50,51).

Techniques for Implanting or Placing the Implant in Humans

Implanted microdialysis probes are still in the early stages of development and they have not been implanted in humans.

Future Horizons

It is likely that future developments of microdialysis probes will focus on experimental models as long-term implantation of these devices in humans may not be feasible in view of the risk of intraocular infection and the probability of dislodging the device during normal daily activities. However, potential human applications could include delivery of relatively short courses of intraocular chemotherapy for retinoblastoma or ocular lymphoma.

MICROELECTROMECHANICAL SYSTEMS DRUG DELIVERY DEVICES

Description of Drug Delivery System

Microfabrication technology has enabled the development of active devices incorporating micrometer scale pumps, valves, and flow channels to deliver liquids (59,60). A solid-state silicon microchip has been developed that can provide controlled release of single or multiple chemicals on demand in laboratory experiments (61). The microchip devices have no moving parts and consist of a standard silicon wafer containing reservoirs with a volume of 25 nL that extend completely through the wafer. The reservoirs are square pyramidal in shape and the surface of the wafer with the smaller square ends of the reservoirs are covered with a gold membrane anode 3-µm thick. After filling of the reservoirs using conventional inkjet printing techniques coupled with a computer-controlled alignment apparatus, the reverse face of the chip is closed with a thin layer of plastic and sealed with a waterproof epoxy (Fig. 13) (61). Furthermore, a microbattery, microcircuitry, and memory could also be contained in the chip. The smallest chip size available at present is 1-cm square.

Spectrum of Diseases for Which This Delivery System
Might Be Appropriate

Microelectromechanical systems (MEMS) devices can potentially administer complex dosing patterns using very small amounts of drugs. This system could be used to deliver protein molecules intraocularly such as neuroprotectors or anti-angiogenic molecules. Delivery of chemotherapeutic agents using the MEMS may also be a future possibility.

(A)

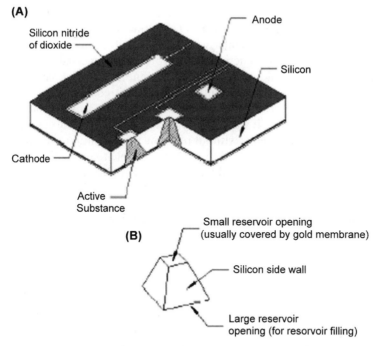

Figure 13 Diagrams of a prototype of a MEMS drug delivery device incorporating multiple sealed compartments that can be opened on demand to deliver a drug dose. (**A**) Prime grade silicon wafers are sandwiched between two layers of silicon nitride and each device contained reservoirs that extended completely through the wafer. The devices also contain a cathode and an anode between which small electric potentials can be passed. (**B**) Each reservoir is square pyramidal in shape with one large and one small square opening. The reservoirs have a volume of approximately 25 nL and are sealed on the small square end with the anode which is a 0.3-μm thick gold membrane. *Abbreviation*: MEMS, microelectromechanical systems.

Animal Models Used to Investigate the Applicability of the Delivery System

MEMS devices are still in the early stages of development and their use against specific diseases have not been investigated in any animal models.

Pharmacokinetic and Pharmacodynamic Studies Using the Delivery System

In vitro release experiments were performed which consisted of immersing the devices in a buffer solution formulated to mimic the body's pH and chloride concentration. Passage of a small electrical potential along the gold anode covering a single reservoir in the presence of chloride ions results in soluble gold chloride complexes, dissolution of the gold membrane, and release of the marker chemicals (Fig. 14). Gold has the advantage of being a biocompatible material and is consistently soluble after passage of the current, in contrast to other metals such as copper and titanium (62).

Further release experiments on the microchips have demonstrated that multiple reservoirs could be opened at different times in a single device. Passage of a current

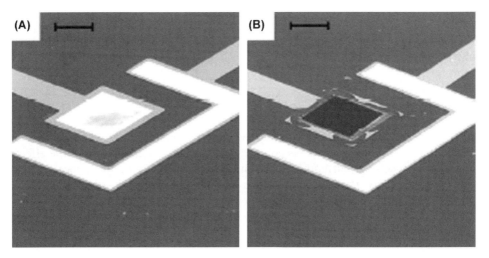

Figure 14 Scanning electromicrograph of a single reservoir in a prototypical MEMS drug delivery device. (**A**) The gold anode is in place over the small square end of the reservoir. (**B**) After passage of a small electrical potential the gold anode has dissolved. *Abbreviation*: MEMS, microelectromechanical systems.

along the electrode over one reservoir resulted in release of marker chemicals in a controlled fashion, without activating adjacent reservoirs. This individual control of multiple reservoirs creates the possibility of achieving complex release patterns of multiple drugs from one device.

Results of Animal Model Studies

Biocompatibility and biofouling studies of a MEMS device have been undertaken in a rodent model. Biocompatibility of the MEMS components which include metallic gold, silicon nitride, silicon dioxide, silicon, and SU-8 TM photoresist were evaluated using a stainless steel cage system measuring 3.5×1.0 cm which were implanted subcutaneously in the rodent. The inflammatory response, measured using leukocyte concentrations extracted from cage exudates, of the device components was similar to that of control cages over a 3-week period (63). Furthermore, all the components with the exception of silicon demonstrated reduced biofouling as shown by scanning electron microscopy studies of macrophages and foreign body giant cells on the surfaces of the material three weeks after subcutaneous implantation in the rodent. The good biocompatibility and biofouling profiles of the MEMS components suggest that long-term implantation would not interfere with the drug delivery capability of the devices (63).

Previous researchers have implanted microchips into the eye in order to stimulate the retina in patients with neuroretinal degenerations (64,65). While these experiments provide proof of the principal that implantation of intraocular microchips is possible, there are a large number of factors that have yet to be considered (66). The microchips were not implanted in the eye for prolonged periods and the degradation profile of such microchips has not been studied. Other important possible complications of prolonged microchip implantation requiring evaluation include the immune response to the chip, the retinal effects of long-term intraocular electrical stimulation (albeit with very low current levels), and the thermal effects of a long-term intraocular electrical current.

Techniques for Implanting or Placing the Implant in Humans

MEMS devices are still in the early stages of development and they have not been implanted in humans.

Future Horizons

At present MEMS devices are probably too large to be inserted into the eye. Further development of retinal prostheses coupled with clearer understanding of the effects of the microelectrical currents and nonorganic components of the devices on the ocular tissues may result in intraocular applications for MEMS drug delivery devices. The complex delivery patterns possible with these devices may allow treatment of diseases with minimal amounts of conventional medications or proteins either in combination or alone.

REFERENCES

1. Ashton P, Blandford DL, Pearson PA, et al. Review: implants. J Ocul Pharmacol 1994; 10:691–701.
2. Jaffe GJ, Yang CH, Guo H, et al. Safety and pharmacokinetics of an intraocular fluocinolone acetonide sustained delivery device. Invest Ophthalmol Vis Sci 2000; 41:3569–3575.
3. Ip MS, Reichel E, Vieira W, Wu M, Cordahi G, Ashton P. Neuroprotective effect of a sustained intravitreal calcium channel blocker in elevated IOP in the rabbit. Invest Ophthalmol Vis Sci 1998; 39:S275.
4. Musch DC, Martin DF, Gordon JF, Davis MD, Kuppermann BD. Treatment of cytomegalovirus retinitis with a sustained-release ganciclovir implant. The Ganciclovir Implant Study Group. N Engl J Med 1997; 337:83–90.
5. Velez G, Robinson MR, Durbin T, Yuan P, Sung C, Whitcup SM. Thalidomide sustained release devices for choroidal neovascularization: an in-vitro analysis. Invest Ophthalmol Vis Sci 1999; 40:S84 (Abstract no 448).
6. Jaffe GJ, Yang C-S, Wang X-C, Cousins SW, Gallemore RP, Ashton P. Intravitreal sustained-release cyclosporine in the treatment of experimental uveitis. Ophthalmology 1998; 105:46–56.
7. Jaffe GJ, Ben-nun J, Guo H, Dunn JP, Ashton P. Fluocinolone acetonide sustained drug delivery device to treat severe uveitis. Ophthalmology 2000; 107:2024–2033.
8. Jaffe GJ, Pearson PA, Ashton P. Dexamethasone sustained drug delivery implant for the treatment of severe uveitis. Retina 2000; 20:402–403.
9. Tao W, Wen R, Goddard MB, Sherman SD, et al. Encapsulated cell-based delivery of CTNF reduces photoreceptor degeneration in animal models of retinitis pigmentosa. Invest Ophthalmol Vis Sci 2002; 43:3292–3298.
10. Cheng-Kuo C, Berger A, Pearson PA, Ashton P, Jaffe GJ. Intravitreal sustained-release dexamethasone device in the treatment of experimental uveitis. Invest Ophthalmol Vis Sci 1995; 36:442–453.
11. Pearson PA, Jaffe GJ, Martin DF, et al. Evaluation of a delivery system providing long-term release of cyclosporine. Arch Ophthalmol 1996; 114:311–317.
12. Enyedi LB, Pearson PA, Ashton P, Jaffe GJ. An intravitreal device providing sustained release of cyclosporine and dexamethasone. Curr Eye Res 1996; 15:549–557.
13. Gilger BC, Molok E, Stewart T, et al. Effect of an intravitreal cyclosporine implant on experimental uveitis in horses. Vet Immunol Immunopathol 2000; 76:239–255.
14. Bellhorn RW. Analysis of animal models of macular edema. Surv Ophthalmol 1984; 28(suppl):520–524.

15. Aiello LP, Bursell SE, Clermont A, et al. Vascular endothelial growth factor-induced retinal permeability is mediated by protein kinase C in vivo and suppressed by an orally effective beta-isoform-selective inhibitor. Diabetes 1997; 46:1473–1480.

16. Robinson MR, Yuan P, Sung C, Nussenblatt RB, Whitcup SM. Sustained-release delivery system for anti-neoplastic agents for the treatment of intraocular lymphoma [ARVO Abstract]. Invest Ophthalmol Vis Sci 1998; 39:S275 (Abstract no. 1254).

17. Durbin TD, Robinson MR, Yuan P, Gogolak L, Sung C, Whitcup SM. Sustained-release devices with trimetrexate for the treatment of intraocular lymphoma: an in-vitro study [ARVO Abstract]. Invest Ophthalmol Vis Sci 1999; 40:S85 (Abstract no. 451).

18. Robinson MR, Baffi J, Yuan P, et al. Safety and pharmacokinetics of intravitreal 2-methoxyestradiol implants in normal rabbit and pharmacodynamics in a rat model. Exp Eye Res 2002; 74:309–317.

19. Tojo K, Isowaki A. Pharmacokinetic model for in vivo/in vitro correlation of intravitreal drug delivery. Adv Drug Deliv Rev 2001; 52:17–24.

20. Baker R. Controlled release of biologically active agents. New York: John Wiley, 1987.

21. Cekic O, Ohjii M. Intraocular gas tamponades. Semin Ophthalmol 2000; 15:3–14.

22. Sanborn GE, Anand R, Torti RE, et al. Sustained-release ganciclovir therapy for treatment of cytomegalovirus retinitis. Use of an intravitreal device. Arch Ophthalmol 1992; 110:188–195.

23. Martin DF, Parks DJ, Mellow SD, et al. Treatment of cytomegalovirus retinitis with an intraocular sustained-release ganciclovir implant. A randomized controlled clinical trial. Arch Ophthalmol 1994; 112:1531–1539.

24. Marx JL, Kapusta MA, Patel SS, et al. Use of the ganciclovir implant in the treatment of recurrent cytomegalovirus retinitis. Arch Ophthalmol 1996; 114:815–820.

25. Perkins SL, Yang C-H, Ashton PA, Jaffe GJ. Pharmacokinetics of the ganciclovir implant in the silicone-filled eye. Retina 2001; 21:10–14.

26. McGuire DE, McAuliffe P, Heinemann M-H, Rahhal FM. Efficacy of the ganciclovir implant in the setting of silicone oil vitreous substitute. Retina 2000; 20:520–523.

27. Perkins SL, Gallemore RP, Yang C-H, Guo H, Ashton P, Jaffe GJ. Pharmacokinetics of the fluocinolone/5-fluorouracil codrug in the gas-filled eye. Retina 2000; 20:514–519.

28. Martin DF, Ferris FL, Parks DJ, et al. Ganciclovir implant exchange. Timing, surgical procedure, and complications. Arch Ophthalmol 1997; 115:1389–1394.

29. Nussenblatt RB, Rodrigues MM, Wacker WB, Cevario SJ, Salinas-Carmona MC, Gery I. Cyclosporin A. Inhibition of experimental autoimmune uveitis in Lewis rats. J Clin Invest 1981; 67:1228–1231.

30. Striph G, Doft B, Rabin B, Johnson B. Retina S antigen-induced uveitis. The efficacy of cyclosporine and corticosteroids in treatment. Arch Ophthalmol 1986; 104:114–117.

31. Nussenblatt RB, Palestine AG, Chan CC. Cyclosporin A therapy in the treatment of intraocular inflammatory disease resistant to systemic corticosteroids and cytotoxic agents. Am J Ophthalmol 1983; 96:275–282.

32. Nussenblatt RB, de Smet MD, Rubin B, et al. A masked, randomized, dose–response study between cyclosporine A and G in the treatment of sight-threatening uveitis of noninfectious origin. Am J Ophthalmol 1993; 115:583–591.

33. Hartzer MK, Blumenkranz MS, Hajek AS, Dailey WA, Cheng M, Margherio AR. Selection of therapeutic agents for intraocular proliferative disease 3. Effects of fluoropyrimidines on cell-mediated contraction of human fibroblasts. Exp Eye Res 1989; 48:321–328.

34. Gilger BC, Malok E, Stewart T, et al. Long-term effect on the equine eye of an intravitreal device used for sustained release of cyclosporine A. Vet Ophthalmol 2000; 3: 105–110.

35. Gilger BC, Wilkie DA, Davidson MG, Allen JB. Use of an intravitreal sustained-release cyclosporine delivery device for treatment of equine recurrent uveitis. Am J Vet Res 2001; 62:1892–1896.

36. Velez G, Yuan P, Sung C, et al. Pharmacokinetics and toxicity of intravitreal chemotherapy for primary intraocular lymphoma. Arch Ophthalmol 2001; 119:1518–1524.

37. Martin DF, Dunn JP, Davis JL. Use of the ganciclovir implant for the treatment of cytomegalovirus retinitis in the era of potent antiretroviral therapy: recommendations of the International AIDS Society–USA panel. Am J Ophthalmol 1999; 127:329–339.
38. Browning DJ. Dislocated ganciclovir implant from use of a nylon fixation suture. Retina 2003; 23:723–724.
39. Lim JI, Wolitz RA, Dowling AH, Bloom HR, Irvine AR, Schwartz DM. Visual and anatomic outcomes associated with posterior segment complications after ganciclovir implant procedures in patients with AIDS and cytomegalovirus retinitis. Am J Ophthalmol 1999; 127:288–293.
40. Taguri AH, Dhillon B, Wharton SB, Kamal A. Foreign body reaction with delayed extrusion of ganciclovir implant in a patient with immune recovery vitritis syndrome. Am J Ophthalmol 2002; 133:147–149.
41. Hatton MP, Duker JS, Reichel E, Morley MG, Puliafito CA. Treatment of relapsed cytomegalovirus retinitis with the sustained-release ganciclovir implant. Retina 1998; 18: 50–55.
42. Roth DB, Feuer WJ, Blenke AJ, Davis JL. Treatment of recurrent cytomegalovirus retinitis with the ganciclovir implant. Am J Ophthalmol 1999; 127:276–282.
43. Guembel HO, Krieglsteiner S, Rosenkranz C, Hattenbach LO, Koch FH, Ohrloff C. Complications after implantation of intraocular devices in patients with cytomegalovirus retinitis. Graefes Arch Clin Exp Ophthalmol 1999; 237:824–829.
44. Anand R, Nightingale SD, Fish RH, Smith TJ, Ashton P. Control of cytomegalovirus retinitis using sustained release of intraocular ganciclovir. Arch Ophthalmol 1993; 111:223–227.
45. Morley MG, Duker JS, Ashton P, Robinson MR. Replacing ganciclovir implants. Ophthalmology 1995; 102:388–392.
46. Boyer DS, Posalski J. Potential complication associated with removal of ganciclovir implants. Am J Ophthalmol 1999; 127:349–350.
47. Ungerstedt U. Measurement of neurotransmitter release by intracranial analysis. In: Marsden CA, ed. Measurement of neurotransmitter release in vivo. New York: John Wiley & Sons Ltd, 1984:81–105.
48. Waga J, Ehringer B. Passage of drugs through different intraocular microdialysis membranes. Graefes Arch Clin Exp Ophthalmol 1995; 233:31–37.
49. Waga J, Ohta A, Ehinger B. Intraocular microdialysis with permanently implanted probes in rabbit. Acta Ophthalmol (Copenh) 1991; 69:618–624.
50. Waga J, Ehringer B. Intravitreal concentrations of some drugs administered with microdialysis. Acta Ophthalmol Scand 1997; 75:36–40.
51. Waga J. Ganciclovir delivery through an intravitreal microdialysis probe in rabbit. Acta Ophthalmol Scand 2000; 78:369–371.
52. Waga J, Ehringer B. NGF administered by microdialysis into rabbit vitreous. Acta Ophthalmol Scand 2000; 78:154–155.
53. Kehr J. A survey on quantitative microdialysis; theoretical models and practical implications. J Neurosci Methods 1993; 48:251–261.
54. Gunnarson G, Jakobsson AK, Hamberger A, Sjostrand J. Free amino acids in the preretinal vitreous space. Effect of high potassium and nipecotic acid. Exp Eye Res 1987; 44:235–244.
55. Ben-Nun J, Cooper RL, Cringle SJ, Constable IJ. Ocular dialysis: a new technique for in vivo intraocular pharmacokinetic measurements. Arch Ophthalmol 1988; 106:254–259.
56. Ben-Nun J, Cooper RL, Cringle SJ, Constable IJ. A new method for continuous intraocular drug delivery. Aust N Z J Ophthalmol 1989; 17:185–190.
57. Ben-Nun J, Joyce DA, Cooper RL, Cringle SJ, Constable IJ. Pharmacokinetics of intravitreal injection. Assessment of a gentamicin model by ocular dialysis. Invest Ophthalmol Vis Sci 1989; 30:1055–1061.
58. Stempels N, Tassignon M-J, Sarre S. A removable ocular microdialysis system for measuring vitreous biogenic amines. Graefes Arch Clin Exp Ophthalmol 1993; 231:651–655.

59. Gravensen P, Branebjerg J, Jensen OS. Microfluidics—a review. J Micromech Microeng 1993; 3:168–182.
60. Shoji S, Esashi M. Microflow devices and systems. J Micromech Microeng 1994; 4: 157–171.
61. Santini JT, Cima MJ, Langer R. A controlled-release microchip. Nature 1999; 397: 335–338.
62. Merchant B. Gold, the noble metal and the paradoxes of its toxicology. Biologicals 1998; 26:49–59.
63. Voskerician G, Shive MS, Shawgo RS, et al. Biocompatibility and biofouling of MEMS drug delivery devices. Biomaterials 2003; 24:1959–1967.
64. Humayun MS. Intraocular retinal prosthesis. Trans Am Ophthalmol Soc 2001; 99: 271–300.
65. Humayun MS, de Juan E Jr, Weiland JD, et al. Pattern electrical stimulation of the human retina. Vision Res 1999; 39:2569–2576.
66. Margalit E, Maia M, Weiland JD, et al. Retinal prosthesis for the blind. Surv Ophthalmol 2002; 47:335–356.

15

Photodynamic Therapy in Human Clinical Studies: Age-Related Macular Degeneration

Ivana K. Kim and Joan W. Miller
Department of Ophthalmology, Harvard Medical School, Massachusetts Eye and Ear Infirmary, Boston, Massachusetts, U.S.A.

INTRODUCTION

The current use of photodynamic therapy (PDT) is founded on evidence from a series of well-designed clinical trials. Work in animal models clearly demonstrated the therapeutic potential of PDT with verteporfin for choroidal neovascularization (CNV) (see Chapter 9). These preclinical studies proved that verteporfin PDT could occlude experimental laser-induced CNV, confirmed by cessation of angiographic leakage on fluorescein angiography and thrombosis of vessels on histologic examination (1–3). Treatment parameters for maximum selectivity were refined by manipulating drug and light doses and the timing of light irradiation, resulting in minimal effect on the choroid and surrounding retina. Although damage was noted at the level of the choriocapillaris and retinal pigment epithelium (RPE), recovery of these structures was observed after both single and repeated PDT (4,5). The preclinical data provided the rationale for a clinical Phase I and II study to assess the safety of PDT and to determine the maximum tolerated dose of PDT using verteporfin for treatment of CNV. Based on encouraging results from these Phase I and II investigations, large Phase III trials followed, providing guidelines for the application of PDT in clinical practice today.

PHASE I/II DESIGN AND METHODOLOGY

Single Treatments

The study was a nonrandomized, multicenter, open-label trial using five different dose regimens with inclusion and exclusion criteria as shown in Table 1 (6). The primary requirement consisted of subfoveal CNV with some classic component. Standardized protocol refractions, visual acuity determinations, complete ophthalmic examinations, color fundus photography, and fluorescein angiography were performed at baseline (between one and seven days prior to the day of treatment), and

Table 1 Eligibility Criteria for Phase I and II Studies

Inclusion criteria
Clinical signs of CNV due to any cause
CNV under the geometric center of the foveal avascular zone (subfoveal)
Some classic CNV (occult CNV could, but need not be present)
Greatest linear dimension of entire CNV ≤5400 µm diameter
Nasal side of CNV ≥500 µm from temporal border of optic nerve
For CNV lesions recurring after standard laser therapy, foveal center must not have been
  included in area treated by laser
Best-corrected visual acuity of 20/40 or worse
≥50 yrs of age
Exclusion criteria
Tears of the RPE at screening
Vitelliform-like detachment retinal pigment epithelium
Central serous retinopathy
Drusenoid pigment epithelium detachment alone
Additional retinovascular diseases compromising visual acuity of study eye
Use of investigational drugs, systemic steroids, cytokines or photosensitive drugs
  in past 3 mos
Significant hepatic, renal or neurologic disease
Class III or IV cardiovascular disease (New York Heart Association functional status
  criteria)
Porphyria, porphyrin sensitivity, hypersensitivity to sunlight or bright artificial light
Any malignancy treatment
Any acute illness during screening or fever on day of treatment prior to verteporfin
  infusion
Uncontrolled hypertension
Ocular surgery within 3 mos prior to study treatment

Abbreviations: CNV, choroidal neovascularization; RPE, retinal pigment epithelium.
Source: From Ref. 6.

at weeks 1, 4, and 12 of follow-up. Repeat visual acuity testing without refraction was also performed on the day of treatment and one day after treatment. Additional testing was carried out at week 2 if any ocular adverse event was noted at week 1, or at any time between visits, if judged to be clinically indicated. In addition, each patient had a physical examination with measurement of vital signs and an electrocardiogram at baseline and week 12. Laboratory testing, including a complete blood count, serum cholesterol, triglycerides, blood urea nitrogen, creatinine, electrolytes, calcium, phosphorus, glucose, protein, albumin, bilirubin, liver function tests, and urinalysis was performed at baseline and at weeks 1 and 12.

The size of the treatment spot was calculated according to the baseline fluorescein angiogram. The greatest linear dimension of the neovascular lesion was measured and this dimension was divided by 2.5 to account for the magnification of the camera systems, yielding the size of the lesion on the retina. A 600-µm border was added to the dimension on the retina to provide at least a 300-µm border at all edges. In some cases, the resulting treatment border at a given edge was >300 µm, but was always < 500 µm. This border provided a margin beyond the CNV detected angiographically and also allowed for adequate treatment despite small errors in the size calculation due to the optics of individual eyes and any small movement by the patient or treating physician during irradiation.

Table 2 Treatment Regimens 1–5 and Duration of Follow-Up

Treatment regimen	Verteporfin dose (mg/m^2)	No. of minutes over which infusion was planned	Light doses (J/cm^2)	Time of light application after start of verteporfin infusion (min)	No. of patients	No. with 4-week follow-up before any retreatment	No. with 12-week follow-up before any retreatment
1	6	10	50, 75, 100, 150	30[a]	22	22	19[b]
2	6	10	50, 75, 100, 150	20	37	28[c]	17[d]
3	12	10	50, 75, 100, 150	30	19	19	19
4	6	10	50, 75, 100	15	22	22	11[e]
5	6	5	12.5, 25, 50	10	28	27[f]	25[g]

[a]Includes two patients in whom light administration was 39 and 55 min late.
[b]One patient withdrew after week 4 assessment; two patients received multiple PDT treatments after week 4, before week 12.
[c]Nine patients received multiple PDT treatments after week 1, before week 4.
[d]Eleven patients received multiple treatments after week 4, before week 12.
[e]Nine patients received multiple treatments after week 4 but before week 12: two additional patients missed the week-12 follow-up visit.
[f]One patient withdrew before week 1.
[g]Two patients withdrew after week 4.
Source: From Ref. 6.

Five treatment regimens as shown in Table 2 were applied to determine whether variations in dosimetry could achieve more persistent closure of CNV. In the first three treatment regimens, the light dose (fluence) was escalated using 50, 75, 100, and 150 J/cm^2, with a minimum of three patients at each light dose. In regimen 4, the 150 J/cm^2 light dose was not included because of nonselective retinal vessel occlusion seen with this light dose in regimens 2 and 3. In regimen 5, lower light doses of 12.5 and 25 J/cm^2 were included, in addition to 50 J/cm^2.

Retreatment Regimens

As preliminary results showed that the effect of verteporfin PDT on angiographic leakage from CNV was only temporary, additional regimens involving retreatment at two- and four-week intervals were designed (7). Prior studies of repeated PDT treatments in normal monkey eyes had suggested the safety of retreatment, demonstrating drug dose-dependent recovery of the retinal pigment epithelium (RPE) and choriocapillaris with minimal damage to the photoreceptors. Two sets of patients were studied, the first group undergoing retreatment two to four weeks after the initial PDT treatment. A second set of patients from the single treatment protocol was reenrolled in a retreatment protocol sometime beyond 12 weeks but < 6 months after their initial treatment, if they met the criteria for retreatment at that time. Criteria for retreatment or re-enrollment included the following: (i) evidence of fluorescein leakage from classic or occult CNV, (ii) greatest linear dimension of leakage from CNV of < 6400 μm, (iii) no adverse event judged to be due to PDT, and (iv) no additional ocular abnormality associated with visual loss identified since the first PDT. Criteria for re-enrollment also required fluorescein leakage from classic CNV.

The first set of patients was retreated two to four weeks after the first PDT treatment. Follow-up fluorescein angiography was obtained and evaluated for CNV leakage one and four weeks after retreatment. If leakage was observed, an additional course of retreatment was applied at two- or four-week intervals, and a final evaluation was performed 12 weeks after the last retreatment. For the set of patients who were re-enrolled, up to three additional courses of retreatment could be performed, if indicated. In this group of patients, the second and third retreatments were scheduled at four-week intervals, and final evaluation was performed 12 weeks after the last retreatment. Treatment regimen 2 with a light dose of 100 J/cm^2 and regimen 4 with light doses of 50, 75, and 100 J/cm^2 (Table 2) were used in the retreatment protocols. The other aspects of the protocol, refraction, ophthalmic examination, PDT protocol, and follow-up were the same as for the single treatment protocol.

Outcome Measures

Two measures were used to assess short-term efficacy and safety: (i) the extent of fluorescein leakage from the CNV (classic and occult), and (ii) stabilization of the best-corrected visual acuity at the 12-week follow-up compared with baseline. A grading system was devised to semi-quantitatively assess the effect of PDT on the extent (area) of fluorescein leakage from the CNV lesions at each follow-up visit compared with that seen at baseline (Table 3). Fluorescein leakage from classic and occult CNV was assessed without any knowledge of the PDT dosage. Other fundus characteristics were graded from fundus photographs and angiograms, including the extent of subretinal hemorrhage, retinal pigment epithelium (RPE) atrophy, as well as retinal vascular nonperfusion.

Table 3 Grading of Extent of Fluorescein Leakage Following PDT

Grade	Description
Absence of leakage	Absence of leakage from 100% of area of CNV noted at baseline and no progression
Minimal leakage	Area of CNV $< 50\%$ of the area of leakage noted at baseline and no progression
Moderate leakage	Area of CNV $\geq 50\%$ of the area of leakage noted at baseline and no progression
Progression of leakage	Leakage from CNV beyond area of CNV noted at baseline, regardless of amount of leakage noted within area of leakage seen at baseline

Abbreviations: CNV, choroidal neovascularization; PDT, photodynamic therapy.
Source: From Ref. 6.

Best-corrected visual acuity was measured primarily as an indicator of safety, not efficacy, given the absence of a control group, the short period of follow-up, and the known variability in the natural history of the disease. However, an exploratory analysis using changes in visual acuity and angiographic leakage compared results of the various regimens and helped to determine the treatment regimen for the subsequent randomized, placebo-controlled trial. Further comparisons using analysis of variance methods were performed to assess the relationship between visual acuity outcomes and baseline characteristics such as visual acuity, lesion size, lesion composition, and lesion status. Adverse events, both ocular and systemic were captured by the treating investigator and by Reading Center review of fundus photographs and angiograms.

PHASE I/II RESULTS

Single Treatment

A total of 128 patients with subfoveal CNV due to age-related macular degeneration (AMD) were treated with at least a single course of PDT, including 31 patients who later received multiple courses of PDT with regimen 2 or 4. The majority of patients (80%) had a baseline visual acuity better than or equal to 20/200 in the treatment eye, with both the mean and median baseline visual acuities measuring 20/125. The mean baseline lesion size was 5.1 Macular Photocoagulation Study (MPS) disc areas (DAs), ranging from 1 to 16 MPS DAs, which was larger than the lesions included in the MPS. The area of classic CNV was $\geq 50\%$ of the area of the lesion in 63% of eyes which presented with classic CNV. Of these lesions, 32% had no occult CNV. Ten (8%) of the baseline lesions were judged to have no classic CNV on review by the Reading Center. In 93% of the lesions, fibrosis was judged to be $\leq 50\%$ of the lesion.

Visual Outcomes

Overall, PDT with verteporfin had no short-term effect on vision, with a mean change at week 1 of +0.7 line. Substantial vision improvement of three or more lines was noted in a small number of patients (14%) at week 1. The mean change in visual acuity at week 4 was +0.2 line and –0.5 line at week 12. There was a small but statistically significant difference in the mean visual acuity change at week 4 between regimen 3 (–1.6 lines) and regimen 4 (+1.1 lines), which contributed to the selection of regimen 4 for subsequent trials. Exploratory analyses also demonstrated that patients with smaller lesions (≤4 MPS DA) and worse baseline vision (≤20/200) had better visual acuity outcomes.

Additionally, patients with lesions composed of purely classic CNV had a better visual acuity outcome at 12 weeks (+0.6 line) than those with lesions containing some occult component (−0.8 line). Some of these early findings were later corroborated in the Phase III clinical trials with two-year follow-up and placebo-controlled, randomized design.

Angiographic Outcomes

Verteporfin PDT was effective at reducing angiographic leakage from CNV. One week after treatment, all patients had decreased leakage as demonstrated by fluorescein angiography. Complete absence of fluorescein leakage from classic CNV at week 1 was achieved in 52–100% of patients depending on the regimen used. By week 4, leakage recurred in some portion of the CNV in most patients in all regimens. Cases with leakage at week 4 were more likely to show progression at week 12. Regimen 4 resulted in the largest proportion of cases with absence of leakage at week 1 (100%) and week 4 (29%). Within regimen 4, the 50 J/cm^2 light dose was associated with the highest percentage of patients with complete absence of leakage at week four and the lowest percentage of patients with progression of classic CNV at 12 weeks. These slight advantages, combined with the visual acuity differences observed, led to the selection of regimen 4 using 50 J/cm^2 for the Phase III clinical trials. Otherwise, there was no clear light-dose/response relationship for inhibiting angiographic leakage from classic CNV. However, a minimally effective dose was identified in that the lowest light doses of 12.5 and 25 J/cm^2 were ineffective in eliminating angiographic leakage. PDT with verteporfin was also effective in reducing leakage from occult CNV although it was qualitatively more difficult to grade.

The extent of post-treatment fluorescein leakage from CNV relative to baseline appeared to be correlated with visual acuity outcome. The 33 patients with lesions that showed a reduced area of leakage from both classic and occult CNV at 12 weeks, compared with baseline, had a mean change in visual acuity of +0.8 line. In comparison, the 49 patients who showed progression of either classic or occult CNV by 12 weeks lost 0.8 line of vision. This difference was statistically significant ($P = 0.012$), suggesting that patients in whom stabilization of CNV leakage is achieved with PDT may have a beneficial visual outcome.

Safety

Thirty-eight (29.7%) of the 128 patients receiving a single course of PDT had an adverse event in the treatment eye and 31 of the cases involved events that were judged to be possibly, or probably, treatment-related. Adverse events included increased subretinal hemorrhage (8.6%), increased fibrosis associated with CNV (8.6%), increased RPE atrophy (3.9%), new subretinal hemorrhage (3.1%), eye pain (3.1%), fibrosis (3.1%), branch retinal arteriolar/venular nonperfusion (2.3%), choroidal vessel staining (1.6%), increased hemorrhage (1.6%), and vitreous hemorrhage (1.6%).

At the highest light dose used [150 J/cm^2] in regimens 2 and 3, PDT-related vision losses were observed in three patients. Two of the three patients treated using the 150 J/cm^2 light dose in regimen 2 had significant vision loss (five and nine lines) at 12 weeks. Two of five patients treated with 150 J/cm^2 in regimen 3 had retinal vessel closure. One developed occlusion of branch retinal arteries and veins in the posterior pole with 12 lines of vision loss at week one. The other had extensive capillary nonperfusion (>30% of the treated area) but only minimal vision loss. Subsequently, patients were not treated at this light dose and the maximally tolerated light dose for verteporfin treatment of CNV should be considered to be $100 \, J/cm^2$.

Retreatments

Visual Outcomes

Thirty-six patients with subfoveal CNV secondary to age-related macular degeneration (AMD) were treated with multiple courses of PDT using verteporfin. Up to four treatments in total were performed, with two PDT regimens and two different retreatment intervals (2 and 4 weeks). The mean visual acuity remained stable even after two or three PDT treatments. Twelve weeks after the last retreatment, the average change in visual acuity was +0.1 line from baseline for the 16 participants in regimen 2. The average visual acuity decreased by one line from baseline 12 weeks after the last retreatment for patients in regimen 4.

Angiographic Outcomes

Prolonged reduction of angiographic leakage was not achieved with the retreatment protocols studied. However, cessation of leakage from classic CNV was achieved 1 week after each treatment, independent of the regimen. Recurrent leakage was typically observed 4 weeks after each PDT course in a portion of the classic CNV lesions. Recurrence with progression beyond the original borders ranged from 17% to 25% in regimen 2, and was as high as 30% in regimen 4. Although this rate was somewhat higher than noted in the single treatment study, the numbers were small and patients were not randomly assigned to the different regimens, making comparisons difficult. On final follow-up 12 weeks after the last treatment, progression rates of classic CNV were 44% in regimen 2, and 89% in regimen 4. While fluorescein leakage eventually recurred following each treatment and often involved areas beyond the borders of the baseline lesion, the total area of leakage typically decreased over the retreatment schedule and was typically less intense at the last follow-up compared with baseline.

A correlation between angiographic results and visual acuity was also observed in the retreatment study. Those patients in whom complete absence of leakage was achieved after each retreatment experienced an increase in mean visual acuity. Additionally, relatively smaller lesions with only classic CNV seemed to respond best to retreatments, as demonstrated by both angiographic and visual outcomes.

Safety

The highest light dose of $150 \, J/cm^2$ was not utilized in the retreatment study and no arteriolar or venular nonperfusion was observed. Retinal capillary nonperfusion (<30% of the treated area) was seen in four patients and retinal vascular leakage in three patients. Increased RPE atrophy was noted in 11 patients (39%), as was increased fibrosis. A correlation between the number of retreatments and progression of RPE atrophy was not noted, but the number of patients receiving more than one retreatment ($n = 14$) was probably too small.

The phase I/II studies confirmed that verteporfin PDT could effectively inhibit angiographic leakage from CNV for approximately four weeks, demonstrated that retreatments were safe, and identified appropriate light and drug doses. Additionally, these studies suggested that reducing angiographic leakage would result in improved vision and that certain subgroups of patients, namely those with smaller lesions composed of purely classic CNV, would experience the most benefit from this new treatment. Thus, two large randomized, placebo-controlled trials were instituted to clarify the potential benefits of PDT with verteporfin.

PHASE III DESIGN/METHODOLOGY

The efficacy of verteporfin PDT in reducing the risk of vision loss in patients with subfoveal CNV due to AMD was investigated in multicenter, double-masked, placebo-controlled, randomized trials. The Treatment of Age-Related Macular Degeneration with Photodynamic Therapy (TAP) study consisted of two identically designed trials conducted in 22 ophthalmology practices in Europe and North America and evaluated PDT for new and recurrent subfoveal CNV with some classic component (8–11). The Verteporfin in Photodynamic Therapy (VIP) study was carried out in 28 practices across Europe and North America and involved patients with only occult CNV or classic CNV with good visual acuity as well as a subset with CNV caused by pathologic myopia (12–14).

Aside from the eligibility criteria, the study design and treatment protocol were the same for both trials. Certification of all study personnel and clinic monitoring assured maximum adherence to the protocol and all photographs were graded at the Wilmer Photograph Reading Center at Johns Hopkins Medical Institutions. An independent Data and Safety Monitoring committee conducted biannual reviews.

The eligibility criteria for the TAP and VIP investigations are listed in Tables 4 and 5. The area of CNV as identified angiographically was required to occupy >50% of the entire lesion, thus excluding patients with large areas of subretinal hemorrhage or RPE detachments. For the TAP study, CNV lesions had to include some classic component and the visual acuity in the study eye needed to fall between approximately 20/40 and 20/200. The VIP criteria identified two subgroups of patients with CNV related to AMD. Those patients with only occult CNV were required to show some evidence of recent disease progression, and participants with some classic CNV were restricted to those with relatively good visual acuity (better than approximately 20/40), presumptively indicating early onset disease.

Patients were randomized in a ratio of 2:1 to verteporfin PDT or placebo. Verteporfin was administered at a dose of $6 \, mg/m^2$ and infused over 10 minutes. Patients assigned to placebo received an infusion of 30 mL of 5% dextrose over 10 minutes. Laser light at 689 nm was applied to the lesion in all patients 15 minutes after the start of the infusion at $600 \, mW/cm^2$ for 83 seconds, giving a total light dose of $50 \, J/cm^2$. The spot size was determined by adding 1000 μm to the greatest linear dimension of the lesion on the fundus, which was calculated by dividing the greatest linear dimension of the lesion on the fluorescein angiogram by 2.5.

Patients were interviewed by telephone two to four days after each treatment to detect any adverse events. The patient was asked to return promptly for reexamination if such an event was suspected. Otherwise, patients were examined three months after each treatment. A protocol refraction, visual acuity and contrast threshold measurement, ophthalmoscopy, stereoscopic color fundus photography, and fluorescein angiography were performed at each scheduled visit. Retreatment was recommended if any leakage from classic or occult CNV was noted.

Table 4 Eligibility Criteria for the TAP Study

Inclusion criteria

CNV secondary to age-related macular degeneration

CNV under the geometric center of the foveal avascular zone

Evidence of classic CNV on fluorescein angiography

Area of CNV at least 50% of the area of the total neovascular lesion

Greatest linear dimension of lesion $\leq$5400 µm (not including any area of prior laser photocoagulation)

Best-corrected TAP protocol visual acuity of 73–34 letters (Snellen equivalent approximately 20/40–20/200)

Age $\geq$50 yrs

Willing and able to provide written informed consent

Exclusion criteria

Tear (rip) of the retinal pigment epithelium

Any significant ocular disease (other than CNV) that has compromised or could compromise vision in the study eye and confound analysis of the primary outcome

Inability to obtain photographs to document CNV, including difficulty with venous access

History of treatment for CNV in study eye other than nonfoveal confluent laser photocoagulation

Participation in another ophthalmic clinical trial of use or any other investigational new drugs within 12 weeks prior to the start of study treatment

Active hepatitis or clinically significant liver disease

Porphyria or other porphyrin sensitivity

Prior photodynamic therapy for CNV

Intraocular surgery within last 2 mos or capsulotomy within last month in study eye

---

Abbreviation: CNV, choroidal neovascularization.
Source: From Ref. 8.

Outcome Measures

The primary measure of efficacy was the proportion of eyes with loss of fewer than 15 letters (moderate vision loss) compared with baseline. Secondary measures included the proportion of eyes with fewer than 30 letters lost (severe vision loss), mean changes in visual acuity and contrast threshold, and angiographic outcomes such as progression of CNV and size of lesion. Angiographic leakage was graded as described previously in Table 3.

Results: TAP Study

A total of 609 patients were enrolled in the TAP study between December 1996 and October 1997 and randomly assigned to verteporfin therapy (402) or placebo (207). Ninety-four percent of patients in each group completed the 12-month follow-up examination while 87% of the verteporfin group and 86% of the placebo group completed the 24-month follow-up.

Vision Outcomes

The proportion of eyes with moderate and severe vision loss was lower in the verteporfin-treated group than placebo at each follow-up visit. At the 12-month point, 39% of verteporfin-treated eyes versus 54% of placebo eyes had lost 15 or more letters (moderate vision loss). Most of the vision loss in both groups was

Table 5 Eligibility Criteria for the VIP–AMD Study

Inclusion criteria

CNV under the geometric center of the foveal avascular zone

Area of CNV at least 50% of the area of the total neovascular lesion

Greatest linear dimension of lesion $\leq 5400\,\mu m$ (not including any area of prior laser photocoagulation)

If evidence of classic CNV, then visual acuity letter score more than 70

If no evidence of classic CNV, then presumed to have recent disease progression because of deterioration (visual or anatomical) within last 3 mos or evidence of hemorrhage from CNV

Best-corrected protocol visual acuity letter score of at least 50 (Snellen equivalent approximately 20/100) or better

Willing and able to provide written informed consent

Exclusion criteria

Features of any condition other than age-related macular degeneration associated with CNV in the study eye

Tear (rip) of the retinal pigment epithelium

Any significant ocular disease (other than CNV) that has compromised or could compromise vision in the study eye and confound analysis of the primary outcome

Inability to obtain photographs to document CNV, including difficulty with venous access

History of treatment for CNV in study eye other than nonfoveal confluent laser photocoagulation

Participation in another ophthalmic clinical trial of use or any other investigational new drugs within 12 wks prior to the start of study treatment

Active hepatitis or clinically significant liver disease

Porphyria or other porphyrin sensitivity

Prior photodynamic therapy for CNV

Intraocular surgery within last 2 mos or capsulotomy within last month in study eye

Abbreviation: CNV, choroidal neovascularization.
Source: From Ref. 13.

observed in the first six months of follow-up. At the 24-month point, 47% of verteporfin-treated eyes and 62% of placebo eyes had lost 15 or more letters, indicating an 8% rate of moderate vision loss in both groups over the second year (Fig. 1). However, the beneficial effect of verteporfin PDT was sustained. Similarly, fewer patients treated with verteporfin experienced severe vision loss (≥ 30 letters lost) compared with those in the placebo group at both 12 (15% vs. 24%) and 24 months (18% vs. 30%). While a definite beneficial effect of treatment was reflected in the prevention of vision loss, few patients experienced improvement. Only 9% of treated and 5% of placebo eyes gained 15 letters or more at the 24-month follow-up.

Secondary measures showed that the proportion of eyes with a visual acuity of 20/200 or less was smaller in the verteporfin-treated group at both 12 months (35% vs. 48%) and 24 months (41% vs. 55%). Significant benefits of verteporfin treatment were also noted with respect to mean change in visual acuity and contrast sensitivity scores at both the one- and two-year follow-up points.

Angiographic Outcomes

Angiographic assessments confirmed the beneficial effects of verteporfin PDT. However, the data suggest that the visual benefits of treatment described above were achieved in spite of some degree of continued growth and leakage of CNV in both

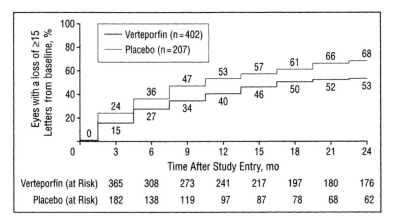

Figure 1 TAP study. Kaplan–Meier estimate of the cumulative proportion of eyes with moderate vision loss (≥15 letters) in verteporfin-treated and placebo groups at each three-month study visit. *Abbreviation*: TAP, Treatment of Age-Related Macular Degeneration with Photodynamic Therapy. *Source*: From Ref. 9.

groups. The proportion of eyes with progression of classic CNV was significantly lower in the verteporfin-treated group than the placebo group. At 12 months, 46% of treated eyes versus 71% of control eyes had progression. By 24 months, the proportion of eyes with progression was lower, 23% of treated and 54% of control, suggesting some regression of CNV in both groups, but still demonstrating an advantage to treatment. Although infrequent at 12 months, absence of leakage from classic CNV was observed in more treated than control eyes (19% vs. 9%). The fraction with no leakage increased by 24 months to 53% of treated eyes and 29% of placebo eyes. Lesion size was also significantly smaller in the verteporfin group at the end of the 24-month study period.

This trend in angiographic leakage was also reflected in an analysis of rates of retreatment. At each follow-up, the percentage of patients receiving retreatment for leakage from CNV was consistently higher in the placebo group, but the overall percentage of patients requiring retreatment decreased with each follow-up. At the month 21 examination, 45% of the verteporfin group versus 65.2% of the control group received retreatment compared with 64% and 79% at month 12. The verteporfin group received an average of 5.6 treatments over the study period out of total possible of eight, compared with an average of 6.5 for the placebo group.

Subgroup Analyses

Of the various lesion characteristics analyzed, only the baseline lesion composition was noted to have a significant relationship to the magnitude of treatment benefit. No subgroups were identified in which treatment had a harmful effect. Lesions were classified according to baseline angiography findings into three groups based on the area of the total CNV lesion occupied by classic CNV components. The lesions was determined to be *predominantly classic*, when classic CNV occupied ≥50% of the entire CNV lesion area and *minimally classic* when classic CNV occupied < 50% of the entire CNV lesion area. The distribution of lesions was similar between the verteporfin and placebo groups: 40% of eyes in each group had predominantly classic and 50% had minimally classic. Nine percent of eyes in each group were judged to contain no classic CNV upon review by the reading center, and 1% of eyes could not be graded.

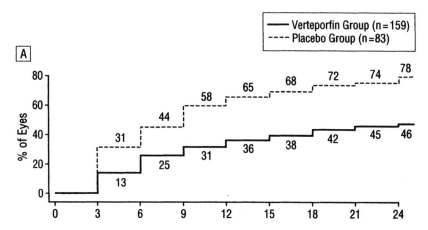

Figure 2 TAP study. Kaplan–Meier estimate of the cumulative proportion of eyes with moderate vision loss (≥15 letters) at each three-month study visit for patients with predominantly classic CNV treated with verteporfin or placebo. *Abbreviations*: CNV, choroidal neovascularization; TAP, Treatment of Age-Related Macular Degeneration with Photodynamic Therapy. *Source*: From Ref. 10.

A greater treatment benefit was demonstrated for those patients with predominantly classic CNV at both the 12- and 24-month time points (Fig. 2). Moderate vision loss of 15 letters or more occurred in 41% of treated versus 69% of control eyes with predominantly classic CNV at the 24-month follow-up compared with 47% versus 62% for the entire study population. The highest magnitude of treatment benefit was seen for the subgroup of patients with only classic and no occult CNV, with 30% of treated patients experiencing moderate vision loss compared with 71% of control patients.

There appeared to be no significant impact of verteporfin treatment on eyes with minimally classic CNV based on the primary visual acuity outcome. Moderate vision loss of 15 letters or more was seen in 44% of treated and 45% of control eyes at 12 months, and 52% versus 56% at 24 months ($P = 0.58$). However, a significant benefit at both 12- and 24-month follow-up was seen for the minimally classic subgroup in terms of contrast sensitivity and angiographic outcomes, including lesion size, progression of classic CNV, and absence of classic CNV leakage.

Additional analyses of the subgroups based on lesion composition revealed that eyes with predominantly classic lesions tended to have a lower mean visual acuity and smaller lesion size than those with minimally classic lesions (10). Similarly, predominantly classic lesions without any occult component also were generally smaller and associated with lower visual acuity compared with predominantly classic lesions with occult CNV. This difference in lesion size and baseline acuity may account for some of the additional treatment benefit described, particularly in light of similar data implicating lesion size as a factor in the VIP trial as discussed in the next section. Furthermore, the placebo eyes in the minimally classic subgroup were more likely to have blood as a component of the lesion, which may have influenced the natural history toward spontaneous resolution, diminishing the potential benefit of verteporfin therapy.

Results: VIP–AMD Study

Of 339 patients enrolled in the VIP–AMD study between March 1998 and September 1998, 225 were assigned to verteporfin treatment and 114 to placebo. The majority of the cases enrolled consisted of occult CNV with no classic component (74% of verteporfin-treated and 81% of placebo). The subgroups containing some classic CNV were considered too small for analysis. Predominantly classic CNV lesions comprised 7% of verteporfin-treated eyes and 3% of placebo eyes while minimally classic CNV was seen in 17% of verteporfin-treated and 16% of placebo-treated eyes. Other baseline lesion characteristics were statistically balanced between treatment and placebo groups for both the entire study population and for the subgroup with occult and no classic CNV.

Rates of follow-up in the VIP study were similar to those seen in the TAP. Ninety-three percent of patients in the verteporfin group and 91% in the control group completed the month 12 examination while 86% and 87%, respectively, completed the 24-month examination.

Vision Outcomes

A statistically significant benefit of verteporfin treatment was demonstrated for the primary vision outcome by the second year of follow-up. At 12 months, the proportion of moderate vision loss ($\geq$15 letters) in the treated group was 51% compared with 54% in the placebo group. Over the second year of treatment, verteporfin eyes tended to stabilize while placebo eyes continued to deteriorate so that by 24 months the treatment benefit had increased to reach statistical significance. Moderate vision loss occurred in 54% of treated eyes versus 67% of control ($P = 0.023$). The subgroup with occult but no classic CNV experienced a similar treatment benefit: 55% treated versus 68% control eyes had lost 15 or more letters of vision at the 24-month follow-up ($P = 0.032$) (Fig. 3). There was also a significant reduction in severe vision loss of 30 letters or more at 24 months for verteporfin-treated compared with placebo eyes in both the entire study population (30% treated vs. 47% placebo, $P = 0.001$) and the subgroup with occult and no classic CNV (29% treated vs. 47% placebo, $P = 0.004$). Other secondary visual outcomes including mean change in visual acuity, visual acuity distribution, percentage of eyes with vision < 34 letters ($\leq 20/200$) and

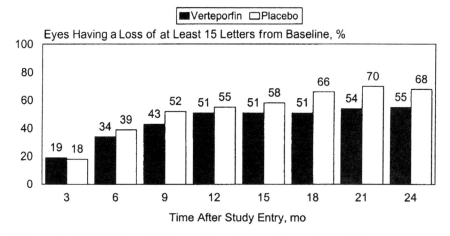

Figure 3 VIP–AMD study. Proportion of eyes with moderate vision loss ($\geq$15 letters) at each three-month study visit for patients with occult with no classic CNV treated with verteporfin or placebo. *Abbreviations*: VIP-AMD, Verteporfin in Photodynamic Therapy-Age-Related Macular Degeneration; CNV, choroidal neovascularization. *Source*: From Ref. 13.

contrast sensitivity significantly favored the verteporfin-treated group at both 12- and 24-month follow-up for the subgroup with occult but no classic CNV.

Angiographic Outcomes

Fluorescein angiography also demonstrated a clinically and statistically significantly treatment benefit at both the 12- and 24-month examinations for the entire study population as well as the subgroup with occult but no classic CNV. At both time points, verteporfin-treated eyes with occult and no classic CNV at baseline were nearly half as likely to develop classic CNV either within or beyond the original lesion boundary. Progression of occult CNV in this subgroup was seen in 55% of treated and 73% of control eyes at the 12-month follow-up ($P = 0.004$), decreasing to 46% of treated and 57% of control eyes by 24 months ($P = 0.12$). Absence of leakage from either classic or occult CNV was infrequent at the 12-month follow-up (14% treated vs. 4% control, $P = 0.02$) but increased by the 24-month follow-up (35% treated vs. 14% control, $P < 0.001$). Lesions in placebo eyes were 2.5 times as likely to be greater than 9 DAs in size compared with verteporfin-treated eyes after both the first and second year of the study.

Again, the trend toward decreased leakage as the study progressed was reflected in the proportion of patients requiring retreatment at each visit. Thirty-three percent of verteporfin-treated eyes and 45% of placebo eyes received treatment at the last treatment visit (21-month follow-up) compared with 61% and 73%, respectively, at the 12-month visit. Fewer verteporfin-treated eyes were assessed by the treating ophthalmologist as requiring treatment at each follow-up visit. These proportions were identical for the subgroup with occult and no classic CNV, with an average of 4.9 treatments (3.1 in the first year and 1.8 in the second year) out of a total possible of 8 given to the verteporfin group and 5.9 for the control group.

Subgroup Analyses

Prospectively planned analyses of the subgroup of occult with no classic CNV revealed that both initial visual acuity and baseline lesion size influenced the magnitude of treatment benefit. Patients with either smaller lesions ($\leq$4 DAs) or lower presenting visual acuity (approximately 20/50 or worse) were found to have a greater benefit (Fig. 4). These patients represented 72% of the total subgroup with occult and no classic CNV. In this group of eyes, moderate vision loss at 24 months occurred in 49% of treated versus 75% of control eyes ($P < 0.001$) and severe vision loss in 21% of treated versus 48% of control eyes. An exploratory analysis also suggested that verteporfin may not be beneficial in eyes with both large occult lesions and relatively good visual acuity. However, the number of patients with these baseline characteristics was small, and this finding should be interpreted cautiously.

This type of data suggesting that factors other than baseline lesion composition could affect vision outcomes from verteporfin PDT, in addition to the lack of demonstrated treatment benefit for minimally classic lesions, raised the issue of whether another parameter might better predict vision outcomes. A retrospective analysis of the AMD populations from the TAP and VIP trials was conducted in order to evaluate the effect of baseline lesion size, visual acuity, and lesion composition on vision outcomes. This analysis revealed that baseline lesion size influenced the treatment benefit for occult with no classic CNV and minimally classic CNV lesions, but not for predominantly classic lesions. When data from all lesion types were combined, lesion size at presentation was shown to be a more important predictor of treatment benefit than either lesion composition or visual acuity (15).

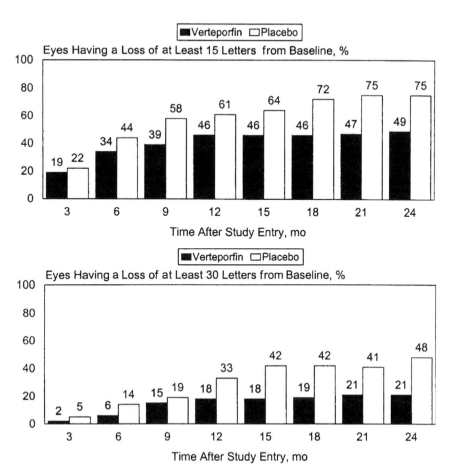

Figure 4 VIP–AMD study. Proportion of eyes with moderate vision loss (≥15 letters) in verteporfin-treated and placebo groups at each three-month study visit for subgroup of patients with occult with no classic CNV with either smaller lesions (≤4 DAs) or lower levels of visual acuity (less than approximate Snellen equivalent 20/50^{-1}). *Abbreviations*: VIP-AMD, Verteporfin in Photodynamic Therapy-Age-Related Macular Degeneration; CNV, choroidal neovascularization. *Source*: From Ref. 13.

Safety

Photodynamic therapy with verteporfin through 24 months of follow-up and treatment in both the TAP and VIP studies appeared quite safe. The treatment was well tolerated with minimal adverse events attributable to treatment. Photosensitivity reactions were rare and lower in incidence in the VIP study (1%) than the TAP study (3%) despite the shorter recommended protection period of 24 hours compared with 48 hours. Allergic reactions were also rare and more frequent in the placebo groups. The only other nonocular adverse event related to treatment was lower back pain during verteporfin infusion, which occurred in a total of 15 people across both studies.

Some visual disturbance following treatment was common in both verteporfin-treated and placebo eyes. However, the incidence of severe decrease in vision, defined as a loss of 20 letters (four lines) or more within seven days of treatment was low. The risk appeared to be higher in eyes with occult but no classic CNV, occurring in 10 (4.4%) patients in the VIP study compared with three (0.75%) in the TAP study.

Of the cases in the VIP trial, the visual loss was attributed to the development of extensive subretinal fluid with choroidal hypofluorescence in one case and sub-retinal pigment epithelial hemorrhage in three cases. No obvious cause was detected in six cases. Vision recovered to < 20 letters lost in five of the 10 patients at three months after the events. Although preclinical studies demonstrated some damage to the RPE with PDT, the Phase III data did not suggest any increase in RPE atrophy in verteporfin-treated patients. For both groups in the TAP study, the distribution of lesion sizes with the inclusion of surrounding atrophy did not differ from the distribution of lesion sizes without atrophy.

Results: VIP–Myopia

While AMD is a major cause of CNV, other ocular conditions also result in blindness due to CNV. Phase I and II trials suggested the safety and efficacy of verteporfin PDT in 13 patients with pathologic myopia. Based on these data and early data from the TAP trial confirming the safety of the treatment, one arm of the VIP study was designed to evaluate the benefit of verteporfin PDT for subfoveal CNV due to pathologic myopia (12,14). The inclusion criteria are shown in Table 6.

A total of 120 patients were enrolled between February and September 1998. Of these patients, 81 were randomized to verteporfin and 39 to placebo with 95% of the treatment group and 92% of the placebo group completing the two-year follow-up. Greater than 90% of patients in each group had evidence of classic CNV at baseline, with the majority of cases containing predominantly classic CNV. Only 15% of the verteporfin group and 13% of the placebo group had evidence of any occult CNV.

Vision Outcomes

The primary outcome for this group of patients was selected as the proportion of eyes with vision loss of fewer than eight letters. A treatment benefit was apparent at the 12-month follow-up with 72% of the verteporfin-treated patients losing fewer than eight letters compared with 44% of the placebo group ($P=0.01$). Moderate visual acuity loss (≥ 15 letters) was decreased in the treated group, occurring in 14% versus 33% of placebo eyes after one year. At the month 24 examination, 64% of verteporfin-treated eyes compared with 49% of placebo eyes lost fewer than eight letters, but this difference did not reach statistical significance (Fig. 5). Similarly, a significant benefit was not demonstrated in the rate of moderate visual loss, which occurred in 21% of treated eyes and 28% of control eyes at 24 months ($P=0.38$). However, the distribution of change in visual acuity continued to show a treatment benefit ($P=0.05$). Patients treated with verteporfin had a median gain of 0.2 lines while those receiving placebo lost a median of 1.6 lines. These figures reflect the finding that 40% of verteporfin-treated eyes had visual acuity improvement of at least five letters versus 13% of placebo.

Angiographic Outcomes

The proportion of patients with progression of classic CNV beyond the baseline borders of the lesion was lower in the verteporfin-treated group than the placebo group at both the one and two year examinations. Similarly, more treated patients had complete absence of leakage than control patients at both time points. Although these differences were not statistically significant at 24 months, differences in lesion

Table 6 Eligibility Criteria for the VIP–Pathologic Myopia Study

---

Inclusion criteria

CNV secondary to pathologic myopia (distance correction of at least –6.0 D, spherical equivalent, or less myopic than –6.0 D with retinal abnormalities consistent with pathologic myopia, such as lacquer cracks, and an axial length at least 26.5 mm

CNV under the geometric center of the foveal avascular zone

Area of CNV at least 50% of the area of the total neovascular lesion

Greatest linear dimension of lesion ≤5400 μm (not including any area of prior laser photocoagulation)

Best-corrected protocol visual acuity letter score of at least 50 (Snellen equivalent approximately 20/100) or better

Willing and able to provide written informed consent

Exclusion criteria

Features of any condition other than pathologic myopia (such as large drusen or multifocal choroiditis) associated with CNV in the study eye

Tear (rip) of the retinal pigment epithelium

Any significant ocular disease (other than CNV) that has compromised or could compromise vision in the study eye and confound analysis of the primary outcome

Inability to obtain photographs to document CNV, including difficulty with venous access

History of treatment for CNV in study eye other than nonfoveal confluent laser photocoagulation

Participation in another ophthalmic clinical trial of use or any other investigational new drugs within 12 weeks prior to the start of study treatment

Active hepatitis or clinically significant liver disease

Porphyria or other porphyrin sensitivity

Prior photodynamic therapy for CNV

Intraocular surgery within last 2 mos or capsulotomy within last month in study eye

Pregnancy

Abbreviations: CNV, choroidal neovascularization; VIP, Verteporfin in Photodynamic Therapy.
Source: From Ref. 12.

size were significant. Verteporfin-treated lesions were more likely to be one DA or smaller in size (55% vs. 36%, $P = 0.05$) and less likely to be more than three DAs in size (9% vs. 28%, $P = 0.01$).

Other Causes of CNV

Based on the success of verteporfin photodynamic for CNV in AMD and pathologic myopia, the efficacy of PDT for subfoveal CNV in the ocular histoplasmosis syndrome was investigated in an open-label, three-center prospective case series (16). Twenty-six patients with classic or occult CNV not larger than 5400 μm in greatest linear dimension and extending under the geometric center of the foveal avascular zone whose vision ranged between approximately 20/40 and 20/200 were treated with verteporfin PDT as described in the TAP and VIP studies. These patients received an average of 2.9 treatments over 12 months and experienced a median improvement from baseline acuity of seven letters. Fifty-six percent of patients gained seven or more letters of vision while 16% lost eight or more letters and only 8% lost 15 or more letters.

Angiographic evaluation revealed absence of fluorescein leakage from classic CNV in 43% of patients at the month 12 visit. Progression of classic CNV was noted

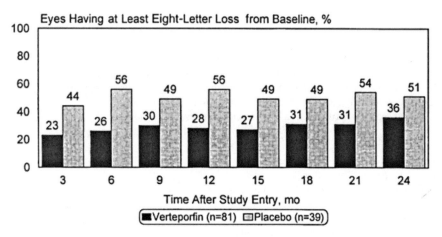

Figure 5 VIP–myopia study. Proportion of eyes with vision loss of at least eight letters in verteporfin-treated and placebo groups at each three-month study visit. *Abbreviations*: VIP, verteporfin in photodynamic therapy. *Source*: From Ref. 14.

in 26% of patients. No cases of severe vision loss within seven days of treatment occurred in this series. Although limited by a small sample size and lack of a control group, these initial results suggest that verteporfin PDT is likely effective for CNV due to ocular histoplasmosis. Two-year results were pending at the time of this writing.

The successful use of verteporfin PDT for CNV secondary to other conditions including angioid streaks, multifocal choroiditis, juxtafoveolar telangiectasis, retinal degenerations, and choroidal osteoma has been reported in various case series (17–24). Beyond CNV, other potential applications for PDT include ocular tumors. Encouraging results have been reported in the treatment of choroidal hemangiomas as well as retinal capillary angiomas (25–30). With future improvements in the delivery and specificity of photosensitizers, expanding indications for PDT in a variety of ocular conditions may follow.

CONCLUSIONS

Based on the safety, efficacy, and dosimetry data from Phase I and II studies of PDT with verteporfin for subfoveal CNV, large, multicenter, randomized, placebo-controlled trials were conducted, providing the evidence for the widespread clinical use of this treatment modality in patients with AMD and other ocular conditions. The key findings of the Phase III trials of verteporfin PDT can be summarized as follows:

1. Verteporfin PDT is a safe and effective treatment for subfoveal CNV due to AMD. The treatment benefit of verteporfin PDT in patients with subfoveal CNV due to AMD is influenced by baseline lesion composition.
2. Patients with predominantly classic lesions had a greater treatment benefit, particularly those with only classic and no occult CNV. Verteporfin PDT is recommended in cases of subfoveal predominantly classic CNV, due to AMD, with or without the presence of occult CNV.
3. A treatment benefit for eyes with occult and no classic CNV and recent evidence of progression was demonstrated after two years of follow-up, particularly in

those eyes with lesions ≤4 MPS DAs or with visual acuity less than approximately 20/50. Verteporfin PDT should be considered for these cases.

4. Treatment benefits have not been proven for minimally classic subfoveal lesions due to AMD. However, a retrospective analysis suggests that smaller lesions may benefit from treatment. A clinical trial addressing this issue is ongoing.

5. Verteporfin PDT is a safe and effective treatment for subfoveal CNV due to pathologic myopia with visual benefits demonstrated through two years of follow-up. Verteporfin PDT is recommended for subfoveal CNV resulting from pathologic myopia regardless of lesion composition.

Based on data from the TAP and VIP trials, the Food and Drug Administration approved vertepofin PDT to treat AMD. The Centers for Medicare and Medicaid Services initially approved Medicare insurance reimbursement on July 1, 2001 for verteporfin PDT to treat predominantly classic CNV associated with AMD. On April 1, 2004, Medicare insurance reimbursement for verteporfin PDT was approved for small (four DAs or less) minimally classic and pure occult CNV lesions associated with AMD that show recent progression. Recent progression was defined as a decrease in visual acuity of five or more letters along with lesion growth (an increase of at least one DA), or the appearance of blood associated with the lesion, within the preceding three months (31).

While the results of the TAP and VIP investigations provide excellent guidelines for the clinical application of PDT, it is likely that advances in the understanding of the pathobiology of AMD and CNV, enhanced imaging capabilities, and the development of improved PDT regimens and adjuvant treatments will alter current clinical strategies. For example, PDT has been recently combined with pharmacological therapy such as intravitreal triamcinolone and pegaptanib (see Chapter 16). Nevertheless, the data obtained from these studies will provide the framework for the evaluation of newer treatment modalities, which will lead to improved visual outcomes for many patients.

REFERENCES

1. Miller JW, Walsh AW, Kramer M, et al. Photodynamic therapy of experimental choroidal neovascularization using lipoprotein-delivered benzoporphyrin. Arch Ophthalmol 1995; 113:810–818.

2. Kramer M, Miller JW, Michaud N, et al. Liposomal benzoporphyrin derivative verteporfin photodynamic therapy. Selective treatment of choroidal neovascularization in monkeys. Ophthalmology 1996; 103:427–438.

3. Husain D, Miller JW, Michaud N, Connolly E, Flotte TJ, Gragoudas ES. Intravenous infusion of liposomal benzoporphyrin derivative for photodynamic therapy of experimental choroidal neovascularization. Arch Ophthalmol 1996; 114:978–985.

4. Husain D, Kramer M, Kenny AG, et al. Effects of photodynamic therapy using verteporfin on experimental choroidal neovascularization and normal retina and choroid up to 7 weeks after treatment. Invest Ophthalmol Vis Sci 1999; 40:2322–2331.

5. Reinke MH, Canakis C, Husain D, et al. Verteporfin photodynamic therapy retreatment of normal retina and choroid in the cynomolgus monkey. Ophthalmology 1999; 106: 1915–1923.

6. Miller JW, Schmidt-Erfurth U, Sickenberg M, et al. Photodynamic therapy with verteporfin for choroidal neovascularization caused by age-related macular degeneration: results of a single treatment in a Phase I and II study. Arch Ophthalmol 1999; 117: 1161–1173.

7. Schmidt-Erfurth U, Miller JW, Sickenberg M, et al. Photodynamic therapy with verteporfin for choroidal neovascularization caused by age-related macular degeneration: results of retreatments in a Phase I and II study. Arch Ophthalmol 1999; 117:1177–1187.

8. Treatment of Age-Related Macular Degeneration with Photodynamic Therapy (TAP) Study Group. Photodynamic therapy of subfoveal choroidal neovascularization in age-related macular degeneration with verteporfin: one-year results of 2 randomized clinical trials—TAP report 1. Arch Ophthalmol 1999; 117:1329–1345.

9. Bressler NM. Photodynamic therapy of subfoveal choroidal neovascularization in age-related macular degeneration with verteporfin: two-year results of 2 randomized clinical trials—TAP report 2. Arch Ophthalmol 2001; 119:198–207.

10. Bressler NM, Arnold J, Benchaboune M, et al. Verteporfin therapy of subfoveal choroidal neovascularization in patients with age-related macular degeneration: additional information regarding baseline lesion composition's impact on vision outcomes—TAP report no. 3. Arch Ophthalmol 2002; 120:1443–1454.

11. Rubin GS, Bressler NM. Effects of verteporfin therapy on contrast on sensitivity: results from the treatment of age-related macular degeneration with photodynamic therapy (TAP) investigation—TAP report no 4. Retina 2002; 22:536–544.

12. Verteporfin in Photodynamic Therapy Study Group. Photodynamic therapy of subfoveal choroidal neovascularization in pathologic myopia with verteporfin. 1-year results of a randomized clinical trial—VIP report no. 1. Ophthalmology 2001; 108:841–852.

13. Verteporfin in Photodynamic Therapy Study Group. Verteporfin therapy of subfoveal choroidal neovascularization in age-related macular degeneration: two-year results of a randomized clinical trial including lesions with occult with no classic choroidal neovascularization—verteporfin in photodynamic therapy report 2. Am J Ophthalmol 2001; 131:541–560.

14. Blinder KJ, Blumenkranz MS, Bressler NM, et al. Verteporfin therapy of subfoveal choroidal neovascularization in pathologic myopia: 2-year results of a randomized clinical trial—VIP report no. 3. Ophthalmology 2003; 110:667–73.

15. Miller JW. Photodynamic therapy for AMD: expanded indications. American Academy of Ophthalmology, Retina Subspecialty Day, Orlando, FL, October 18–19, 2002.

16. Saperstein DA, Rosenfeld PJ, Bressler NM, et al. Photodynamic therapy of subfoveal choroidal neovascularization with verteporfin in the ocular histoplasmosis syndrome: one-year results of an uncontrolled, prospective case series. Ophthalmology 2002; 109: 1499–1505.

17. Battaglia Parodi M, Da Pozzo S, Toto L, Saviano S, Ravalico G. Photodynamic therapy for choroidal neovascularization associated with choroidal osteoma. Retina 2001; 21: 660–661.

18. Coquelet P, Postelmans L, Snyers B, Verougstraete C. Successful photodynamic therapy combined with laser photocoagulation in three eyes with classic subfoveal choroidal neovascularisation affecting two patients with multifocal choroiditis: case reports. Bull Soc Belge Ophthalmol 2002; 283:69–73.

19. Dantas MA, Slakter JS, Negrao S, Fonseca RA, Kaga T, Yannuzzi LA. Photodynamic therapy with verteporfin in mallatia leventinese. Ophthalmology 2002; 109:296–301.

20. Karacorlu M, Karacorlu S, Ozdemir H, Mat C. Photodynamic therapy with verteporfin for choroidal neovascularization in patients with angioid streaks. Am J Ophthalmol 2002; 134:360–366.

21. Potter MJ, Szabo SM, Chan EY, Morris AH. Photodynamic therapy of a subretinal neovascular membrane in type 2A idiopathic juxtafoveolar retinal telangiectasis. Am J Ophthalmol 2002; 133:149–151.

22. Shaikh S, Ruby AJ, Williams GA. Photodynamic therapy using verteporfin for choroidal neovascularization in angioid streaks. Am J Ophthalmol 2003; 135:1–6.

23. Spaide RF, Freund KB, Slakter J, Sorenson J, Yannuzzi LA, Fisher Y. Treatment of subfoveal choroidal neovascularization associated with multifocal choroiditis and panuveitis with photodynamic therapy. Retina 2002; 22:545–549.

24. Valmaggia C, Niederberger H, Helbig H. Photodynamic therapy for choroidal neovascularization in fundus flavimaculatus. Retina 2002; 22:111–113.
25. Atebara NH. Retinal capillary hemangioma treated with verteporfin photodynamic therapy. Am J Ophthalmol 2002; 134:788–790.
26. Porrini G, Giovannini A, Amato G, Ioni A, Pantanetti M. Photodynamic therapy of circumscribed choroidal hemangioma. Ophthalmology 2003; 110:674–680.
27. Jurklies B, Anastassiou G, Ortmans S, et al. Photodynamic therapy using verteporfin in circumscribed choroidal haemangioma. Br J Ophthalmol 2003; 87:84–89.
28. Robertson DM. Photodynamic therapy for choroidal hemangioma associated with serous retinal detachment. Arch Ophthalmol 2002; 120:1155–1161.
29. Schmidt-Erfurth UM, Michels S, Kusserow C, Jurklies B, Augustin AJ. Photodynamic therapy for symptomatic choroidal hemangioma: visual and anatomic results. Ophthalmology 2002; 109:2284–2294.
30. Schmidt-Erfurth UM, Kusserow C, Barbazetto IA, Laqua H. Benefits and complications of photodynamic therapy of papillary capillary hemangiomas. Ophthalmology 2002; 109:1256–1266.
31. Medlearn Medicare Matters #MM3191, on internet at URL http://www.cms.hhs.gov/medlearn/matters/mmarticles/2004/mm3191.pdf

16

Age-Related Macular Degeneration Drug Delivery

Kourous A. Rezaei

Department of Ophthalmology, Rush University Medical Center, University of Chicago, Chicago, Illinois, U.S.A.

Sophie J. Bakri and Peter K. Kaiser

The Cole Eye Institute, Cleveland Clinic Foundation, Cleveland, Ohio, U.S.A.

Age-related macular degeneration (AMD) is the leading cause of blindness in patients older than 60 years of age in the United States (1). Clinically, AMD is divided into non-neovascular (dry) and neovascular (wet) forms. Although the non-neovascular form is more prevalent, severe vision loss is most commonly encountered in patients with neovascularization (2). The term neovascularization refers to the growth of new vessels, and in the case of AMD neovascularization originates from the choroid [choroidal neovascularization (CNV)].

TREATMENT MODALITIES FOR NEOVASCULAR AMD

Although several treatment modalities have been proposed for the treatment of CNV, the visual outcome after therapy remains modest at best. Thermal laser photocoagulation treatment was extensively studied in the Macular Photocoagulation Study (MPS) (3–15). The effectiveness of photodynamic therapy (PDT), using selective laser ablation of CNV was assessed in the TAP and VIP multicenter trials (16–20). In PDT a photosensitizing dye is preferentially activated within the CNV by a sensitizing laser beam of a specific wavelength, with relative sparing of adjacent tissue (see Chapters 9 and 15). To further reduce neuronal damage associated with thermal laser treatment, the effect of transpupillary thermotherapy (TTT) on neovascular AMD is being evaluated (21–27). In TTT, a low-irradiance infrared laser is used with a large spot size and prolonged exposure (long-pulse). The low temperature and long-pulse photocoagulation is a potential strategy for decreasing neural retinal damage. Radiation therapy in the form of teletherapy with photons or irradiation with proton beam has also been evaluated in clinical trials for the treatment of CNV in AMD (28–33).

Various surgical techniques have also been proposed for the treatment of CNV. The submacular surgical trials are investigating the outcome for surgical removal of the choroidal neovascular membranes in patients with AMD (34). Furthermore, macular translocation has also been proposed as a surgical technique in the management of some neovascular AMD lesions (35–40).

Pharmacologic intervention for the treatment of CNV in AMD is an attractive treatment approach, as it avoids tissue damage induced by laser irradiation or surgical trauma. Furthermore, it may address the pathogenesis of angiogenesis. Interferon alpha-2a was one of the first anti-angiogenic agents studied to suppress or stabilize the growth of subretinal neovascularization (41–45). In this chapter, human clinical trials of intraocular drug delivery to treat neovascular AMD are discussed.

CLINICAL TRIALS OF DRUG DELIVERY DEVICES FOR THE TREATMENT OF NEOVASCULAR AMD

Nonselective Drugs

Triamcinolone Acetonide

Penfold et al. (46) published in 1995 a pilot study of neovascular AMD treatment with intravitreal triamcinolone acetonide. Their preliminary data evaluating 30 eyes treated with intravitreal triamcinolone injection demonstrated decreased leakage by fluorescein angiography and increased visual acuity. In an 18-month follow-up to this trial, of the 20 eyes with initial visual acuity of 20/200 or better, the vision was maintained (± 1 Bailey-Lovie lines) in 11 eyes (55%), while six eyes (30%) suffered severe visual loss (six or more lines). The visual acuity improved by five to six lines in three of 10 eyes with initial vision of 3/60 or worse.

Jonas et al. (47) performed an uncontrolled study of intravitreal triamcinolone acetonide to treat exudative AMD. Of 71 treated eyes, 68 had predominantly or totally occult CNV, as determined by fluorescein angiography. With a mean follow-up of seven months, the visual acuity increased from a preinjection mean of 0.16 to a maximum of 0.23 ($P < 0.001$). The maximal visual acuity was attained at 1–3 months postinjection. However, there was no significant visual acuity difference by 7.5 months, when compared with pretreatment visual acuity. The average intraocular pressure (IOP) increased from a baseline of 15.1 to 23.0 mmHg. There were no significant postoperative complications such as endophthalmitis and retinal detachment.

A small, randomized clinical trial was conducted to evaluate the safety and effectiveness of a single 4 mg intravitreal triamcinolone acetonide injection for neovascular AMD (48). At both the 3- and 6-month follow-up visits, the treated group had statistically significant better visual acuity than the control group. The angiographic appearance was also better in the treated group compared with the control group. Adverse events from the injection in the treated group included IOP elevation (seen in 25%) and cataract progression.

The largest randomized trial to date was performed by Gillies et al. (49). This study was a randomized, double-masked, placebo-controlled trial of triamcinolone acetonide intravitreal injection in patients with classic CNV associated with AMD. Patients were randomized to receive a single injection of triamcinolone acetonide, 4 mg ($n = 75$) or to receive a sham injection ($n = 76$). At 12 months the risk of severe visual loss (30 letters) was 35% for both the treated group and the placebo group. Although the visual acuity did not differ between the treated and control groups, at three months, the choroidal neovascular complex appeared to be smaller in the treated group. At 12 months, however, there was no difference in lesion size between the groups. The smaller lesion size at three months suggested an anti-angiogenic effect of intravitreal triamcinolone acetonide over the three months following injection, which may have diminished as the drug was cleared. It is not known whether repeated injections would provide a sustained benefit to stabilize visual acuity and lesion size. However, there was a statistically significant elevation in IOP in the treated group compared with the placebo group, which may be one of the factors (among others) that may limit a long-term reinjection protocol for the treatment of CNV.

Triamcinolone has been combined with laser photocoagulation to treat CNV associated with AMD (50). Following intravitreal injection, the mean visual acuity remained stable; these results were comparable to the laser retreatment group in the MPS (51).

Verteporfin ocular PDT has been combined with intravitreal triamcinolone acetonide to treat CNV associated with AMD. The rationale for this approach is to decrease the number of photodynamic treatments (as well as the cost associated with these multiple treatments) by combining the angiostatic properties of triamcinolone acetonide with the vascular occlusion induced by ocular PDT. In a pilot study of 26 eyes (13 eyes naïve to PDT and 13 eyes previously treated with PDT), the retreatment rate at 3-month follow-up was 7.7% (two of 26 eyes). At the 6-month follow-up visit, no eyes required retreatment (52).

In another trial, triamcinolone acetonide was injected into the vitreous cavity of 14 eyes within six weeks of PDT (53). Eleven received one initial combined treatment and three received an additional combined treatment after six months. Median follow-up was 18 months (range 12–25 months). Overall, 7% gained 30 or more letters, 50% maintained stable vision, 14% lost 15–29 letters, and 29% lost 30 or more letters. The mean number of PDT treatments during the first year was 2.57. Side effects were mild and included IOP elevation in 28.5% and cataract progression in 50% of phakic eyes.

Multicenter, randomized trials are currently underway to compare verteporfin ocular PDT alone to ocular PDT with intravitreal triamcinolone acetonide for neovascular AMD. The main outcome measures include visual acuity improvement and the number of verteporfin treatments.

Anecortave Acetate

Anecortave acetate is a synthetic steroid derivative which is thought to inhibit blood vessel growth by inhibiting the proteases necessary for vascular endothelial cell migration (54,55). It inhibits both urokinase-like plasminogen activator and matrix metalloproteinase-3 (56). Anecortave acetate has been specifically modified to eliminate its in vivo corticoid activity (57). It is administered posterior to the eye as a juxtascleral depot injection onto the outer surface of the sclera near the macula, using a specially designed cannula (see Chapter 5) (57).

Currently anecortave acetate is being evaluated in an ongoing multicenter trial as monotherapy for treatment of subfoveal AMD. The first 6- and 12-month clinical safety and efficacy data following a single treatment were recently reported (57,58). This double-masked Phase II trial investigated the efficacy of anecortave acetate versus a placebo for maintenance of vision and inhibition of CNV growth in AMD. Between April 1999 and May 2001, 128 patients were enrolled in the study at 18 participating sites. Of these 128 patients, 80% (102 patients) had predominantly classic lesions and 20% (26 patients) had minimally classic lesions. Predominantly classic lesions were defined as those in which classic CNV, as determined by fluorescein angiography, occupied at least 50% of the total lesion area. Of the CNV area, 50% had to have been classic CNV or the area of the classic CNV must have been at least 0.75 MPS disc areas.

Follow-up examinations included detailed ophthalmic examinations (best-corrected log minimum angle of resolution (MAR) visual acuity evaluation, query of patient as to double vision, external examination of the eye(s), routine screen for changes in extraocular motility and/or restriction of gaze, routine screen for pupil responsiveness, slit-lamp examination of anterior segment and lens, dilated fundus examination, IOP measurement). On scheduled visits, fluorescein angiography and indocyanine green angiography were performed. Annual general physical examinations with electrocardiogram, interim physical examinations and periodic examinations of blood and urine, and pharmacokinetic sampling were performed at scheduled visits.

The baseline characteristics of the patients were similar to those reported in the verteporfin TAP trial (other than the higher number of patients with predominantly classic patients) (80% vs. 40%; see Chapter 15) (16). Patients with a log MAR visual acuity of 0.3 (20/40 Snellen equivalent) to 1.2 (20/320 Snellen equivalent) and primary or recurrent subfoveal CNV secondary to AMD with a lesion size up to 30.48 mm^2 (12 disc areas) were enrolled. The visual acuity was measured according to Early Treatment Diabetic Retinopathy Study (ETDRS) guidelines. The patients were equally randomized into four groups: anecortave acetate sterile suspension for injection 30 mg ($n = 33$), 15 mg ($n = 33$), or 3 mg ($n = 32$) or to placebo (vehicle, $n = 30$). Upon enrollment anecortave acetate or placebo was administered behind the eye as a 0.5-mL posterior juxtascleral injection onto the outer surface of the sclera near the macula using a specially designed cannula.

The primary efficacy variable was mean change from baseline in best-corrected log MAR visual acuity. Secondary efficacy variables included the percentage of patients with stable vision (< 3 log MAR lines of visual acuity decrease from baseline) and assessments of size of both the CNV and classic CNV lesion components. Clinical data was assessed at days 1–2, week 2, week 6, month 3, and month 6 following therapy. A 6-month retreatment interval was established based on the laboratory data confirming therapeutic levels of the drug in the retina and choroid for up to six months. The data reported in the 6-month interim study, however, reflects a single administration.

The analysis of mean change in log MAR visual acuity indicated that the 15 mg dose of anecortave acetate was statistically superior to placebo treatment at six months ($P = 0.003$). Trends also favored treatment with 30 and 3 mg over placebo, although statistical significance was not reached. Of the four groups, the 15 mg dose of anecortave acetate exhibited the greatest vision stabilization. As a secondary visual outcome, preservation of vision was defined as a decrease of <3 log MAR lines of visual acuity from baseline values. Although there was a greater preservation of vision for treatment with anecortave acetate 15 mg, it was not statistically significant when compared with placebo. However, in the subgroup of predominantly classic patients this treatment did reach statistical significance ($P = 0.021$) when compared with placebo. Eighteen percent of patients treated with anecortave acetate 15 mg improved at least 2 log MAR lines, compared with 3% of patients in the 30 mg group, 6% in the 3 mg group, and 0% in the placebo group. The difference between 15 mg and placebo was significant ($P = 0.025$). As anecortave acetate is an angiostatic agent, the change in CNV surface area was also analyzed. Total lesion area, CNV area, and total classic CNV area were measured and compared between the groups. The inhibition of surface area was significant only for the 15 mg group.

Of the 128 patients enrolled and treated, 76 patients (59.4%) completed their month 12 visit. The most common reason for exiting was AMD progression (24 of the 52 discontinued patients), with fewer patients in the anecortave acetate (15 mg) group than in the other three treatment groups discontinued from the study. More than 75% of the eyes in each of the four treatment groups had predominantly classic lesions at baseline. The 24-month data from this study confirm and support the 12-month data (Anecortave Acetate Clinical Study Group, 2003). Anecortave acetate 15 mg for depot suspension was statistically superior to placebo treatment for long-term stabilization of vision (< 3 log MAR line change from baseline values) ($P = 0.032$), prevention of vision loss ($P = 0.033$), and suppression of growth of both the CNV and classic CNV lesion components ($P < 0.05$). However, each of the three concentrations of anecortave acetate tested in this study was numerically superior to placebo treatment.

As part of this completed study, more than 300 posterior juxtascleral administration procedures were performed, and some patients received up to seven injections into the superotemporal quadrant. The most frequently reported ocular adverse events were cataract (32% after anecortave acetate vs. 40% after placebo), decreased visual acuity (26% after anecortave acetate vs. 43% after placebo), ptosis, eye pain, visual abnormalities, and subconjunctival hemorrhage. Only visual abnormalities (e.g., flashes, floaters, hazy vision, photophobia, blind spot enlargement) occurred more frequently following administration of anecortave acetate compared with placebo. Ptosis of the study eye, untreated fellow eye or both eyes occurred in 26 of the patients, and was observed in all four treatment groups. The ptosis was mild to moderate, and transient in all untreated cases. Nine patients had transient IOP changes of 1–13 mmHg from baseline.

The most frequently reported systemic adverse events were hypertension, arthritis, urinary tract infection, and hypercholesterolemia. Of the 36 patients reporting a serious adverse event, 11 exited the study for that reason. Serious adverse events were reported by 23 of the 98 anecortave acetate-treated patients and 13 of the 30 placebo-treated patients. Five deaths from lung carcinoma, heart failure/cerebrovascular accident, accidental injury, or myocardial infarction were reported. However, none of these serious adverse events or deaths was assessed as related to study treatment. An independent safety committee concluded that there were no clinically relevant medication-related or administration-related safety concerns.

Phase III Clinical Trials. In view of the positive long-term safety and efficacy outcomes from this study, a Phase III comparison of Anecortave Acetate 15 mg for Depot Suspension (RETAANE™ 15 mg Depot; Alcon Research, Ltd) to verteporfin (Visudyne® Novartis) PDT for treatment of subfoveal CNV was completed.

In the Phase III prospective, randomized multicenter trial, the 15-mg anecortave acetate dose was compared with the standard protocol for administration of verteporfin PDT. Patients were 50 years of age or older and had subfoveal predominantly classic CNV. The lesions were <5400 μm in greatest linear dimension. The visual acuity ranged from 20/40 to 20/400 Snellen equivalent. Of the 511 patients enrolled in the study, 255 patients received treatment with anecortave acetate 15 mg every six months with a corresponding sham PDT treatment every three months, while 256 patients received PDT every three months if there was any leakage, according to the standard PDT guidelines, with corresponding sham juxtascleral depot.

The primary end point was the percentage of patients with fewer than three lines of visual acuity loss. The goal was to demonstrate the noninferiority of anecortave acetate to verteporfin PDT at month 12 after the start of treatment. There was a

10% or less dropout rate over the 12-month time frame in the two groups of patients; results of 214 patients in the anecortave acetate group and 220 patients in the PDT group were analyzed.

There was no statistical difference between the anecortave acetate treatment and the PDT treatment outcomes regarding the primary end point of loss of fewer than three lines of visual acuity. In the anecortave acetate group 45% had less than a three-line loss of visual acuity, and in the PDT group, 49% of patients had less than a three-line loss of visual acuity. The statistical end point, however, was not reached regarding noninferiority of anecortave acetate compared with PDT.

Two factors that are potentially controllable that may have contributed to the failure of anecortave to reach the primary end point of noninferiority to PDT are: treatment interval and drug reflux. Future trials are being designed with this in mind.

Future Clinical Trials. Study C-04-59 is designed to evaluate the dose and administration frequency of anecortave acetate depot every three months (15 mg) or six months (15 or 30 mg). Inclusion criteria are similar to those of the previous trials.

Anecortave Acetate Risk Reduction Trial

Trial C-02-60 is a 48-month study of anecortave acetate (15 or 30 mg) versus sham (1:1:1) administered every six months, to determine whether anecortave reduces the risk of CNV developing in eyes with non-neovascular AMD. Approximately 2500 patients will be enrolled having neovascular AMD in the nonstudy eye, and presence of the following characteristics in the study eye: five or more intermediate or larger soft drusen, and/or confluent drusen with 3000 μm of the foveal center, and hyperpigmentation. Best-corrected ETDRS log MAR visual acuity must be 0.5 (20/62.5 Snellen equivalent) or better in the study eye.

The primary outcome measure is the development of sight-threatening CNV, defined as any CNV within 2500 μm of the center of the fovea. The greatest linear diameter (GLD) of classic CNV must be 100 μm or more, and of occult, 500 μm or greater, unless associated with subretinal hemorrhage or lipid, in which case 100 μm meets the criteria. Subretinal hemorrhage >500 μm GLD is also considered sight-threatening CNV.

Selective Drug Delivery Systems

There are currently more than 30 angiogenesis inhibitors in clinical trials and new candidates are under investigation in animal studies or in vitro (59). Many of these anti-angiogenesis inhibitors are currently being investigated for the treatment of CNV in AMD.

Squalamine, a broad-spectrum aminosterol antibiotic was originally isolated from the dogfish squalus acanthias (60). In various animal models it inhibits ocular neovascularization (61,62). Its anti-angiogenic activity is ascribed, at least in part, to its blockage of mitogen-induced proliferation and migration of endothelial cells (62). Early results from a Mexican Phase I–II clinical trial of squalamine, in patients with AMD indicated that squalamine induced CNV shrinkage in some patients, and lesion stabilization in others. In addition, in early study results, visual acuity was improved greater than three lines in some patients, and was stabilized in all patients. Some patients have been followed for up to four months after initiation of therapy. The responses observed include each angiographic subtype of AMD lesion. Further follow-up evaluations on all of the patients enrolled in the study, and additional Phase II and III trials are currently underway.

AG3340 (Prinomastat), a selective inhibitor of matrix metalloproteases, also inhibits retinal neovascularization in an animal model (63). Subsequently, a Phase II, randomized, double-masked, placebo-controlled study of the matrix metalloprotease inhibitor AG3340 in patients with subfoveal CNV associated with AMD was conducted. The outcome of this study, however, was not released and the company decided not to proceed with a Phase III trial.

Anti-Vascular Endothelial Growth Factor Agents

Vascular endothelial growth factor (VEGF) is an important molecule in angiogenesis development. Thus, anti-VEGF therapy is an attractive approach to treat AMD (64–66). In human studies, high VEGF concentrations are present in the vitreous in angiogenic retinal disorders but not in inactive or non-neovascularization-associated disease states (65,66). Further, VEGF is preferably localized within the cytoplasm of retinal pigment epithelial cells in the highly vascularized regions of surgically excised CNV membranes in humans and in animal models (64,67). Recent preclinical and clinical studies have demonstrated that blocking VEGF may have potential importance in the treatment of CNV secondary to AMD (68,69). In addition, anti-VEGF therapy may address the destructive effects caused by leakage secondary to CNV. VEGF, also known as vascular permeability factor, increases vascular leakage 50,000 times more potently than does histamine (70). Recent laboratory work suggested that anti-VEGF therapy may inhibit diabetes-induced blood–retinal barrier breakdown in animals (71).

Pegaptanib Sodium (Anti-VEGF Aptamer, Macugen^{TM})

Pegaptanib sodium is a pegylated anti-VEGF aptamer (see Chapter 5). The drug product is preservative-free and intended for single use by intravitreous injection using a sterile 27-gauge needle

Fifteen patients were entered into a Phase IA safety study (68). This was a multicenter open-label dose-escalation study of a single intravitreal injection of pegaptanib sodium in eyes with subfoveal CNV due to AMD and visual acuity worse than 20/200 as tested with early treatment diabetic retinopathy study (ETDRS) protocol-charts. Doses tested varied from 0.25 to 3.0 mg per eye. Visual acuity at three months was stable (unchanged) or improved in 80% of eyes, and 26.7% had a three-line or greater increase in ETDRS acuity. Eleven of the 15 patients experienced adverse events including mild intraocular inflammation, scotoma, visual distortion, hives, eye pain, and fatigue. There were no signs of retinal or choroidal toxicity on color photos or fluorescein angiography. This Phase IA study has demonstrated that intravitreal pegaptanib sodium is safe in doses of up to 3.0 mg/eye.

A multicenter, open-label, repeat-dose Phase IIA study of pegaptanib sodium (3.0 mg/eye) was performed in patients with subfoveal CNV secondary to AMD (72). The ophthalmic criteria included best-corrected visual acuity in the study eye worse than 20/100 on the ETDRS chart, best-corrected visual acuity in the fellow eye equal to or better than 20/400, subfoveal CNV with active CNV (either classic and/or occult) of less than 12 total disc areas in size secondary to AMD, clear ocular media and adequate pupillary dilation to permit good quality stereoscopic fundus photography, and IOP of 21 mmHg or less. A cohort scheduled to receive PDT with verteporfin prior to their first dose of pegaptanib sodium had to have equal to, or more than a 50% classic component (predominantly classic lesion).

If three or more patients experienced dose-limiting toxicity, the dose was reduced to 2 mg and then 1 mg, if necessary. The intended number of patients to be treated was 20; 10 patients with pegaptanib sodium alone and 10 patients with both anti-VEGF therapy and PDT. Eleven sites in the United States were selected for the studies. Hundred microliters of intravitreal pegaptanib sodium (3 mg/injection) was administered on three occasions at 28-day intervals.

PDT with verteporfin was given with pegaptanib sodium only in cases with predominantly classic (> 50%) CNV. The standard requirements and procedures for PDT administration were used. PDT was required to be given 5–10 days prior to administration of pegaptanib sodium.

Patients were clinically evaluated by the ophthalmologist two and eight days after each injection and again one month later just prior to the next injection. ETDRS visual acuities, color fundus photography, and fluorescein angiography were performed monthly for the first four months. Blood samples were drawn prior to and one week after each injection for routine hematologic and biochemical analyses and at additional time points to monitor the circulating levels of pegaptanib sodium.

Visual acuity in both patient groups remained stable throughout their study participation. One patient died prior to the final visit. No dose decrease was required for any patients in the study.

Of those patients ($n = 8$) who completed the 3-month treatment regimen of pegaptanib sodium alone, 87.5% had stabilized or improved visual acuity and 25.0% had a three-line improvement on the ETDRS chart at three months. Eleven patients were treated with both pegaptanib sodium and PDT. In this group of patients ($n = 10$) who completed the three months treatment regimen, 90% had stabilized or improved vision and 60% showed a three-line improvement of visual acuity on the ETDRS chart at three months.

Of the remaining patients who did not show a three-line gain, only one showed a loss of vision at three months and this patient lost only one line of vision at this time point. No patient in this group lost more than one line of visual acuity at three months. Repeat PDT treatment at three months (whose need was solely determined by the investigator) was performed in four of 10 eyes (40%) that participated for the complete study duration.

Although there were no serious adverse events that were directly attributed to the pegaptanib sodium injection, one patient in the Phase II clinical trial suffered two myocardial infarctions. As circulating plasma levels of pegaptanib sodium have been documented in the pharmacokinetic studies, these myocardial infarctions are of concern because of the critical role VEGF plays in cardiovascular angiogenesis. The patient had her first myocardial infarction 11 days after injection with pegaptanib sodium and her second, fatal, myocardial infarction 16 days after injection. Despite this patient not having elevated plasma drug levels, only further study will fully delineate the role these circulating plasma drug levels have on the cardiovascular system in these elderly patients.

In patients treated with pegaptanib sodium alone, ocular adverse events considered likely to be associated with intravitreal injection of pegaptanib sodium included vitreous floaters or haze, mild transient anterior chamber inflammation, ocular irritation, increased IOP, intraocular air, subconjunctival hemorrhage, eye pain, lid edema/erythema, dry eye, and conjunctival injection. In patients treated with pegaptanib sodium and PDT, adverse events probably associated included ptosis (due to the contact lens), mild anterior chamber inflammation, corneal abrasion, eye pain, foreign body sensation, chemosis, subconjunctival hemorrhage, and vitreous prolapse.

The results of this Phase IIA multiple intravitreal injection clinical study of anti-VEGF therapy expanded the favorable safety profile reported in the Phase IA single-injection study. Specifically, the Phase IIA study showed that three consecutive anti-pegaptanib sodium intravitreal injections given monthly did not cause serious ocular or systemic adverse events. The adverse events encountered appeared to be unrelated to study drug and were generally minor. In most cases they were probably related to the intravitreal injection procedure or to the PDT therapy. These results provided the basis for the Phase III pegaptanib sodium trial described below.

Phase III (VISION) Trial—VEGF Inhibition Study in Ocular Neovascularization. In the Phase III (VISION) clinical trial, patients were randomized to intravitreal injection or sham injection given every six weeks for 54 weeks (73). Two separate trials were conducted, one in North America and the other in Europe. Patients received either 0.3, 1.0, or 3.0 mg of pegaptanib sodium or sham injection in 1:1:1:1 randomization. Inclusion criteria included subfoveal CNV secondary to AMD (< 12 MPS disc areas in size, including lesion components) with any lesion composition and ETDRS visual acuity between 20/40 and 20/320 in the study eye and better or equal to 20/800 in the fellow eye. For patients with minimally classic or purely occult CNV, subretinal hemorrhage had to be present (but comprising no more than 50% of the lesion) and/or lipid and/or documented evidence of three or more lines of vision loss (ETDRS or equivalent) during the previous 12 weeks. Patients with predominantly classic CNV could receive combination treatment with verteporfin and ocular PDT based on investigator discretion.

A total of 1186 patients were included in efficacy analyses; 7545 intravitreous injections of pegaptanib sodium and 2557 sham injections were administered. Approximately 90% of the patients in each treatment group completed the study. An average of 8.5 injections were administered per patient out of a possible total of nine injections.

Efficacy was demonstrated, without a dose–response relationship, for all three doses of pegaptanib sodium. In the 0.3 mg group, 70% of patients lost fewer than 15 letters of visual acuity, compared with 55% of controls ($P < 0.001$). The risk of severe visual acuity loss (loss of 30 letters or more) was reduced from 22% in the sham group to 10% in the group receiving 0.3 mg of pegaptanib sodium ($P < 0.001$). More patients receiving pegaptanib sodium 0.3 mg, compared with sham injection, maintained or gained visual acuity (33% vs. 23%; $P = 0.003$). At all subsequent points from six weeks after beginning therapy, the mean visual acuity among those receiving 0.3 mg of pegaptanib sodium was better than in those receiving sham injections ($P < 0.002$). There was no evidence that any angiographic lesion subtype, the lesion size or the visual acuity level at baseline precluded a treatment benefit. In the study, 78% of patients never received PDT. A slightly higher proportion of patients receiving sham injections than those receiving pegaptanib sodium received PDT after baseline, suggesting a possible bias against pegaptanib sodium.

The most common adverse events were endophthalmitis (in 1.3% of patients— 0.16% per injection), traumatic injury to the lens (in 0.7% patients), and retinal detachment (in 0.6% patients).

Based on the results described above, Eyetech received Food and Drug Administration approval to market pegaptanib sodium for treatment of patients with neovascular AMD. This drug is the first anti-angiogenic agent to receive approval to treat AMD.

Patients who were initially enrolled in the VISION study will be re-randomized after 54 weeks of treatment to either continue or discontinue therapy for a further 48 weeks. Patients who show improved vision in the first year and deteriorate to baseline vision after stopping treatment may receive previously assigned (active) treatment.

Ranibizumab^{TM} (Rhufab V2, Lucentis)

Rhufab V2 (Lucentis®) is a humanized anti-VEGF antibody fragment which binds to VEGF, thus blocking CNV and vascular leakage (74). RhuFab is the Fab portion (the antigen-binding portion) of anti-VEGF monoclonal antibody (74). It is a recombinant antibody that consists of two parts: a nonbinding human sequence and a high-affinity binding epitope derived from the mouse, which serves to bind the antigen. Its molecular weight, 48,000, makes it much smaller molecule than the full-length monoclonal antibody which has a molecular weight of 148,000. Unlike the full-length antibody, after intravitreal ranibizumab injection, it can penetrate the internal limiting membrane and gain access to the subretinal space. In primates intravitreal rhuFab injection prevented the formation of clinically significant CNV and decreased the leakage of already formed CNV with no significant toxicity (see Chapter 5) (75).

Ranibizumab^{TM} has been studied in three Phase I/II trials in humans and is currently undergoing two pivotal Phase III clinical trials in patients with neovascular AMD and subfoveal CNV.

Study FVF1770g was a Phase I, open-label, dose-escalation trial of a single intravitreal injection of ranibizumab in subjects with new or recurrent CNV caused by exudative AMD.

Study FVF2128g, a Phase IB/II randomized, controlled, single-agent trial of two different rhuFabV doses given as multiple intravitreal injections was conducted (76). Sixty-four patients were enrolled in a single-agent, multicenter trial. The mean age of the treated patients was 78 years (range 63–87), 56% were female, and 92% were Caucasian. In the drug-treated group ($n = 53$), intravitreal rhuFab V2 injections (either 300 or 500 µg) were administered to one eye every four weeks for four weeks. Control eyes ($n = 11$) were treated with standard of care (no ranibizumab). Three different groups of subjects were enrolled in the study based on disease pattern as determined by fluorescein angiography (predominantly classic or minimally classic) and prior treatment: minimally classic (48% of treated patients), predominantly classic (28% of treated patients), and patients previously treated with PDT (24% of treated patients). Patients were monitored for safety and visual acuity. Visual acuity was defined as change from baseline in total number of letters read correctly (gained or lost) on the ETDRS chart. Baseline visual acuity of the enrolled patients ranged from 20/50 to 20/400, with a median of 20/125. There were no drug-related serious adverse events, and only two of 25 had transient vitreal inflammation. By day 98, after four injections, the visual acuity increased by three lines or greater in eight of 24, was stable in 14 of 24, and decreased by three lines or greater in two of 24 patients. The results from the first cohort of treated patients suggest that ranibizumab is well tolerated and visual acuity results were promising.

Study FVF 2425g was a Phase I, open-label, randomized study of three escalating multiple-dose regimens of intravitreal ranibizumab administered to subjects with primary or recurrent CNV caused by AMD. The goal was to ascertain whether a dose higher than 500 µg (up to 2000 µg) was safe to inject every two or four weeks, through 20 weeks. Mean visual acuity improved in each dose group, and no serious ocular adverse events were encountered. Overall, visual acuity was stable in 48%, improved by at least 15 letters in 44%, and decreased by 15 letters or more in 7%. The study concluded that the more frequent and higher doses of ranibizumab were well-tolerated.

The FOCUS trial was a Phase II trial for patients with predominantly classic subfoveal CNV due to AMD, designed to evaluate the efficacy of intravitreal ranibizumab in combination with PDT, versus PDT alone (2:1 randomization). All patients receive PDT every three months if indicated by leakage from CNV. In addition, one group receives 13 monthly 500 µg intravitreal ranibizumab injections, and the other group receives 13 monthly sham injections, for two years.

There are two pivotal Phase III clinical trials that enrolled patients with neovascular AMD and subfoveal CNV. In each Phase III trial, the primary end point is the proportion of patients losing 15 or more letters of vision. In the Minimally classic/occult trial of Anti-VEGF antibody RhuFab V2 in the treatment of Neovascular AMD Trial (MARINA, FVF2598g; Genentech), patients were randomized (1:1:1) to receive Ranibizumab™ (300 or 500 µg) versus a sham injection for 24 months, for primary, minimally classic/occult, subfoveal CNV. Similarly, the Anti-VEGF Antibody for the Treatment of Predominantly Classic Choroidal Neovascularization in AMD trial (ANCHOR, FVF2587g; Genentech) is evaluating ranibizumab versus PDT for the treatment of primary or recurrent, predominantly classic, subfoveal CNV. Patients were randomized 1:1:1 to receive 24-monthly intravitreal ranibizumab injections of 300 or 500 µg or PDT. Patients in the ranibizumab group were eligible to receive additional PDT every three months if they show leakage from CNV on fluorescein angiography.

The PIER trial was a Phase IIIB trial in which AMD patients with subfoveal minimally classic, predominantly classic, or occult-only CNV were randomized 1:1:1 to receive 300 or 500 µg of ranibizumab or a sham injection. Three monthly injections were followed by injections every three months thereafter.

In summary, the pharmacological treatment of AMD is a promising method to stop the progression of this devastating disease. The route of drug delivery plays a crucial role in the success of this treatment modality. A less invasive approach may also make the prophylactic treatment of AMD possible.

REFERENCES

1. Ferris FL III. Senile macular degeneration: review of epidemiologic features. Am J Epidemiol 1983; 118:132–151 (Review).
2. Hyman LG, Lilienfeld AM, Ferris FL III, Fine SL. Senile macular degeneration: a case–control study. Am J Epidemiol 1983; 118:213–227.
3. Macular Photocoagulation Study Group. Argon laser photocoagulation for neovascular maculopathy. Three-year results from randomized clinical trials. Arch Ophthalmol 1986; 104:694–701.
4. Macular Photocoagulation Study Group. Persistent and recurrent neovascularization after krypton laser photocoagulation for neovascular lesions of age-related macular degeneration. Arch Ophthalmol 1990; 108(6):825–831.
5. Macular Photocoagulation Study Group. Argon laser photocoagulation for neovascular maculopathy. Five-year results from randomized clinical trials. Arch Ophthalmol 1991; 109:1109–1114.
6. Macular Photocoagulation Study Group. Laser photocoagulation of subfoveal neovascular lesions in age-related macular degeneration. Results of a randomized clinical trial. Arch Ophthalmol 1991; 109:1220–1231.
7. Macular Photocoagulation Study Group. Laser photocoagulation of subfoveal recurrent neovascular lesions in age-related macular degeneration. Results of a randomized clinical trial. Arch Ophthalmol 1991; 109:1232–1241.
8. Macular Photocoagulation Study Group. Subfoveal neovascular lesions in age-related macular degeneration. Guidelines for Evaluation and Treatment in the Macular Photocoagulation Study. Arch Ophthalmol 1991; 109:1242–1257.

9. Macular Photocoagulation Study Group. Five-year follow-up of fellow eyes of patients with age-related macular degeneration and unilateral extrafoveal choroidal neovascularization. Arch Ophthalmol 1993;111:1189–1199.

10. Macular Photocoagulation Study Group. Laser photocoagulation of subfoveal neovascular lesions of age-related macular degeneration. Updated findings from two clinical trials. Arch Ophthalmol 1993; 111:1200–1209.

11. Macular Photocoagulation Study Group. Persistent and recurrent neovascularization after laser photocoagulation for subfoveal choroidal neovascularization of age-related macular degeneration. Arch Ophthalmol 1994; 112:489–499.

12. Macular Photocoagulation Study Group. Laser photocoagulation for juxtafoveal choroidal neovascularization. Five-year results from randomized clinical trials. Arch Ophthalmol 1994; 112:500–509.

13. Macular Photocoagulation Study Group. Evaluation of argon green vs. krypton red laser for photocoagulation of subfoveal choroidal neovascularization in the macular photocoagulation study. Arch Ophthalmol 1994; 112:1176–1184.

14. Macular Photocoagulation Study Group. The influence of treatment extent on the visual acuity of eyes treated with Krypton laser for juxtafoveal choroidal neovascularization. Arch Ophthalmol 1995; 113:190–194.

15. Macular Photocoagulation Study Group. Occult choroidal neovascularization. Influence on visual outcome in patients with age-related macular degeneration. Arch Ophthalmol 1996; 114:400–412.

16. Photodynamic therapy of subfoveal choroidal neovascularization in age-related macular degeneration with verteporfin: one-year results of 2 randomized clinical trials—TAP report. Treatment of age-related macular degeneration with photodynamic therapy (TAP) Study Group. Arch Ophthalmol 1999; 117:1329–1345.

17. Bressler NM. Photodynamic therapy of subfoveal choroidal neovascularization in age-related macular degeneration with verteporfin: two-year results of 2 randomized clinical trials—TAP report 2. Arch Ophthalmol 2001; 119:198–207.

18. Bressler NM, Arnold J, Benchaboune M, et al. Verteporfin therapy of subfoveal choroidal neovascularization in patients with age-related macular degeneration: additional information regarding baseline lesion composition's impact on vision outcomes—TAP report no. 3. Arch Ophthalmol 2002; 120:1443–1454.

19. Rubin GS, Bressler NM. Effects of verteporfin therapy on contrast on sensitivity: results from the treatment of age-related macular degeneration with photodynamic therapy (TAP) investigation—TAP report no 4. Retina 2002; 22:536–544.

20. Blumenkranz MS, Bressler NM, Bressler SB, et al. Verteporfin therapy for subfoveal choroidal neovascularization in age-related macular degeneration: three-year results of an open-label extension of 2 randomized clinical trials—TAP report no. 5. Arch Ophthalmol 2002; 120:1307–1311.

21. Ip M, Kroll A, Reichel E. Transpupillary thermotherapy. Semin Ophthalmol 1999; 14:11–18 (Review).

22. Reichel E, Berrocal AM, Ip M, et al. Transpupillary thermotherapy of occult subfoveal choroidal neovascularization in patients with age-related macular degeneration. Ophthalmology 1999; 106:1908–1914.

23. Mainster MA, Reichel E. Transpupillary thermotherapy for age-related macular degeneration: long-pulse photocoagulation, apoptosis, and heat shock proteins. Ophthalmic Surg Lasers 2000; 31:359–373.

24. Newsom RS, McAlister JC, Saeed M, McHugh JD. Transpupillary thermotherapy (TTT) for the treatment of choroidal neovascularisation. Br J Ophthalmol 2001; 85(2):173–178.

25. Ciulla TA, Harris A, Kagemann L, et al. Transpupillary thermotherapy for subfoveal occult choroidal neovascularization: effect on ocular perfusion. Invest Ophthalmol Vis Sci 2001; 42:3337–3340.

26. Algvere PV, Libert C, Lindgarde G, Seregard S. Transpupillary thermotherapy of pre-dominantly occult choroidal neovascularization in age-related macular degeneration with 12 months follow-up. Acta Ophthalmol Scand 2003; 81:110–117.

27. Salvetti P, Rosen JM, Reichel E. Subthreshold infrared footprinting with indocyanine green for localizing low-intensity infrared photocoagulation. Ophthalmic Surg Lasers Imaging 2003; 34:44–48.

28. Berson AM, Finger PT, Sherr DL, Emery R, Alfieri AA, Bosworth JL. Radiotherapy for age-related macular degeneration: preliminary results of a potentially new treatment. Int J Radiat Oncol Biol Phys 1996; 36:861–865.

29. Sasai K, Murata R, Mandai M, et al. Radiation therapy for ocular choroidal neovascu-larization (Phase I/II study): preliminary report. Int J Radiat Oncol Biol Phys 1997; 39:173–178.

30. Bergink GJ, Hoyng CB, van der Maazen RW, Vingerling JR, van Daal WA, Deutman AF. A randomized controlled clinical trial on the efficacy of radiation therapy in the con-trol of subfoveal choroidal neovascularization in age-related macular degeneration: radiation versus observation. Graefes Arch Clin Exp Ophthalmol 1998; 236:321–325.

31. Ciulla TA, Danis RP, Klein SB, et al. Proton therapy for exudative age-related macular degeneration: a randomized, sham-controlled clinical trial. Am J Ophthalmol 2002; 134:905–906.

32. Hart PM, Chakravarthy U, Mackenzie G, et al. Visual outcomes in the subfoveal radio-therapy study: a randomized controlled trial of teletherapy for age-related macular degeneration. Arch Ophthalmol 2002; 120:1029–1038.

33. Kirwan JF, Constable PH, Murdoch IE, Khaw PT. Beta irradiation: new uses for an old treatment: a review. Eye 2003; 17:207–215 (Review).

34. Submacular surgery trials randomized pilot trial of laser photocoagulation versus surgery for recurrent choroidal neovascularization secondary to age-related macular degenera-tion: I. Ophthalmic outcomes submacular surgery trials pilot study report number 1. Am J Ophthalmol 2000; 130:387–407.

35. Ninomiya Y, Lewis JM, Hasegawa T, Tano Y. Retinotomy and foveal translocation for surgical management of subfoveal choroidal neovascular membranes. Am J Ophthalmol 1996; 122:613–621.

36. Machemer R. Macular translocation. Am J Ophthalmol 1998; 125:698–700.

37. Wolf S, Lappas A, Weinberger AW, Kirchhof B. Macular translocation for surgical man-agement of subfoveal choroidal neovascularizations in patients with AMD: first results. Graefes Arch Clin Exp Ophthalmol 1999; 237:51–57.

38. Eckardt C, Eckardt U, Conrad HG. Macular rotation with and without counter-rotation of the globe in patients with age-related macular degeneration. Graefes Arch Clin Exp Ophthalmol 1999; 237:313–325.

39. de Juan E Jr, Vander JF. Effective macular translocation without scleral imbrication. Am J Ophthalmol 1999; 128:380–382.

40. Lewis H, Kaiser PK, Lewis S, Estafanous M. Macular translocation for subfoveal chor-oidal neovascularization in age-related macular degeneration: a prospective study. Am J Ophthalmol 1999; 128:135–146.

41. Guyer DR, Adamis AP, Gragoudas ES, Folkman J, Slakter JS, Yannuzzi LA. Systemic antiangiogenic therapy for choroidal neovascularization. What is the role of interferon alfa? Arch Ophthalmol 1992; 110(10):1383–1384.

42. Thomas MA, Ibanez HE. Interferon alfa-2a in the treatment of subfoveal choroidal neo-vascularization. Am J Ophthalmol 1993; 115(5):563–568.

43. Gillies MC, Sarks JP, Beaumont PE, et al. Treatment of choroidal neovascularisation in age-related macular degeneration with interferon alfa-2a and alfa-2b. Br J Ophthalmol 1993; 77:759–765.

44. Chan CK, Kempin SJ, Noble SK, Palmer GA. The treatment of choroidal neovascular membranes by alpha interferon. An efficacy and toxicity study. Ophthalmology 1994; 101:289–300.

45. Pharmacological Therapy for Macular Degeneration Study Group. Interferon alfa-2a is ineffective for patients with choroidal neovascularization secondary to age-related macular degeneration. Results of a prospective randomized placebo-controlled clinical trial. Arch Ophthalmol 1997; 115:865–872.
46. Penfold PL, Gyory JF, Hunyor AB, Billson FA. Exudative macular degeneration and intravitreal triamcinolone. A pilot study. Aust N Z J Ophthalmol 1995; 23:293–298.
47. Jonas JB, Kreissig I, Hugger P, Sauder G, Panda-Jonas S, Degenring R. Intravitreal triamcinolone acetonide for exudative age related macular degeneration. Br J Ophthalmol 2003; 87:462–468.
48. Danis RP, Ciulla TA, Pratt LM, Anliker W. Intravitreal triamcinolone acetonide in exudative age-related macular degeneration. Retina 2000; 20:244–250.
49. Gillies MC, Simpson JM, Luo W, et al. A randomized clinical trial of a single dose intravitreal triamcinolone acetonide for neovascular age-related macular degeneration: one-year results. Arch Ophthalmol 2003; 121:667–673.
50. Ranson NT, Danis RP, Ciulla TA, Pratt L. Intravitreal triamcinolone in subfoveal recurrence of choroidal neovascularization after laser treatment in macular degeneration. Br J Ophthalmol 2002; 86:527–529.
51. Macular Photocoagulation Study Group. Five-year follow-up of fellow eyes of patients with age-related macular degeneration and unilateral extrafoveal choroidal neovascularization. Arch Ophthalmol 1993; 111:1189–1199.
52. Spaide RF, Sorenson J, Maranan L. Combined photodynamic therapy with verteporfin and intravitreal triamcinolone acetonide for choroidal neovascularization. Ophthalmology 2003; 110:1517–1525.
53. Rechtman E, Danis RP, Pratt LM, Harris A. Intravitreal triamcinolone with photodynamic therapy for subfoveal choroidal neovascularisation in age related macular degeneration. Br J Ophthalmol 2004; 88:344–347.
54. DeFaller JM, Clark AF. A new pharmaceutical treatment for angiogenesis. In: Taylor HR, ed. Pterygium. The Hague: Kugler Publications, 2000:159–181.
55. Penn JS, Rajaratnam VS, Collier RJ, Clark AF. The effect of an angiostatic steroid on neovascularization in a rat model of retinopathy of prematurity. Invest Ophthalmol Vis Sci 2001; 42:283–290.
56. Blei, F, Wilson, EL, Mignatti, P, Rifkin, DB. Mechanism of action of angiostatic steroids: suppression of plasminogen activator activity via stimulation of plasminogen activator inhibitor synthesis. J Cell Physiol 1993; 155:568–578.
57. D'Amico DJ, Goldberg MF, Hudson H, et al. Anecortave acetate as monotherapy for the treatment of subfoveal lesions in patients with exudative age-related macular degeneration (AMD): interim (month 6) analysis of clinical safety and efficacy. Retina 2003; 23:14–23.
58. D'Amico DJ, Goldberg MF, Hudson H, et al. Anecortave Acetate Clinical Study Group. Anecortave acetate as monotherapy for treatment of subfoveal neovascularization in age-related macular degeneration: twelve-month clinical outcomes. Ophthalmology 2003; 110:2372–2383; Discussion 2384–2385.
59. Hagedorn M, Bikfalvi A. Target molecules for anti-angiogenic therapy: from basic research to clinical trials. Crit Rev Oncol Hematol 2000; 34:89–110.
60. Moore KS, Wehrli S, Roder H, et al. Squalamine: an aminosterol antibiotic from the shark. Proc Natl Acad Sci U S A 1993; 90:1354–1358.
61. Higgins RD, Sanders RJ, Yan Y, Zasloff M, Williams JI. Squalamine improves retinal neovascularization. Invest Ophthalmol Vis Sci 2000; 41:1507–1512.
62. Genaidy M, Kazi AA, Peyman GA, et al. Effect of squalamine on iris neovascularization in monkeys. Retina 2002; 22:772–778.
63. Garcia C, Bartsch DU, Rivero ME, et al. Efficacy of Prinomastat (AG3340), a matrix metalloprotease inhibitor, in treatment of retinal neovascularization. Curr Eye Res 2002; 24:33–38.
64. Lopez PF, Sippy BD, Lambert HM, et al. Transdifferentiated retinal pigment epithelial cells are immunoreactive for vascular endothelial growth factor in surgically excised

age-related macular degeneration-related choroidal neovascular membranes. Invest Ophthalmol Vis Sci 1996; 37:855–868.

65. Adamis AP, Miller JW, Bernal MT, et al. Increased vascular endothelial growth factor levels in the vitreous of eyes with proliferative diabetic retinopathy. Am J Ophthalmol 1994; 118:445–450.

66. Aiello LP, Avery RL, Arrigg PG, et al. Vascular endothelial growth factor in ocular fluid of patients with diabetic retinopathy and other retinal disorders. N Engl J Med 1994; 331:1480–1487.

67. Kwak N, Okamoto N, Wood JM, Campochiaro PA. VEGF is major stimulator in model of choroidal neovascularization. Invest Ophthalmol Vis Sci 2000; 41:3158–3164.

68. The Eyetech Study Group. Pre-clinical and Phase IA clinical evaluation of an anti-VEGF pegylated aptamer (EYE001) for the treatment of exudative age-related macular degeneration. Retina 2002; 22:143–152.

69. Eyetech Study Group. Anti-vascular endothelial growth factor therapy for subfoveal choroidal neovascularization secondary to age-related macular degeneration: Phase II study results. Ophthalmology 2003; 110:979–986.

70. Senger DR, Galli SJ, Dvorak AM, et al. Tumor cells secrete a vascular permeability factor that promotes accumulation of ascites fluid. Science 1983; 219:983–985.

71. Qaum T, Xu Q, Joussen AM, et al. VEGF-initiated blood–retinal barrier breakdown in early diabetes. Invest Ophthalmol Vis Sci 2001; 42:2408–2413.

72. Eyetech Study Group. Anti-vascular endothelial growth factor therapy for subfoveal choroidal neovascularization secondary to age-related macular degeneration: Phase II study results. Ophthalmology 2003; 1105:979–986.

73. Gragoudas ES, Adamis AP, Cunningham ET Jr, Feinsod M, Guyer DR. VEGF Inhibition Study in Ocular Neovascularization Clinical Trial Group. Pegaptanib for neovascular age-related macular degeneration. N Engl J Med 2004; 351:2805–2816.

74. Mordenti J, Cuthbertson RA, Ferrara N, et al. Comparisons of the intraocular tissue distribution, pharmacokinetics, and safety of 125I-labeled full-length and Fab antibodies in rhesus monkeys following intravitreal administration. Toxicol Pathol 1999; 27:536–544.

75. Krzystolik MG, Afshari MA, Adamis AP, et al. Prevention of experimental choroidal neovascularization with intravitreal anti-vascular endothelial growth factor antibody fragment. Arch Ophthalmol 2002; 120:338–346.

76. Heier JS, SY JP, McCluskey ER. RhuFab V2 (anti-VEGF antibody) for treatment of exudative AMD. Combined Vitreous Society and Retina Society Meeting, San Francisco, CA, 2002.

17

Intraocular Sustained-Release Drug Delivery in Uveitis

Mark T. Cahill and Glenn J. Jaffe
Duke University Eye Center, Durham, North Carolina, U.S.A.

INTRODUCTION

The main goal in the treatment of uveitis is to eliminate intraocular inflammation and thereby relieve discomfort and prevent visually significant complications such as cataract, cystoid macular edema, and hypotony. When anti-inflammatory drugs are given systemically, it is often necessary to give high doses over long time periods to achieve an effective ocular anti-inflammatory effect. Corticosteroids are the mainstay of uveitis therapy, however treatment may not be fully effective, or side effects may be treatment-limiting. It is then often necessary to switch to alternative drugs. These agents can be broadly described as steroid-sparing drugs as they can either reduce the amount of corticosteroids needed in a given patient, or they can replace corticosteroids altogether. Broadly speaking these steroid-sparing medications can be classified as immunosuppressives and immunomodulators. Immunosuppressives include antimetabolites, such as methotrexate, azathioprine and mycophenolate mofetil and alkylating agents, such as cyclophosphamide and chlorambucil. Immunomodulators include calcineurin inhibitors such as cyclosporin A (CsA) and tacrolimus (FK506), and cytokine inhibitors including etanercept and infliximab.

Uveitis treatments can be delivered topically, periocularly, intraocularly or systemically, and there are problems common to all delivery techniques and specific to each delivery method. Compliance with any form of regular medication can be a problem particularly if its administration is associated with discomfort or if its side effects are unpleasant. Some medications, particularly hydrophobic compounds, may cross the blood–retinal barrier poorly, which is an important consideration for all delivery systems except intraocular injections. Topical medications, which have the least side effects, do not penetrate into the posterior segment and are unsuitable for posterior uveitis, which is often sight-threatening.

Intraocular injections are associated with significant complications and often must be repeated at regular intervals in patients with a chronic disease such as uveitis. Similar difficulties are associated with periocular injections although the complication rate is lower and those that do occur are usually less severe. The main advantage of intraocular or periocular injections is that there is a relatively high dose of drug delivered to the eye with few or no systemic side effects.

In severe, sight-threatening uveitis systemic immunosuppression may be the only treatment that can adequately control the disease. However, all the previously outlined medications have significant systemic side effects particularly if they are used for prolonged periods of time and these side effects can be treatment-limiting. Many immunosuppressive drugs are teratogenic and are contraindicated during pregnancy or may prevent conception. Most side effects become apparent during treatment and can damage specific organs such as the liver and kidneys, the functioning of which need to be monitored during treatment. However, some side effects including osteoporosis and lymphoproliferative malignancies may not become apparent until many years after treatment has stopped. Osteoporosis is particularly related to corticosteroid use even at low doses. Rapid and extensive bone loss has been associated with corticosteroid doses >5 mg for three months or more. Long-term immunosuppression may also increase the risk of developing cancer, particularly solid tumors and lymphomas. This increased cancer risk is probably the result of reduction of normal immunosurveillance or direct effects of the medication on the patient's DNA.

An increasing number of sustained-release drug devices using different mechanisms and containing a variety of immunosuppressive agents have been developed to treat uveitis. All these devices try to maximize the time that an effective amount of a given drug for a given disease remains in the eye while minimizing any side effects associated with device insertion, prolonged exposure of the eye to components of the medication or the device itself, and systemic absorption of the medication.

CORTICOSTEROID DEVICES

Although corticosteroids are the first-line treatment for uveitis and can be delivered topically, local injection or systemically, they are associated with a number of side effects and recurrence of inflammation frequently occurs after cessation of treatment (1). Topical corticosteroids do not penetrate the posterior segment well and can cause poor wound healing, corneal toxicity, and elevated intraocular pressure (2). Local therapy given as periocular corticosteroid injection can be effective. However, multiple periocular corticosteroid injections may be necessary for disease control and each injection carries a number of risks including localized toxic drug delivery vehicle reactions, extraocular muscle fibrosis, and inadvertent globe injury (3). The side effects of chronic systemic corticosteroid administration have been well documented and include changes in appearance, hypertension, hyperglycemia, gastritis, opportunistic infections, and life-threatening psychosis (Fig. 1) (4).

Sustained intraocular corticosteroid delivery can overcome systemic side effects associated with oral topical and periocular therapy while at the same time provide effective suppression of intraocular inflammation (5–9). However, sustained intraocular therapeutic corticosteroid levels are required to adequately treat uveitis that typically has a chronic and recurrent course. Some corticosteroids such as dexamethasone phosphate are less suitable for treatment of chronic intraocular inflammation as they have half-lives of <4 hours when administered intravitreally and are rapidly removed from the eye (5,9).

In contrast 4 mg of triamcinolone acetonide, a minimally water-soluble steroid, injected intravitreally has a mean elimination half-life of 18 days in nonvitrectomized

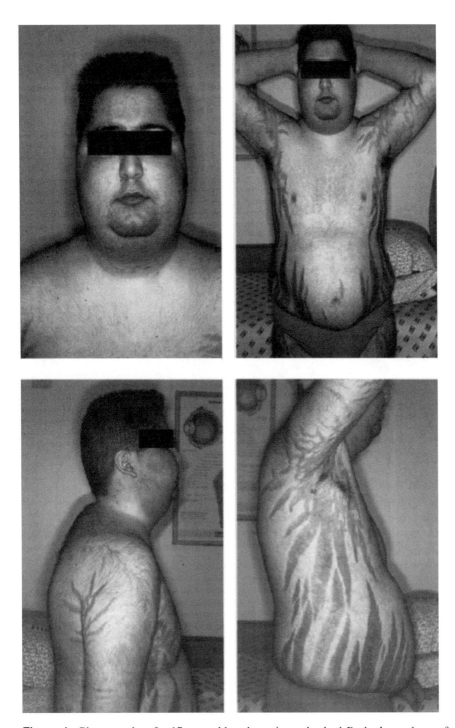

Figure 1 Photographs of a 17-year-old male patient who had Bechet's syndrome for three years which required treatment with long-term systemic corticosteroids as well as immuno-suppressive drugs. Note the pronounced corticosteroid-induced side effects including skin striations and Cushingoid appearance.

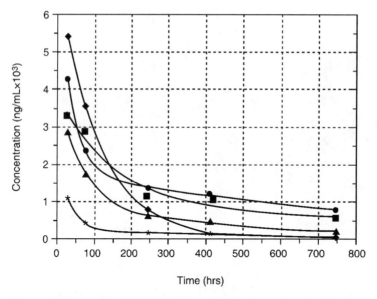

Figure 2 Intravitreal triamcinolone concentrations after a single intravitreal triamcinolone acetonide injection in five eyes. Concentration–time data for each vitreous sample for all patients are shown along with two-compartment model-derived pharmacokinetic curves.

eyes and measurable concentrations of triamcinolone acetonide would be expected to last for up to three months in such eyes (Fig. 2) (10). Furthermore, a measurable concentration of triamcinolone acetonide has been documented to last as long as 1.5 years after a larger 25 mg intravitreal injection (11). Intravitreal triamcinolone acetonide may be an effective short-term treatment for cystoid macular edema associated with intraocular inflammation secondary to idiopathic uveitis, HLA B27-associated uveitis, intermediate uveitis, birdshot retinochoroidopathy, and sympathetic ophthalmitis (12–16). However, even with a possible therapeutic effect of three months using a standard 4 mg dose, long-term use of triamcinolone acetonide to treat chronic uveitis would still require multiple intraocular injections. In contrast both biodegradable and nondegradable devices have been used to deliver intraocular steroids over sustained periods without the need for repeated intraocular injections.

Biodegradable Dexamethasone Device

A biodegradable polymer matrix containing dexamethasone has been developed to treat postcataract uveitis. The device, called Surodex®, is 1 mm long and 0.5 mm wide and can be designed to release dexamethasone for periods ranging from days to months (Fig. 3) (17). In a Phase I human clinical trial the device effectively suppressed postoperative inflammation in six patients who underwent cataract surgery without any clinically significant adverse safety problems.

A subsequent Phase II, multicenter, randomized, double-masked, placebo-controlled clinical trial was performed in patients undergoing phacoemulsification and intraocular lens implantation (18). Only one eye per patient was eligible for treatment and exclusion criteria included previous uveitis, concurrent anterior segment disease or intraoperative surgical complications. Patients were randomized in a 2:1 ratio into an active treatment group or a control group. Patients in the two

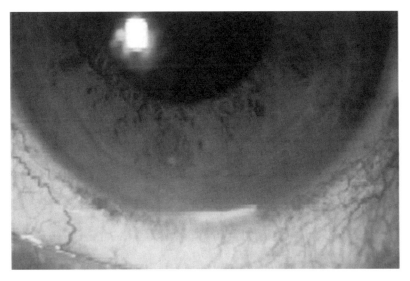

Figure 3 (*See color insert*) Slit-lamp photograph of an isolated incidence in which two Surodex devices are visible in the inferior angle. In most cases Surodex pellets are sited deeper within the angle and are not visible on slit-lamp examination.

treatment groups received either one or two biodegradable devices containing 60 µg of dexamethasone, respectively. There were two different control groups; one received no device and the other received a placebo device that contained no drug. All devices were placed in the posterior chamber between the iris and the anterior surface of the intraocular lens. The device was positioned in the posterior chamber to ensure that postoperative examiners were masked as to whether any given eye had received an implant (18). Ocular or systemic steroidal or nonsteroidal anti-inflammatory medications were not allowed for two weeks before surgery or two days after surgery. On the third postoperative day the masked examiner could start any anti-inflammatory medications required, while antibiotics and pressure-lowering medications were allowed at any point in the postoperative period.

Ninety patients were randomized into the four groups, of which 89 completed the study follow-up period of 60 days. Thirty eyes received two drug devices, 29 eyes received one drug device, 15 eyes received a placebo device, and the remaining 15 eyes received no device at all. A significantly higher proportion of eyes in the control groups (80.0% at week 2) required topical corticosteroid medication to control intraocular inflammation when compared with the treatment groups (7.0% at week 2). Furthermore, eyes in the control groups required rescue anti-inflammatory medications sooner than eyes in the treatment groups. Control patients had significantly more objective signs of intraocular inflammation than treatment patients, while there was no difference in mean intraocular pressure in either treatment or control groups during the study period.

On the first postoperative day eyes that received one device had significantly higher cell and flare scores than eyes that received two devices. However, there was no statistically significant difference in these objective measurements of inflammation between eyes that received one or two drug devices at any of the remaining follow-up examinations. Similarly, no significant difference was found between the cell and flare counts at any of the follow-up examinations in eyes that received a placebo device or no device at all. Comparative data on the need for rescue medications and intraocular pressure between eyes with one or two devices and eyes with a placebo device or no device was not available in the published study report (18).

A separate randomized, double-masked study compared postcataract anterior segment inflammation in eyes that received the same biodegradable device containing 60 µg of dexamethasone with eyes that were treated with 0.1% topical dexamethasone (17). In total 60 eyes in 60 patients were studied of which 32 eyes received a drug device. Eyes were followed for four weeks after surgery. After the fourth postoperative day anterior segment flare as measured by a laser flare meter was significantly lower in eyes that received a drug device when compared with eyes that were treated with drops alone. Anterior chamber cell counts at the slit lamp were also lower in eyes that had a drug implant when compared with those that did not but this difference was not significant. A third study by the same authors subsequently demonstrated that the drug device was equally effective if it was placed in the anterior segment or the ciliary sulcus and that insertion in the anterior chamber did not affect corneal endothelial cell counts (19).

A biodegradable dexamethasone sustained-release device called Posurdex® which is inserted into the posterior segment has been developed to treat macular edema secondary to diabetes, retinal vascular occlusion, cataract surgery, and uveitis (Fig. 4). A Phase II trial randomized eyes with macular edema to one of three groups to study the safety and efficacy of the device. Eyes in the two treatment groups received a device containing either 700 or 350 µg of dexamethasone while a third, control group received no device and was observed only. Although over 300 patients

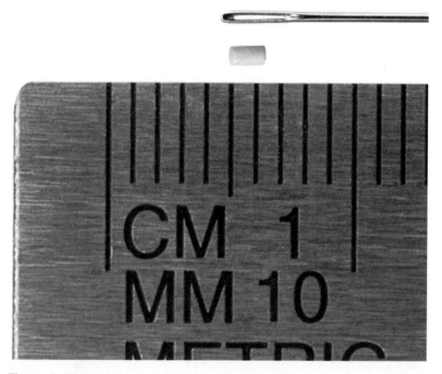

Figure 4 Photograph of a Posurdex, a biodegradable dexamethasone sustained-release device which is inserted into the posterior segment.

were enrolled in the study only a small percentage of eyes had cystoid macular edema secondary to uveitis. However, when the 6-month data of all eyes was analyzed a greater than three-line improvement in visual acuity was seen in a significantly higher proportion of eyes that received a device than eyes that did not. Furthermore, there were no significant adverse advents associated with the device. Based on these results, a Phase III trial of the device has been initiated.

Nondegradable Dexamethasone Device

In view of successful animal experiments (20; see also Chapter 14), a sustained-release dexamethasone device was implanted in one eye of a patient with bilateral severe uveitis associated with multiple sclerosis (20,21). The patient had previously undergone pars plana lensectomy and vitrectomy in the right eye for decreased vision associated with cataract. Despite chronic topical corticosteroids, the patient had persistent bilateral low-grade inflammation and recurrent severe bilateral iridocyclitis. Best corrected visual acuity was 20/400 in both eyes. Systemic corticosteroids and methotrexate controlled the intraocular inflammation but the patient was intolerant of these medications because of systemic side effects. The nondegradable dexamethasone device was inserted into the patient's left eye.

The implant, which consisted of a 5-mg drug core surrounded by ethylene vinyl acetate and polyvinyl alcohol, was inserted through the pars plana after pars plana vitrectomy and lensectomy. The device was well tolerated and the patient's eye remained quiet and did not require any supplemental local corticosteroids. The visual acuity remained at 20/400 in the left eye and the intraocular pressure remained normal. In contrast, the patient had two episodes of recurrent anterior segment inflammation in the right eye, despite intensive topical steroids. After 10 months, the intraocular inflammation recurred in the patient's left eye, presumably because the device and surrounding tissues had become depleted of dexamethasone.

Pars plana vitrectomy with lensectomy has been reported to reduce the severity and intensity of uveitic episodes (22). However despite those procedures, it was not possible to eliminate topical steroid treatment in the right eye which did not become quiet until supplemental systemic immunosuppressants were started. In contrast, the left eye was quiet three months after the device implantation and remained quiet for 10 months. These results suggest that the dexamethasone implant had an anti-inflammatory effect not attributable to the lensectomy or vitrectomy. While the device was effective for 10 months, an even longer duration of drug effect is desirable. The limited dexamethasone device duration of action is a function of the drug solubility and the size of the device. While dexamethasone is relatively insoluble a very large device would be needed to permit drug release over a long period. This large size is a significant disadvantage when dexamethasone used in a device to treat a chronic disease like uveitis.

Nondegradable Fluocinolone Acetonide Device

Fluocinolone acetonide is a lipophilic, synthetic corticosteroid with a potency similar to dexamethasone. However, fluocinolone acetonide is 1/24th as soluble as dexamethasone, which makes it very insoluble. Thus it can be released over a much longer period of time than dexamethasone without an excessively bulky polymer system (23; see also Chapter 14). A prospective, noncomparative interventional case

series of a sustained-release (24) drug delivery device containing fluocinolone acetonide to treat uveitis was undertaken. Pellets containing 2.1 mg of the drug were
coated in a polyvinyl alcohol and silicone laminate and fixed to a polyvinyl alcohol
suture strut (Fig. 5). The devices were designed to release 2 µg of fluocinolone acetonide per day for approximately three years.

Inclusion criteria for patients enrolled in the study included severe uveitis with
posterior segment involvement with or without iridocyclitis, previous favorable
response to oral or periocular corticosteroids, treatment-limiting side effects associated with systemic or periocular corticosteroids or systemic nonsteroidal immunosuppressive agents, intraocular pressure controlled at ≤21 mmHg with no more than
one anti-ocular hypertensive drop and ability to attend follow-up visits. In total,
seven eyes of five patients were included and the patients had a diagnosis of Bechet's
syndrome or idiopathic panuveitis. The mean uveitis duration before device implantation was six years and the mean visual acuity was 20/207.

The device was implanted via the pars plana and a posterior infusion was used
in three eyes that had undergone prior vitrectomy. Cataract surgery was undertaken
at the time of the device implantation in two eyes. There were no intraoperative complications in any of the eyes. The mean follow-up was 10 months. All seven eyes had
stabilized or improved visual acuity while four eyes improved three or more lines.
Mean postoperative visual acuity was 20/57, which was significantly better than
the mean preoperative visual acuity.

All eyes remained quiet for the duration of the follow-up period, with no more
than occasional anterior chamber cells, clear media, and no recurrences of vitritis,

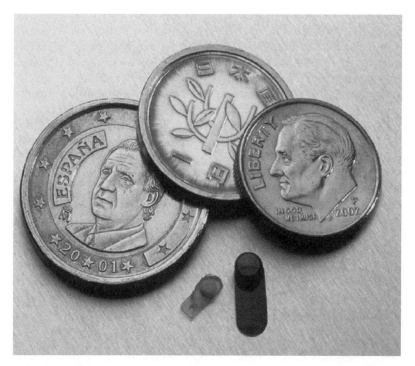

Figure 5 Photograph of a drug delivery device containing 2 mg of fluocinolone acetonide
(*left*). Ganciclovir device is shown on the right for comparison. Note the smaller size of the
fluocinolone device. The drug core is surrounded by a polyvinyl alcohol/silicone laminate.

retinitis or chorioretinitis. Resolution of cystoid macular edema was documented in one eye (Fig. 6). In contrast, three of the four binocular patients had severe inflammation of the fellow eye in the follow-up period. A drug device was subsequently placed in the fellow eye of two of the patients. All eyes had a marked reduction in anti-inflammatory medication use. Before drug device implantation, all seven eyes required periocular corticosteroid injections at regular intervals ranging from every two weeks to every three months. Furthermore, six of seven eyes required intensive topical steroids and one patient required high-dose systemic steroids. Postoperatively, no eye required topical or periocular steroids and one patient was on 10 mg prednisone every other day to permit adrenal gland recovery after chronic corticosteroid use. The mean pre- and postoperative intraocular pressure was 13.1 and 15.7 mmHg, respectively, a difference that was not significant. However, four of the seven eyes had elevation of the intraocular pressure, which was controlled in all eyes with topical medications.

Based on the favorable results obtained from the initial series, additional patients were enrolled into a single-investigator randomized trial in which the 2.1 mg device described above, or a 0.59 mg device designed to release drug at 0.5 mg/day for approximately three years was inserted (25). Enrollment criteria were identical to those used for the initial five patients. The results of this trial that included 36 eyes of 32 patients were similar to the initial series and were maintained

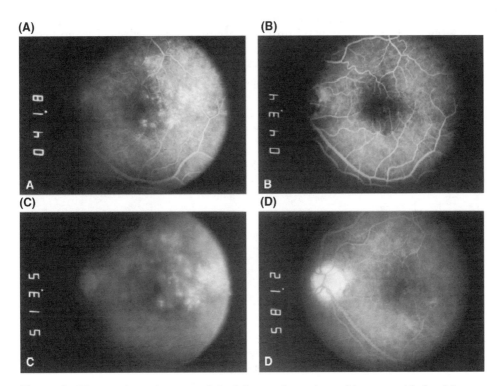

Figure 6 Fluorescein angiograms of the left eye of a patient with pan-uveitis for 5.5 years treated with a fluocinolone acetonide implant. (**A, C**) Mid arteriovenous phase and late phase frames, respectively, before device implantation showing petaloid hyperfluorescence corresponding to cystoid macular edema. (**B, D**) Mid arteriovenous phase and late phase frames, respectively, four months after device implantation showing resolution of the petaloid hyperfluorescence.

over an extended time period. Mean follow-up duration was 683.2 ± 461.1 days (204–1817 days). Inflammation was effectively controlled during the follow-up period. The average number of recurrences in the 12 months preceding device implantation was 2.5 episodes per eye. There were no recurrences during the first two years following device implantation, and there were only five recurrences for the group as a whole during the follow-up period (Fig. 7). Baseline visual acuity for the fluocinolone device-implanted eye was +1.1 log MAR units (20/250) which improved significantly to +0.81 log MAR units (20/125) at 30 months ($P < 0.05$). Additionally, there was a reduction in systemic and local therapy use in the device-implanted eyes. There was a reduction in systemic anti-inflammatory medication use from baseline, 53.1% to 38.8%, 24 months following device implantation. Of those still on systemic medications, the dose was reduced in 68% of the patients postoperatively, and none required systemic medications for the device-implanted eye (they were given for the fellow, nonimplanted eye, or for systemic disease manifestations). The percentage of eyes that required posterior subtenon's capsule injection following fluocinolone device implantation was significantly less than those requiring a posterior subtenon's injection in the 12 months preceding device implantation. In the 12 months preceding injection, the average posterior subtenon's capsule injection rate was 2.2 injections per eye per year. After implantation, no injections were given for the first two years, and only 0.03 injections per eye were given by 36 months. At baseline, 78.1% of eyes required topical anti-inflammatory drops; this percentage continuously declined over the follow-up period. On average at baseline, 6.1 drops per eye per day were given; by 30 months, only 0.3 drops per eye per day were given. The most common adverse event was intraocular pressure rise. At baseline 16.7% of patients used pressure lowering agents compared with 47.6% over the follow-up period ($P < 0.05$). Filtering procedures were performed in seven (19.4%) eyes. Of eight phakic eyes at

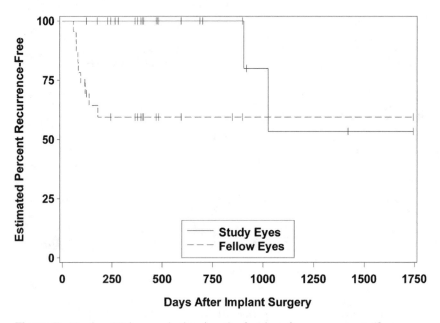

Figure 7 Kaplan–Meier graph showing the freedom from recurrence (first recurrence after implantation) for the device-implanted eyes and fellow eye over time.

baseline, four required cataract extraction during the follow-up period. There were no device explantations or patients lost to follow-up during the investigation.

In view of these promising results a randomized, prospective, multicenter study was undertaken at 27 sites in the United States and one in Singapore to further assess the role of sustained-release fluocinolone in the treatment of chronic uveitis. In this trial, 278 eyes with noninfectious posterior uveitis were randomized to receive a device containing either 0.59 or 2.1 mg of fluocinolone acetonide. In bilateral cases the device was placed in the more severely affected eye. When the aggregate 34-week data were analyzed, eyes that received a drug device had a very low uveitis recurrence rate. The aggregate recurrence rate for the two dose implants decreased from 51.4% in the 34 weeks prior to device implantation to 6.1% in the 34 weeks following implantation. The recurrence rate in the eyes implanted with the 0.59 mg implant did not differ significantly from that observed in eyes implanted with the 2.1 mg device. The vast majority of eyes had stabilized or improved visual acuity following device implantation and 21% of eyes that received an implant had an improvement in visual acuity of three or more lines. Overall the use of systemic corticosteroid and/or immunosuppressive therapy decreased at 34 weeks as did the need for adjunctive topical and periocular corticosteroid treatments. The most common adverse effects included elevation of intraocular pressure and cataract progression. The proportion of eyes that required ocular antihypertensive agents increased from approximately 14% at baseline to 50% at week 34, and 6% of eyes required filtration surgery. Approximately 10% of eyes required cataract extraction during the follow-up period. Based on the individual-investigator data described previously, the proportion of eyes in the multicenter-study population with elevated intraocular pressure requiring topical therapy and filtering surgery, and the proportion that require cataract extraction, is expected to rise, with increased follow-up duration. Based on the favorable efficacy results, and manageable safety concerns observed in the multicenter trial described above, and similar results obtained in a second multicenter pivotal trial performed in the United States, Canada, and Asia, in April, 2005, the United States Food and Drug Administration granted approval to market the implant, called Retisent (Bausch and Lomb Inc., Rochester, NY) for the treatment of severe posterior uveitis. As of this writing, the fluocinolone implant represents the only FDA-approved intraocular sustained drug delivery implant to treat severe uveitis.

CYCLOSPORINE DEVICES

While corticosteroids are often effective in treating uveitis when administered systemically and locally, an alternative treatment may be required in eyes with a history of steroid-induced glaucoma or in a case of uveitis refractory to steroid treatment. Cyclosporin A (CsA) is a naturally occurring hydrophobic macrolide produced by soil fungi and is a potent immunosuppressive agent that selectively suppresses T-cell activation by inhibiting the phosphatase action of calcineurin thereby suppressing transcription of interleukin 2 and other early Phase T-cell activation genes (26,27). This cytokine inhibition prevents clonal expansion of helper and cytotoxic T-cells. CsA also prevents inflammatory cell chemotaxis, particularly eosinophils. Peak CsA blood levels are reached six hours after ingestion, the drug is metabolized by the cytochrome P450 microsomal system in the liver and it is concentrated in the liver.

CsA penetrates the eye poorly when given as a topical agent (28). Accordingly, topical CsA is not generally given to control intraocular inflammation. In contrast, moderate intraocular CsA levels are achieved with oral administration and systemic administration of CsA alone effectively suppresses intraocular inflammation in two uncontrolled case series of pediatric and adult patients with various uveitic diseases (29,30). A randomized controlled trial showed that CsA was as effective as corticosteroids in suppressing intraocular inflammation and an enhanced anti-inflammatory effect was noticed when CsA was used in conjunction with corticosteroids (31).

Side effects of systemic CsA may include oral ulceration and gingivitis, hypertrichosis, malaise, headaches, muscle cramps, and gastrointestinal disturbance. Serious side effects such as nephrotoxicity and hypertension may be treatment-limiting (32). Prolonged use of cyclosporine and, thus, good patient compliance with the treatment is required in order to adequately control intraocular inflammation.

Nondegradable Cyclosporine Device

Direct intraocular injection of cyclosporine has been shown to control intraocular inflammation in an animal model of uveitis (33). However, the half-life of intravitreal cyclosporine is short which limits its effectiveness in a chronic disease such as uveitis (34). A small Phase I pilot study was undertaken by the National Eye Institute in 1998 to evaluate the safety and effectiveness of cyclosporine, delivered directly into the eye, using a sustained-release device. Patients with uveitis with active inflammation and poor vision were eligible to participate in this study and patients were randomly assigned to one of two treatment groups. One group received a 1-mg implant that releases 0.8 µg of drug per day while the second group received a 2-mg implant that delivers 1.4 µg of drug a day. Baseline examination included a complete eye examination and fluorescein angiography. Electroretinography (ERG) was also performed in view of reversible ERG changes seen in rabbits but not primates (34). Patients were reviewed one and two weeks after surgery, then once a month for six months, and then every three months until the implant was either depleted of drug or removed. This study is ongoing and results are not yet available.

CONCLUSIONS

Uveitis is a potentially sight-threatening disease which has traditionally been treated with topical corticosteroids supplemented by periocular, intravitreal, and systemic corticosteroids in more severe cases. Chronic use of local and systemic corticosteroids can result in significant side effects some of which are life threatening. While newer immunosuppressive agents can reduce the amount of systemic steroids needed to control sight-threatening uveitis, most of these agents are poorly absorbed by the eye when applied topically, and have significant associated side effects. The chronic nature of uveitis also requires long-term patient compliance with prescribed medications. Sustained-release devices used to treat uveitis allow local delivery of immunosuppressive agents in adequate concentrations without the systemic side effects and remove the need for patient medication compliance. For selected patients the potential complications associated with the insertion of these devices have not outweighed these advantages to date. A range of anti-inflammatory medications some of which have been used systemically to treat uveitis and some which have not could be used in sustained-release devices in the future. Other future horizons could include devices containing complementary drugs and responsive sustained-release delivery systems.

REFERENCES

1. Stern AL, Taylor DM, Dalbung CA, et al. Pseudophakic cystoid maculopathy: a study of 50 cases. Ophthalmology 1981; 88:942–946.
2. Leopold IH, Gaster RU. Ocular inflammation and anti-inflammatory drugs. In: Kaufmann HE, Barron BA, McDonald MB, Waltman SR, eds. The Cornea. New York: Church Livingstone, 1987:70.
3. Nussenblatt RB, Palestine AG. Uveitis, Fundamentals and Clinical Practice. Chicago: Yearbook Medical, 1989:104–116.
4. Hayes RC Jr, Murad F. Adrenocorticotropic hormone: adrenocortical steroids and their synthetic analogs, inhibitors of adrenocortical steroid biosynthesis. In: Gilman AG, Tall TW, Nies AS, Taylor P, eds. Goodman and Gilman's the pharmacological basis of therapeutics. New York: Pergamon Press, 1990:1431–1462.
5. Graham RO, Peyman GA. Intravitreal injection of dexamethasone: treatment of experimentally induced endophthalmitis. Arch Ophthalmol 1974; 92:149–154.
6. Peyman GA, Herbst R. Bacterial endophthalmitis: treatment with intraocular injection of gentamicin and dexamethasone. Arch Ophthalmol 1974; 91:416–418.
7. Maxwell DP Jr, Brent BD, Diamond JG, et al. Effect of intravitreal dexamethasone on ocular histopathology in a rabbit model of endophthalmitis. Ophthalmology 1991; 98:1370–1375.
8. Keller N, Jamieson L, Olejnik O, et al. Efficacy of intravitreal dexamethasone in experimental uveitis: effect of drug delivery in a biodegradable polymer. Invest Ophthalmol Vis Sci 1986; 27:248 [ARVO abstract].
9. Kwak HW, D'Amico DJ. Evaluation of the retinal toxicity and pharmacokinetics of dexamethasone after intravitreal injection. Arch Ophthalmol 1992; 110:259–266.
10. Beer PM, Bakri SJ, Singh RJ, Liu W, Peters GB, Miller M. Intraocular concentration and pharmacokinetics of triamcinolone acetonide after a single intravitreal injection. Ophthalmology 2003; 110:681–686.
11. Jonas JB. Intraocular availability of triamcinolone acetonide after intravitreal injection. Am J Ophthalmol 2004; 137:560–562.
12. Antcliff RJ, Spalton DJ, Stanford MR, Graham EM, Ffytche TJ, Marshall J. Intravitreal triamcinolone for uveitic cystoid macular edema: an optical coherence tomography study. Ophthalmology 2001; 108:765–772.
13. Martidis A, Duker JS, Puliafito CA. Intravitreal triamcinolone for refractory cystoid macular edema secondary to birdshot retinochoroidopathy. Arch Ophthalmol 2001; 119:1380–1383.
14. Young S, Larkin G, Branley M, Lightman S. Safety and efficacy of intravitreal triamcinolone for cystoid macular oedema in uveitis. Clin Experiment Ophthalmol 2001; 29:2–6.
15. Degenring RF, Jonas JB. Intravitreal injection of triamcinolone acetonide as treatment for chronic uveitis. Br J Ophthalmol 2003; 87:361.
16. Jonas JB. Intravitreal triamcinolone acetonide for treatment of sympathetic ophthalmia. Am J Ophthalmol 2004; 137:367–368.
17. Tan DT, Chee SP, Lim L, Lim AS. Randomized clinical trial of a new dexamethasone delivery system (Surodex) for treatment of post-cataract surgery inflammation. Ophthalmology 1999; 106:223–231.
18. Chang DF, Garcia IH, Hunkeler JD, Minas T. Phase II results of an intraocular steroid delivery system for cataract surgery. Ophthalmology 1999; 106:1172–1177.
19. Tan DT, Chee SP, Lim L, Theng J, Van Ede M. Randomized clinical trial of Surodex steroid drug delivery system for cataract surgery: anterior versus posterior placement of two Surodex in the eye. Ophthalmology 2001; 108:2172–2181.
20. Cheng C, Berger AS, Pearson PA, et al. Intravitreal sustained-release dexamethasone device in the treatment of experimental uveitis. Invest Ophthalmol Vis Sci 1995; 36:442–453.

21. Jaffe GJ, Pearson PA, Ashton P. Dexamethasone sustained drug delivery implant for the treatment of severe uveitis. Retina 2000; 20:402–403.
22. Diamond JG, Kaplan HJ. Uveitis: effect of vitrectomy combined with lensectomy. Ophthalmology 1978; 86:1320–1329.
23. Jaffe GJ, Yang CH, Guo H, Denny JP, Lima C, Ashton P. Safety and pharmacokinetics of an intraocular fluocinolone acetonide sustained delivery device. Invest Ophthalmol Vis Sci 2000; 41:3569–3575.
24. Jaffe GJ, Ben-Nun J, Guo H, Dunn JP, Ashton P. Fluocinolone acetonide sustained drug delivery device to treat severe uveitis. Ophthalmology 2000; 107:2024–2033.
25. Jaffe GJ, McCallum RM, Branchaud B, Skalak C, Butuner Z, Ashton P. Long-term follow-up results of a pilot trial of a fluocinolone acetonide implant to treat posterior uveitis. Ophthalmology 2005; 112:1192–1198.
26. Liu J. FK506 and cyclosporin: molecular probes for studying intracellular signal transduction. Trends Pharmacol Sci 1993; 14:182–188.
27. Manez R, Jain A, Marino IR, et al. Comparative evaluation of tacrolimus (FK506) and cyclosporin A as immunosuppressive agents. Transpl Rev 1995; 9:63.
28. BenEzra D, Maftzir G, deCourten C, Timonen P. Ocular penetration of cyclosporine A. III: The human eye. Br J Ophthalmol 1992; 76:350–352.
29. Nussenblatt RB, Palestine AG, Chan CC. Cyclosporin A therapy in the treatment of intraocular inflammatory disease resistant to systemic corticosteroids and cytotoxic agents. Am J Ophthalmol 1983; 96:275–282.
30. Walton RC, Nussenblatt RB, Whitcup SM. Cyclosporine therapy for severe sight-threatening uveitis in children and adolescents. Ophthalmology 1998; 105:2028–2034.
31. Nussenblatt RB, Palestine AG, Chan CC, et al. Randomized, double-masked study of cyclosporine compared to prednisolone in the treatment of endogenous uveitis. Am J Ophthalmol 1991; 112:138–146.
32. Deray G, Benhmida M, Le Hoang P, et al. Renal function and blood pressure in patients receiving long-term, low-dose cyclosporine therapy for idiopathic autoimmune uveitis. Ann Intern Med 1992; 117:57–83.
33. Nussenblatt RB, Dinning WJ, Fujikawa LS, Chan CC, Palestine AG. Local cyclosporine therapy for experimental autoimmune uveitis in rats. Arch Ophthalmol 1985; 103: 1559–1562.
34. Pearson PA, Jaffe GJ, Martin DF, et al. Evaluation of a delivery system providing long-term release of cyclosporine. Arch Ophthalmol 1996; 114:311–317.

18

Drug Delivery for Proliferative Vitreoretinopathy: Prevention and Treatment

Stephen J. Phillips and Glenn J. Jaffe
Duke University Eye Center, Durham, North Carolina, U.S.A.

RETINAL DETACHMENT/PROLIFERATIVE VITREORETINOPATHY

Despite improvements in surgical technique, proliferative vitreoretinopathy (PVR) remains a common and significant vision-threatening complication of retinal detachment repair and trauma. A clinically significant form of PVR occurs following approximately 5–10% of all rhegmatogenous detachment repairs (1–6). In these eyes, PVR typically develops approximately six weeks after the initial surgical repair. In eyes with established severe PVR at the time of surgery, postoperative reproliferation and redetachment occurs in up to 55% of cases (7).

After retinal detachment or surgical repair, breakdown of the blood–retinal barrier occurs allowing serum components access to the vitreous cavity (8–10). Cells migrate and proliferate on the surface and undersurface of detached retina (Fig. 1) (11). Retinal pigment epithelial (RPE) cells are the key cell type in these proliferative membranes. Other cell types, including glial cells, cells resembling fibroblasts, monocytes, and T-lymphocytes have also been observed (12–16). Extracellular matrix, produced by some of these cells comprises a large portion of the membrane (17). Cell-mediated contraction within this fibrocellular membrane may lead to traction retinal detachment, may cause new tears or reopen old tears and result in loss of vision (Fig. 2).

Cytokines play a critical role in the migration and proliferation of cells, the production of extracellular matrix, and the ultimate development of proliferative membranes. Some of the cytokines that have been implicated in PVR development include platelet-derived growth factor, acidic and basic fibroblast growth factor, epithelial growth factor, interleukin-1, tumor necrosis factor-alpha, transforming growth factor-beta (TGF-β), macrophage colony-stimulating factor and macrophage chemotactic protein-1 (MCP-1) (15,18–23).

Surgery is the primary form of therapy used to manage eyes with retinal detachment and PVR. Epiretinal and occasionally subretinal membranes (Fig. 3) are removed to release retinal traction, followed by reattachment and tamponade with either gas or silicone oil. In a multicenter randomized clinical trial of 340 eyes with

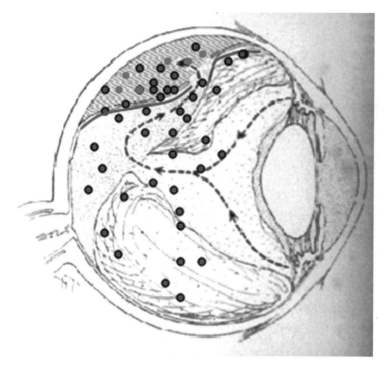

Figure 1 Migrating and proliferating cells in the subretinal space, on the retinal surface and undersurface, and in the vitreous cavity following rhegmatogenous retinal detachment.

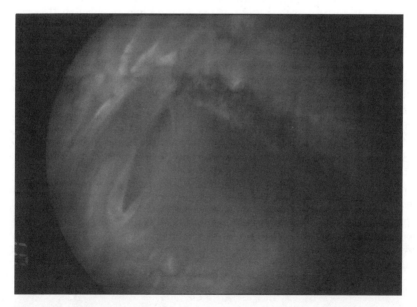

Figure 2 PVR with traction and retinal tear with rolled edges adjacent to cryopexy scar. *Abbreviation*: PVR, proliferative vitreoretinopathy.

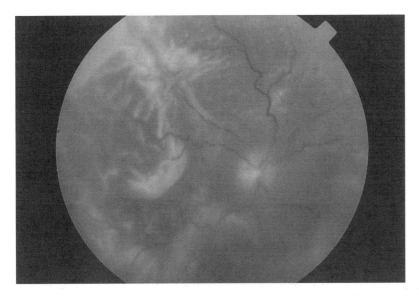

Figure 3 Retinal detachment with epiretinal membranes creating fixed retinal folds.

rhegmatogenous retinal detachment and severe PVR comparing silicone oil with C3F8 tamponade (Silicone Study Report), there was a final macular attachment rate of 77% for cases managed with silicone oil and 79% for those receiving C3F8 gas tamponade (24,25). In a separate study, a final macular attachment rate of 90% for initial PVR surgery and 86% for repeat surgery was reported (26,27). However, visual results remain disappointing. In the Silicone Study Report, only 25% achieved 20/200 acuity or better. In the subsequent study, visual acuity of 20/100 or better was achieved in only 19% of patients after initial surgery and 11% after repeat surgery.

To minimize PVR, care is taken to ensure that all retinal breaks are closed, cryotherapy is minimized to reduce RPE cell dispersion, and intraoperative bleeding is avoided. The risk for PVR is highest in cases of giant retinal tear (16–41%) and in penetrating ocular trauma (10–45%) (28). These cases are usually excluded from series examining PVR incidence following primary retinal detachment repair. Certain features have been consistently reported to be predictive of PVR following surgical repair of primary rhegmatogenous retinal detachment. Those features include anterior uveitis, aphakia, preoperative PVR, preoperative cryotherapy, detachment size, and vitreous hemorrhage (29). Identifying those patients at highest risk for development of PVR has been a critical step in the development and testing of potential prevention strategies. Asaria et al. (29) prospectively tested the accuracy of a risk formula in determining which patients are likely to develop postoperative PVR. Based on the presence or absence of known risk factors for PVR, patients were stratified into either a high-risk or a low-risk group. Of the 130 patients in the low-risk group, 12 (9.2%) developed postoperative PVR. Of the 82 patients in the high-risk group, 23 (28%) developed PVR. The difference was statistically significant. Given the relatively high surgical success rate of rhegmatogenous detachment primary repair, some studies that have failed to identify a treatment effect in prevention of PVR, may have lacked statistical power. Therefore, future studies of prevention will need to focus on enrolling populations at risk for the development of PVR.

While surgical techniques and instruments have improved and anatomic success has remained fairly high, functional improvement has been limited. There is hope that by addressing the underlying disease at the cellular level, need for reoperation will decrease and functional results will improve. Therefore, recent efforts have focused on finding adjuncts to surgical treatment of PVR.

The ideal drug for treating patients at high risk for PVR, or for use as an adjunct to surgery in established PVR would be one that achieves therapeutic levels in the eye for weeks to months, one that addresses a variety of mechanisms of PVR development, that can be delivered with minimal added surgical risk, and that has few to no ocular or systemic side effects.

To date, only a few therapeutic agents have been tested in human clinical trials. Methods of delivery have included systemic (oral), direct injection into the vitreous or subconjunctival space, and as an additive to the infusion fluid at the time of surgery.

SYSTEMIC

While direct delivery of drug affords greater tissue concentration and has been the focus of recent efforts, systemic delivery has the advantage of allowing daily dosing, providing constant levels of drug for extended duration.

Prednisone

Corticosteroids have been investigated as a potential treatment for PVR. It has been hypothesized that they might directly inhibit cellular proliferation and suppress inflammation, and thereby prevent epiretinal and subretinal membrane formation (30). The safety and efficacy of corticosteroids to treat PVR have been demonstrated in animal studies (30–33). Systemic corticosteroids have been used to treat a variety of ocular and nonocular conditions in humans. Systemic side effects of corticosteroids are well-documented and include, among others, cushingoid changes, osteoporosis, elevated serum glucose, hypertension, peptic ulcer disease, and psychiatric disturbance.

Prednisone was given in the first human trial of systemic corticosteroid treatment for PVR (34). A total of 141 patients were randomized to receive either prednisone (100 mg for 5 days, 50 mg for 10 days, and 50 mg every other day for 40 days) or placebo. At the six-month follow-up visit, an examination was done to detect any evidence of PVR. The results suggested an inhibitory effect of corticosteroids on postoperative retinal fibrosis, especially for subtle signs in the posterior pole. There was no difference between the two groups in the incidence of advanced retinal fibrosis, or in the presence of peripheral retinal fibrosis. The authors concluded that higher tissue concentrations and a larger number of patients need to be evaluated before a definitive statement can be made regarding the efficacy of corticosteroids in preventing the development of PVR.

Retinoic Acid

Retinoids, a group of compounds related to vitamin A which have profound effects on DNA transcription, have an inhibitory effect on cellular proliferation and have been implicated in cellular differentiation (35–37). It has been suggested that depletion of retinoic acid (RA) secondary to retinal detachment may result in morphologic and proliferative changes in RPE that ultimately lead to PVR (38). RA also markedly reduces the production of TGF-β. In eyes of patients with PVR, TGF-β is increased (39). It has been hypothesized that decreased levels of TGF-β, mediated by delivery of RA acid, would be beneficial in the treatment of PVR (40).

The efficacy of RA in PVR has been evaluated in animal models. A single intravitreal injection of RA used in conjunction with silicone oil reduces the incidence of traction retinal detachment (40). Use of microsphere encapsulation of RA to prolong the intravitreal half-life reduced the incidence of PVR in the rabbit model by 64% (41). Intravitreal RA suspended in 1% sodium hyaluronate has also been studied in a rabbit model of PVR and has similar inhibitory effect on PVR progression. The authors suggested that in cases in which silicone oil was not necessary, sodium hyaluronate could be used as an RA vehicle, though transient elevations of intraocular pressure could be expected (42).

Retinoic acid, available as an oral commercial preparation for the treatment of severe cystic acne (Accutane) has been one of the great successes in dermatology. However, it is not without side effects, some of which are severe. Oral retinoids are potent teratogens, and approximately one-fourth of all exposed fetuses develop birth defects. Their teratogenic potential has led to strict guidelines governing their use in women of child-bearing potential. Other side effects, such as elevated liver function tests, elevated serum lipids, and nyctalopia have also limited their use (43).

Retinoic acid has been studied in a limited retrospective human trial of PVR. Ten patients who received 40 mg oral 13-cis RA (Accutane) twice daily for four weeks following surgery for PVR detachment were compared with 10 case-matched controls. Though the sample size was small, the redetachment rate of the control group (60%) was greater than that of the study group (10%) (44). Based on the results of this pilot study, a clinical trial was planned, but has not yet been performed. To date, no human trials evaluating local delivery of RA have been reported.

LOCAL DELIVERY

While systemic delivery of medications does not have associated surgical risks, it may be difficult to achieve adequate intraocular penetration without causing systemic side effects. To minimize systemic toxicity and increase retinal tissue drug levels, local delivery methods to treat and prevent PVR have been investigated.

DIRECT INJECTION—SUBCONJUNCTIVAL OR INTRAVITREAL

5-Fluorouracil

5-Fluorouracil (5-FU) was one of the first drugs to be studied in human trials of PVR. 5-FU, a pyrimidine analog that inhibits thymidilate synthetase, and therefore DNA synthesis, has a greater effect on proliferating cells than it does on resting cells. It has been used as an adjuvant in the treatment of breast, pancreas and abdominal neoplasms, and has been used as a topical agent to treat dermatologic disorders, including premalignant keratoses, basal cell carcinomas and viral warts. Parenteral administration has been associated with bone marrow suppression and mucosal alteration. While 5-FU has significant toxic effects if given systemically, local delivery appears to be relatively nontoxic (45,46). 5-FU has been studied in animal models, and has also been studied in humans as adjunct to surgery for retinal detachments with advanced PVR.

Blumenkranz et al. (47) studied the effects of direct 5-FU injection in eyes of 22 consecutive patients undergoing retinal detachment repair for advanced PVR. Eighteen patients had vitrectomy combined with scleral buckle and either air or SF6 gas injection. Three had scleral buckle repair only and one that had previously undergone vitrectomy had fluid–gas exchange alone.

Twelve patients received at least five consecutive 5-FU injections (10 mg per injection) beginning on the day of surgery (average 6.4 injections). Fourteen patients received intravitreal 5-FU injections (1 mg per injection) on one or more occasions (average 1.7 injections). To avoid concentrating drug in the aqueous phase surrounding the intraocular gas used as a tamponade, intravitreal 5-FU was not given at the time of surgery. Instead, patients were followed clinically until the intraocular gas bubble had decreased in size to 50% or less before intravitreal injections were given. Four patients had both intravitreal and subconjunctival injection.

No serious systemic or ocular complications were observed, and retinal reattachment was achieved in 60% of patients postoperatively. The number of patients studied was small. However, the authors believed this therapy resulted in an improved success rate compared with other studies of the reattachment rate following surgery for advanced PVR without use of silicone oil and that the reattachment rate was comparable with the reattachment rate achieved in studies using silicone oil tamponade (47).

Corticosteroids

Dexamethasone alcohol and triamcinolone acetonide have been studied in PVR animal models (30–33). These agents are particularly suited for local delivery as they are both relatively lipophilic, and therefore, may be administered as a suspension. The crystalline drug then acts as a depot, providing relatively long-term intraocular levels of steroid that can be given at high doses without apparent retinal toxicity (33). While direct delivery of corticosteroids has been demonstrated to be both safe and effective in the treatment of PVR, few studies have been performed in humans (31,32).

Jonas et al. (48) examined postoperative inflammation in 16 patients in the first two weeks following vitrectomy to repair rhegmatogenous retinal detachments complicated by PVR. All 16 patients received silicone oil tamponade followed by direct injection of 10–20 mg crystalline triamcinolone acetonide into the silicone oil bubble through closed sclerotomies. At the end of the two-month follow-up period, three of the 16 patients had retinal redetachment. The study did not look at visual outcomes or at anatomic success rates beyond the first two months. The authors believe that those patients receiving intravitreal triamcinolone acetonide had less intraocular inflammation as estimated by slit-lamp biomicroscopy, and clearer fundus appearance in the first two weeks after surgery compared with a similar group of historical controls (48). However, it is difficult to make definitive conclusions regarding corticosteroid efficacy because of the small patient number, the lack of controls, and the short follow-up duration. To date, no human trial has established the efficacy of intravitreal triamcinolone injection to treat PVR.

Infusate

Drugs have been placed in the intraocular infusate during vitrectomy surgery to repair rhegmatogenous retinal detachment with PVR. This method can produce local delivery of a relatively high drug dose, avoid systemic complications, and minimize additional surgical risk.

Daunomycin

Daunorubicin hydrochloride (daunomycin hydrochloride) is an anthracycline antibiotic that inhibits cellular proliferation by a variety of mechanisms, including DNA binding, free radical formation, membrane binding, and metal-ion chelation (49). A number of studies examining efficacy and toxicity in an animal model of PVR have been performed (50–56). Prior to 1998, experience with daunorubicin in human trials was more limited (57–59).

The efficacy and safety of daunorubicin in the infusion fluid was examined in a multicenter, prospective, randomized, controlled human clinical trial of 286 eyes undergoing vitrectomy for stage C2 (Retina Society Classification, 1983) or more advanced PVR. An encircling scleral buckle was placed, followed by a three-port pars plana vitrectomy. Preretinal and subretinal membranes were removed, and a relaxing retinotomy was performed if necessary. Eyes were then randomized to drug treatment or control. Drug-treated eyes received a continuous 10-minute infusion with daunorubicin (7.5 μg/mL in balanced saline solution). After the 10-minute infusion period, the surgery was completed using air or perfluorocarbon reattachment of the retina, photocoagulation and silicone oil exchange. In the control group, surgery was completed similarly, but without a 10-minute waiting period.

Six months after surgery, there was no statistically significant difference between the rate of complete retinal reattachment in the daunorubicin group (62.7%) and the control group (54.1%). There was also no significant difference in best-corrected visual acuity between the two groups. The overall one-year reattachment rate was the same for the two groups (80.2% vs. 81.8%). However, in the control group, 65 of 135 patients (48%) had at least one vitreoretinal reoperation, while in the daunorubicin group, only 50 of 142 patients (35%) required at least one reoperation. The difference was statistically significant. The authors point toward the significant reduction in the number of operations as evidence that PVR is amenable to pharmacologic treatment, but do not advocate the use of daunorubicin in similar PVR cases based on the results of this study (60).

Heparin and 5-Fluorouracil

Low-molecular-weight heparin (LMWH) has been shown to reduce fibrin formation after vitrectomy (61–63). Fibrin has the potential to serve as a scaffold for attachment and proliferation of RPE with subsequent membrane formation (64). Treatment with heparin can prevent fibrin formation but has the potential to increase intraoperative bleeding (62,65). Because LMWH has been shown to produce less hemorrhage for an equivalent antithrombotic effect, a number of investigators have chosen to use it instead of heparin in the infusion (66).

As mentioned previously, 5-FU injections have been shown to improve reattachment rates after surgery for PVR (47). The combination of heparin and 5-FU has the potential to decrease fibrin formation and proliferation. In a large randomized, placebo-controlled trial, Asaria et al. (29) examined the efficacy of 5-FU and LMWH in the infusate to prevent the postoperative development of PVR in patients at high risk. Asaria et al. used a statistical model to predict which eyes were at highest risk for development of PVR after primary vitrectomy repair of rhegmatogenous retinal detachment to identify those patients most likely to benefit from PVR prevention treatment. This model has been used to select patients at highest risk for PVR in order to study the efficacy of interventions aimed at PVR prevention. The treatment group received 5-FU in the infusion bag at a concentration of 200 μg/mL and LMWH at a concentration of 5 IU/mL. After a follow-up period of six months, the rate of postoperative PVR in the placebo group was 26.4% (23/87) and the rate of PVR in the treatment group was 12.6% (11/87). While the reattachment rates with a single operation were not statistically significant between the two groups, the final visual acuity was significantly better in the treatment group than in the control group. The difference in outcome between the two groups led the authors to advocate the use of this drug combination in the infusate in all patients at increased risk for developing postoperative PVR (67).

While the 5-FU, LMWH combination was effective in preventing PVR in high-risk patients, the use of 5-FU and LMWH in the infusate has not been proven to improve postoperative results in patients with established PVR (David Charteris unpublished results).

Dexamethasone and Heparin

The combination of heparin and dexamethasone in the infusate as treatment has been studied in a group of patients with established severe PVR undergoing vitrectomy. In a randomized trial of 59 eyes, 34 eyes received placebo and 25 received heparin (1 U/mL) and dexamethasone (4 µg/mL). After follow-up of at least six months, visual acuities in the two groups were comparable. Recurrent detachment requiring additional surgery developed in 26.5% of the control eyes and 16% of the treated eyes. The differences between the two groups were not statistically significant as the study was underpowered, but a trend to improved reattachment rates with treatment was suggested (68).

SUSTAINED DELIVERY AND CO-DRUGS

While for some intraocular disorders, such as uveitis or CMV retinitis, long-term (months to years) delivery may be necessary, for PVR, therapeutic intraocular drug levels that are present for weeks may be all that is needed. Nonetheless, a number of drugs with proven antiproliferative effect are limited in their usefulness because their short half-life makes repeat postoperative injections necessary. 5-FU is an example of a drug that has limited clinical effectiveness because of its pharmacodynamics. In aphakic, vitrectomized rabbit eyes, therapeutic levels are only maintained for 12–24 hours (69). Extrapolating to human eyes postvitrectomy, intraocular therapeutic levels are unlikely to be maintained through the period when proliferation is most likely to occur. Repeated injections are inconvenient to the patient and may increase the risk of endophthalmitis and retinal detachment. 5-FU delivered to the vitreous cavity in a lactide and glycolide copolymer, however, has the potential for achieving sustained intravitreal concentrations of drug. In an animal model of PVR, this delivery method of 1 mg of 5-FU allowed levels of drug between 1 and 13 mg/L to be maintained for at least 14 days. Animals receiving the 5-FU implant also had a lower rate of retinal detachment (70).

While corticosteroids, proven to reduce the incidence of retinal detachment in a rabbit PVR model, have some limited antiproliferative properties, they may play a more important role in stabilizing the blood–retinal barrier and preventing the access of inflammatory mediators (30,71,72). A combination of corticosteroid and 5-FU has the potential advantage of addressing both the inflammation and the proliferation associated with PVR formation.

Formation of conjugated triamcinolone–5-FU in suspension allows sustained release of both drugs. The efficacy and safety of this combination has been tested in a rabbit PVR model. The two drugs, covalently linked through the carbonate bond makes the drugs relatively insoluble and allows release over a longer period. Delivery of co-drugs, dexamethasone–5-FU as an implantable, sustained-release pellet, and triamcinolone–5-FU as injectable intravitreal sustained-release suspension, each have been effective in inhibiting the progression of PVR in an animal model (73). To date, human studies with this agent have not been conducted.

The challenges in treating PVR lie not only in identifying drugs that are safe and effective but in developing delivery methods that allow for sustained therapeutic levels over several months. The development of effective strategies for preventing PVR is also dependent on identifying and treating those patients who are at the highest risk of development of PVR.

REFERENCES

1. Speicher MA, Fu AD, Martin JP, von Fricken MA. Primary vitrectomy alone for repair of retinal detachments following cataract surgery. Retina 2000; 20:459–464.
2. Greven CM, Sanders RJ, Brown GC, et al. Pseudophakic retinal detachments. Anatomic and visual results. Ophthalmology 1992; 99:257–262.
3. Cowley M, Conway BP, Campochiaro PA, Kaiser D, Gaskin H. Clinical risk factors for proliferative vitreoretinopathy. Arch Ophthalmol 1989; 107:1147–1151.
4. Girard P, Mimoun G, Karpouzas I, Montefiore G. Clinical risk factors for proliferative vitreoretinopathy after retinal detachment surgery. Retina 1994; 14:417–424.
5. Bonnet M, Fleury J, Guenoun S, Yaniali A, Dumas C, Hajjar C. Cryopexy in primary rhegmatogenous retinal detachment: a risk factor for postoperative proliferative vitreoretinopathy? Graefes Arch Clin Exp Ophthalmol 1996; 234:739–743.
6. Heimann H, Bornfeld N, Friedrichs W, et al. Primary vitrectomy without scleral buckling for rhegmatogenous retinal detachment [comment]. Graefes Arch Clin Exp Ophthalmol 1996; 234:561–568.
7. Anonymous. Proliferative vitreoretinopathy. The Silicone Study Group. Am J Ophthalmol 1985; 99:593–595.
8. Ando N, Sen HA, Berkowitz BA, Wilson CA, de Juan E Jr. Localization and quantitation of blood–retinal barrier breakdown in experimental proliferative vitreoretinopathy [erratum appears in Arch Ophthalmol 1994; 112(6):789]. Arch Ophthalmol 1994; 112: 117–122.
9. Clausen R, Weller M, Wiedemann P, Heimann K, Hilgers RD, Zilles K. An immunochemical quantitative analysis of the protein pattern in physiologic and pathologic vitreous. Graefes Arch Clin Exp Ophthalmol 1991; 229:186–190.
10. Grisanti S, Wiedemann P, Heimann K. Proliferative vitreoretinopathy. On the significance of protein transfer through the blood–retina barrier. Ophthalmology 1993; 90:468–471.
11. Glaser BM, Lemor M. Pathobiology of Proliferative Vitreoretinopathy. In: Ryan SJ, ed. Retina: Mosby, St. Louis, 1988:369.
12. Jerdan JA, Pepose JS, Michels RG, et al. Proliferative vitreoretinopathy membranes. An Immunohistochemical Study. Ophthalmology 1989; 96:801–810.
13. Charteris DG, Hiscott P, Robey HL, Gregor ZJ, Lightman SL, Grierson I. Inflammatory cells in proliferative vitreoretinopathy subretinal membranes. Ophthalmology 1993; 100: 43–46.
14. Vinores SA, Campochiaro PA, Conway BP. Ultrastructural and electron-immunocytochemical characterization of cells in epiretinal membranes. Invest Ophthalmol Vis Sci 1990; 31:14–28.
15. Baudouin C, Fredj-Reygrobellet D, Brignole F, Negre F, Lapalus P, Gastaud P. Growth factors in vitreous and subretinal fluid cells from patients with proliferative vitreoretinopathy. Ophthalmic Res 1993; 25:52–59.

16. Nicolai U, Eckardt C. The occurrence of macrophages in the retina and periretinal tissues in ocular diseases. German J Ophthalmol 1993; 2:195–201.

17. Machemer R. Massive periretinal proliferation: a logical approach to therapy. Trans Am Ophthalmol Soc 1977; 75:556–586.

18. Elner S, Strieter R, Elner V, Rollins B, Monte MD, Kunkel S. Monocyte chemotactic protein gene expression by cytokine-treated human retinal pigment epithelial cells. Lab Invest 1991; 64:819.

19. Fredj-Reygrobellet D, Baudouin C, Negre F, Caruelle JP, Gastaud P, Lapalus P. Acidic FGF and other growth factors in preretinal membranes from patients with diabetic retinopathy and proliferative vitreoretinopathy. Ophthalmic Res 1991; 23:154–161.

20. Malecaze F, Mathis A, Arne JL, Raulais D, Courtois Y, Hicks D. Localization of acidic fibroblast growth factor in proliferative vitreoretinopathy membranes. Curr Eye Res 1991; 10:719–729.

21. Planck SR, Andresevic J, Chen JC, et al. Expression of growth factor mRNA in rabbit PVR model systems. Curr Eye Res 1992; 11:1031–1039.

22. Robbins SG, Mixon RN, Wilson DJ, et al. Platelet-derived growth factor ligands and receptors immunolocalized in proliferative retinal diseases [erratum appears in Invest Ophthalmol Vis Sci 1995; 36(3):519]. Invest Ophthalmol Vis Sci 1994; 35:3649–3663.

23. Westra I, Robbins SG, Wilson DJ, et al. Time course of growth factor staining in a rabbit model of traumatic tractional retinal detachment. Graefes Arch Clin Exp Ophthalmol 1995; 233:573–581.

24. McCuen BW II, Azen SP, Stern W, et al. Vitrectomy with silicone oil or perfluoropropane gas in eyes with severe proliferative vitreoretinopathy. Silicone Study Report 3. Retina 1993; 13:279–284.

25. Anonymous. Vitrectomy with silicone oil or sulfur hexafluoride gas in eyes with severe proliferative vitreoretinopathy: results of a randomized clinical trial. Silicone Study Report 1 [comment]. Arch Ophthalmol 1992; 110:770–779.

26. Lewis H, Aaberg TM. Causes of failure after repeat vitreoretinal surgery for recurrent proliferative vitreoretinopathy. Am J Ophthalmol 1991; 111:15–19.

27. Lewis H, Aaberg TM, Abrams GW. Causes of failure after initial vitreoretinal surgery for severe proliferative vitreoretinopathy. Am J Ophthalmol 1991; 111:8–14.

28. Charteris DG, Sethi CS, Lewis GP, Fisher SK. Proliferative vitreoretinopathy—developments in adjunctive treatment and retinal pathology. Eye 2002; 16:369–374.

29. Asaria RH, Kon CH, Bunce C, et al. How to predict proliferative vitreoretinopathy: a prospective study [comment]. Ophthalmology 2001; 108:1184–1186.

30. Chandler DB, Hida T, Sheta S, Proia AD, Machemer R. Improvement in efficacy of corticosteroid therapy in an animal model of proliferative vitreoretinopathy by pretreatment. Graefes Arch Clin Exp Ophthalmol 1987; 225:259–265.

31. Tano Y, Chandler D, Machemer R. Treatment of intraocular proliferation with intravitreal injection of triamcinolone acetonide. Am J Ophthalmol 1980; 90:810–816.

32. Tano Y, Sugita G, Abrams G, Machemer R. Inhibition of intraocular proliferations with intravitreal corticosteroids. Am J Ophthalmol 1980; 89:131–136.

33. McCuen BW II, Bessler M, Tano Y, Chandler D, Machemer R. The lack of toxicity of intravitreally administered triamcinolone acetonide. Am J Ophthalmol 1981; 91:785–788.

34. Koerner F, Merz A, Gloor B, Wagner E. Postoperative retinal fibrosis—a controlled clinical study of systemic steroid therapy. Graefes Arch Clin Exp Ophthalmol 1982; 219: 268–271.

35. Lotan R, Nicolson GL. Inhibitory effects of retinoic acid or retinyl acetate on the growth of untransformed, transformed, and tumor cells in vitro. J Natl Cancer Inst 1977; 59: 1717–1722.

36. Strickland S, Mahdavi V. The induction of differentiation in teratocarcinoma stem cells by retinoic acid. Cell 1978; 15:393–403.

37. Jones-Villeneuve EM, Rudnicki MA, Harris JF, McBurney MW. Retinoic acid-induced neural differentiation of embryonal carcinoma cells. Mol Cell Biol 1983; 3:2271–2279.

38. Campochiaro PA, Hackett SF, Conway BP. Retinoic acid promotes density-dependent growth arrest in human retinal pigment epithelial cells. Invest Ophthalmol Vis Sci 1991; 32:65–72.

39. Connor TB Jr, Roberts AB, Sporn MB, et al. Correlation of fibrosis and transforming growth factor-beta type 2 levels in the eye. J Clin Invest 1989; 83:1661–1666.

40. Araiz JJ, Refojo MF, Arroyo MH, Leong FL, Albert DM, Tolentino FI. Antiproliferative effect of retinoic acid in intravitreous silicone oil in an animal model of proliferative vitreoretinopathy. Invest Ophthalmol Vis Sci 1993; 34:522–530.

41. Giordano GG, Refojo MF, Arroyo MH. Sustained delivery of retinoic acid from microspheres of biodegradable polymer in PVR. Invest Ophthalmol Vis Sci 1993; 34:2743–2751.

42. Takahashi M, Refojo MF, Nakagawa M, Veloso A Jr, Leong FL. Antiproliferative effect of retinoic acid in 1% sodium hyaluronate in an animal model of PVR. Curr Eye Res 1997; 16:703–709.

43. Ellis CN, Krach KJ. Uses and complications of isotretinoin therapy [comment]. J Am Acad Dermatol 2001; 45:S150–S157.

44. Fekrat S, de Juan E Jr, Campochiaro PA. The effect of oral 13-cis-retinoic acid on retinal redetachment after surgical repair in eyes with proliferative vitreoretinopathy. Ophthalmology 1995; 102:412–418.

45. Blumenkranz MS, Ophir A, Claflin AJ, Hajek A. Fluorouracil for the treatment of massive periretinal proliferation. Am J Ophthalmol 1982; 94:458–467.

46. Leon JA, Britt JM, Hopp RH, Mills RP, Milam AH. Effects of fluorouracil and fluorouridine on protein synthesis in rabbit retina. Invest Ophthalmol Vis Sci 1990; 31:1709–1716.

47. Blumenkranz M, Hernandez E, Ophir A, Norton EW. 5-fluorouracil: new applications in complicated retinal detachment for an established antimetabolite. Ophthalmology 1984; 91:122–130.

48. Jonas JB, Hayler JK, Panda-Jonas S. Intravitreal injection of crystalline cortisone as adjunctive treatment of proliferative vitreoretinopathy. Br J Ophthalmol 2000; 84:1064–1067.

49. Myers C. Pharmacologic Principles of Cancer Treatment. In: Chabner B, ed. Philadelphia: WB Saunders, 1982:416.

50. Wiedemann P, Kirmani M, Santana M, Sorgente N, Ryan SJ. Control of experimental massive periretinal proliferation by daunomycin: dose–response relation. Graefes Arch Clin Exp Ophthalmol 1983; 220:233–235.

51. Kirmani M, Santana M, Sorgente N, Wiedemann P, Ryan SJ. Antiproliferative drugs in the treatment of experimental proliferative vitreoretinopathy. Retina 1983; 3:269–272.

52. Khawly JA, Saloupis P, Hatchell DL, Machemer R. Daunorubicin treatment in a refined experimental model of proliferative vitreoretinopathy. Graefes Arch Clin Exp Ophthalmol 1991; 229:464–467.

53. Wiedemann P, Sorgente N, Bekhor C, Patterson R, Tran T, Ryan SJ. Daunomycin in the treatment of experimental proliferative vitreoretinopathy. Effective doses in vitro and in vivo. Invest Ophthalmol Vis Sci 1985; 26:719–725.

54. Wiedemann P, Heimann K. Toxicity of intraocular daunomycin. Lens Eye Toxic Res 1990; 7:305–310.

55. Santana M, Wiedemann P, Kirmani M, et al. Daunomycin in the treatment of experimental proliferative vitreoretinopathy: retinal toxicity of intravitreal daunomycin in the rabbit. Graefes Arch Clin Exp Ophthalmol 1984; 221:210–213.

56. Steinhorst UH, Hatchell DL, Chen EP, Machemer R. Ocular toxicity of daunomycin: effects of subdivided doses on the rabbit retina after vitreous gas compression. Graefes Arch Clin Exp Ophthalmol 1993; 231:591–594.

57. Wiedemann P, Leinung C, Hilgers RD, Heimann K. Daunomycin and silicone oil for the treatment of proliferative vitreoretinopathy. Graefes Arch Clin Exp Ophthalmol 1991; 229:150–152.

58. Wiedemann P, Lemmen K, Schmiedl R, Heimann K. Intraocular daunorubicin for the treatment and prophylaxis of traumatic proliferative vitreoretinopathy. Am J Ophthalmol 1987; 104:10–14.

59. Wiedemann P, Evans PY, Wiedemann R, Meyer-Schwickerath R, Heimann K. A fluorescein angiographic study on patients with proliferative vitreoretinopathy treated by vitrectomy and intraocular daunomycin. Int Ophthalmol 1989; 13:211–216.

60. Wiedemann P, Hilgers RD, Bauer P, Heimann K. Adjunctive daunorubicin in the treatment of proliferative vitreoretinopathy: results of a multicenter clinical trial. Daunomycin Study Group. Am J Ophthalmol 1998; 126:550–559.

61. Iverson DA, Katsura H, Hartzer MK, Blumenkranz MS. Inhibition of intraocular fibrin formation following infusion of low-molecular-weight heparin during vitrectomy. Arch Ophthalmol 1991; 109:405–409.

62. Johnson RN, Blankenship G. A prospective, randomized, clinical trial of heparin therapy for postoperative intraocular fibrin. Ophthalmology 1988; 95:312–317.

63. Blumenkranz MS, Hartzer MK, Iverson D. An overview of potential applications of heparin in vitreoretinal surgery. Retina 1992; 12:S71–S74.

64. Vidaurri-Leal JS, Glaser BM. Effect of fibrin on morphologic characteristics of retinal pigment epithelial cells. Arch Ophthalmol 1984; 102:1376–1379.

65. Murray TG, Stern WH, Chin DH, MacGowan-Smith EA. Collagen shield heparin delivery for prevention of postoperative fibrin. Arch Ophthalmol 1990; 108:104–106.

66. Hirsch J, Ofosu F, Levine M. The development of low molecular weight heparins for clinical use. In: Verstrate M, Vermylen H, Arnout J, eds. Thrombosis and Hemostasis. Leuven University Press, Leuven, Netherlands, 1987.

67. Asaria RH, Kon CH, Bunce C, et al. Adjuvant 5-fluorouracil and heparin prevents proliferative vitreoretinopathy: results from a randomized, double-blind, controlled clinical trial [comment]. Ophthalmology 2001; 108:1179–1183.

68. Williams RG, Chang S, Comaratta MR, Simoni G. Does the presence of heparin and dexamethasone in the vitrectomy infusate reduce reproliferation in proliferative vitreoretinopathy? Graefes Arch Clin Exp Ophthalmol 1996; 234:496–503.

69. Jarus G, Blumenkranz M, Hernandez E, Sossi N. Clearance of intravitreal fluorouracil. Normal and aphakic vitrectomized eyes. Ophthalmology 1985; 92:91–96.

70. Rubsamen P, Davis P, Hernandez E, O'Grady G, Cousins S. Prevention of experimental proliferative vitreoretinopathy with a biodegradable intravitreal implant for the sustained release of fluorouracil. Arch Ophthalmol 1994; 112:407–413.

71. Ruhmann A, Berliner D. Effect of steroids on growth of mouse fibroblasts in vitro. Endocrinology 1965; 76:916–927.

72. Stahl JH, Miller DB, Conway BP, Campochiaro PA. Dexamethasone and indomethacin attenuate cryopexy. Induced breakdown of the blood–retinal barrier. Graefes Arch Clin Exp Ophthalmol 1987; 225:418–420.

73. Berger AS, Cheng CK, Pearson PA, et al. Intravitreal sustained release corticosteroid-5-fluorouracil conjugate in the treatment of experimental proliferative vitreoretinopathy. Invest Ophthalmol Vis Sci 1996; 37:2318–2325.

19

Pharmacologic Treatment in Diabetic Macular Edema

Zeshan A. Rana and P. Andrew Pearson
Department of Ophthalmology and Visual Science, Kentucky Clinic, Lexington, Kentucky, U.S.A.

INTRODUCTION

Diabetes mellitus is a group of metabolic diseases characterized by an increased blood glucose level secondary to defects in insulin secretion and/or action. According to the American Diabetes Association, as of 2002, the United States diabetes prevalence is 18.2 million people (6.3% of the population). Diabetics suffer from acute complications of the disease such as diabetic ketoacidosis and hyperosmolar nonketotic syndrome as well as chronic complications ranging from microvascular disease (nephropathy, neuropathy, retinopathy) to macrovascular disease (1).

Diabetic retinopathy is the leading cause of blindness in people aged 20–74 years in the United States and causes from 12,000 to 24,000 new cases of blindness each year. Manifestations of diabetic retinopathy include retinal microaneurysms, hemorrhages, hard exudates, cotton-wool spots, microvascular abnormalities, growth of abnormal blood vessels and fibrous tissue, and macular edema (2). It was estimated that of the 7.8 million patients affected with diabetes in 1993, 95,000 are expected to develop macular edema each year (3).

Diabetic macular edema (DME) is an important cause of vision loss and is estimated to occur in 29% of patients who have had diabetes for 20 years or more (4–9). Moreover, when thickening involves or threatens the foveal center, the 3-year risk of moderate vision loss (decrease of three lines or more) is 32%. If treated, this risk decreases by 50% (10). Diabetic macular edema is a result of blood–retinal barrier (BRB) breakdown. Endothelial cell tight junction incompetence results in increased vascular permeability causing intraretinal and subretinal fluid accumulation. Retinal microvascular basement membrane thickening and a reduced number of pericytes (smooth muscle cells that help provide vascular stability) further contribute to increased retinal vessel permeability (11). Microaneurysms also play a role in DME by acting as sites of fluid transudation. Microaneurysms are thought to be caused by loss of pericytes and astrocytes, increased capillary pressure, and production of vasoproliferative factors such as vascular endothelial growth factor (VEGF). Increased oxidative stress, accumulation of advanced glycation end-products, and generation of diacylglycerol (DAG) caused by hyperglycemia, activate protein kinase C (PKC), which in turn increases VEGF expression (12,13). There are currently no data to suggest that one racial group or gender develops DME more than others (11).

Diabetic macular edema is defined as retinal thickening within two disc diameters of the center of the macula. Focal edema is associated with hard exudate rings resulting from leakage from microaneurysms, while diffuse edema is the result of breakdown of the BRB with leakage from microaneurysms, retinal capillaries, and arterioles. Clinically significant macular edema (CSME) is defined by the Early Treatment Diabetic Retinopathy Study (ETDRS) in any of the following cases (14):

- retinal thickening within 500 μm of the center of the fovea,
- hard, yellow exudates within 500 μm of the center of the fovea with adjacent retinal thickening,
- at least one disc area of retinal thickening, any part of which is within one disc diameter of the center of the fovea.

TREATMENT

Clinical studies, such as the ETDRS, support the use of laser photocoagulation as standard therapy for DME (8,15). In patients with CSME, focal laser photocoagulation was found to significantly reduce the risk for moderate visual loss and significantly increase the chance for improved visual acuity (VA) when pretreatment VA was worse than 20/40. However, improvement in VA of three lines at 36 months post-treatment was only 3% (8). Laser photocoagulation was less effective in patients with advanced cases of retinopathy or diffuse macular edema, and was not found to be beneficial in patients with non-CSME (8). Furthermore, laser photocoagulation may be complicated by subretinal fibrosis, choroidal neovascularization, and progressive scar expansion. Therefore, although laser photocoagulation is a beneficial treatment for DME, it has significant limitations. Thus, more effective treatments are still required.

Currently, there are three main pharmacologic approaches to the treatment of DME in practice or under study: steroids (the intravitreal injection of triamcinolone acetonide, fluocinolone acetonide sustained drug delivery implant, dexamethasone biodegradable implant), PKC inhibitors, and VEGF inhibitors.

CORTICOSTEROIDS

Corticosteroids have antiangiogenic, antifibrotic, and antipermeability properties that help to stabilize the BRB, aid in exudation resorption, and downregulate inflammatory mediators [including interleukin (IL)-5, IL-6, IL-8, prostaglandins, interferon-gamma (IF-γ), tumor necrosis factor-alpha]. Corticosteroids stabilize cell and lysosomal membranes, reduce prostaglandin release, inhibit cellular proliferation, block macrophage recruitment, inhibit phagocytosis, and decrease neutrophil infiltration into injured tissue (16). It is thought that ischemia associated with diabetic retinopathy elevates levels of VEGF, a potent vasopermeability factor, that compromises vascular endothelial cell intercellular tight junctions and thereby produces macular edema in eyes with diabetic retinopathy (17–21). In animal studies, corticosteroids can reduce levels of growth factors, including VEGF (17,22,23). Collectively, these corticosteroid actions promote resolution of macular edema.

Corticosteroids have been used to treat a variety of ocular diseases. Traditionally, delivery of corticosteroids for posterior-segment eye diseases has been achieved through oral systemic therapy and periocular injections. Oral corticosteroids have not been widely used to treat DME, but when used for posterior inflammatory uveitis, they require high concentrations to reach therapeutic levels in the posterior segment. These high doses often result in systemic side effects (24). Periocular corticosteroid administration often must be repeated and may be associated with complications such as ptosis and inadvertent needle penetration of the globe.

INTRAVITREAL TRIAMCINOLONE ACETONIDE (KENALOG) INJECTION

Topical corticosteroids do not readily penetrate the posterior segment, and posterior sub-Tenon's steroid injections take a long time to diffuse into the posterior segment. Direct placement of corticosteroids into the vitreous cavity may be the best way to deliver corticosteroids to the posterior segment and minimize systemic side effects (16).

Triamcinolone acetonide inhibits basic fibroblast growth factor-induced migration and tube formation in choroidal microvascular endothelial cells. Furthermore, it downregulates metalloproteinase 2, decreases permeability, decreases intercellular adhesion molecule-1 expression, and decreases major histocompatibility complex (MHC)-II antigen expression, all of which are important factors in the inflammatory process (16). Penfold et al. (25,26) showed downregulation of inflammatory modulators and endothelial cell permeability by significantly decreasing MHC-II expression. This group demonstrated that triamcinolone acetonide re-establishes the BRB and downregulates inflammatory mediators. Together, these studies indicate that triamcinolone acetonide favorably influences cellular permeability, including the barrier function of the RPE (16).

Several studies have investigated the efficacy of intravitreal triamcinolone for DME. In a retrospective review study, Martidis et al. (27) injected 4 mg of intravitreal triamcinolone into 16 eyes of patients with CSME who had failed to respond to at least two previous laser photocoagulation treatments. At one- three- and six-month follow-up visits the average VA improved 2.4, 2.4, and 1.3 Snellen lines. Concurrently, central macular thickness [measured by optical coherence tomography (OCT)] decreased by 55%, 57.5%, and 38%, respectively (16). Subsequently, Micelli-Ferrari et al. (28) further confirmed with OCT the effectiveness of intravitreal triamcinolone on macular thickness. In another case series study, 26 eyes given 25 mg of triamcinolone were compared with a control group that underwent only macular grid laser coagulation. The study group was found to have significant improvement in VA from an average of 20/166 to 20/105 ($P < 0.001$) (16,29). In contrast, VA in the control group did not significantly change. In another prospective pilot study termed the ISIS trial, 30 patients received either 2 or 4 mg of intravitreal triamcinolone for CSME. Thirty-three percent of the patients had increase in VA of greater than or equal to three ETDRS lines at three months follow-up, and 21% of patients had similar increases at six month follow-up visits (16,30). The 4 mg group had significantly greater increased VA and decreased macular edema when compared with the 2 mg group. The study also grouped eyes into cystoid and noncystoid foveal edema based on fluorescein angiogram; the group with cystoid foveal edema had a significantly greater VA improvement postintravitreal triamcinolone compared with the group without cystoid foveal edema. In another study, Massin et al. (31) selected 12 patients with bilateral DME and randomly assigned one of the eyes to receive 4 mg of intravitreal triamcinolone while the other eye was simply monitored (control). The study eyes had significantly improved retinal thickness while the control eyes did not change significantly at 12 weeks ($P < 0.001$).

The appropriate intravitreal triamcinolone dose has not yet been established. The ISIS study compares 2 mg versus 4 mg of triamcinolone in patients with macular edema. Preliminary data indicates that the 4 mg dose is more effective at treating macular edema (30). Two studies by Jonas et al. (32,33) in Europe, demonstrated measurable vitreous triamcinolone for seven to eight months and 1.5 years, respectively, after an injection of 20–25 mg of the drug. Prospective trials that compare the 20–25 mg dose to lower doses will be necessary to clarify the optimal dosing regimen.

Intravitreal injections to deliver corticosteroids minimize systemic side effects; however, they may be associated with complications such as retinal detachment, retinal tears, vitreous hemorrhage, endophthalmitis, increased intraocular pressure (IOP), cataract formation, and, with repeated use (required for successful treatment), fibrosis and ptosis. The most common side effect is increased IOP, which has been found on rare occasion to increase drastically (up to 50 mmHg in one case report by Detry-Morel et al.) (16,34,35). Close IOP monitoring is crucial following intravitreal injection.

INTRAVITREAL FLUOCINOLONE ACETONIDE IMPLANT (RETISERT)

Several drug delivery systems for the sustained release of medication within the posterior segment are either under investigation or already in clinical use (36–38). These systems offer a promising approach to the treatment of ocular diseases in cases where systemic drug administration may be associated with unacceptable toxicity and where repeated periocular and/or intravitreal injection carries unacceptable risk (36,37,39). A nonbiodegradable intravitreal implant (Vitrasert) was approved in 1996 to deliver ganciclovir to the posterior segment to treat cytomegalovirus infection (38). Jaffe et al. (40) have reported their experience with an implant based on similar technology that delivers fluocinolone acetonide (Retisert) in a sustained and linear fashion (41) to treat posterior uveitis. In seven eyes, five patients with severe uveitis, the device stabilized or improved VA, and virtually eliminated clinically detectable inflammation (24).

Recent results have been released of a Phase III study of the fluocinolone acetonide implant to treat DME (42). This study, CDS FL-002, was a multicenter, randomized, masked, controlled trial involving 80 patients with DME. At 24 months, there was statistically significant data showing that 0.5 mg fluocinolone had better results on DME than the standard of care (SOC) control group who received either macular grid laser or observation. The study demonstrated that retinal edema at the center of the macula had resolved completely in 53.7% of eyes in the fluocinolone group compared with 28.6% of eyes in the SOC group. There was a greater than two grade improvement in retinal thickness at the center of the macula in 46.2% of the study group compared with only 14.8% in the SOC group. The diabetic retinopathy severity score remained stable or improved in 87.2% of the study group versus 62.9% in the SOC group. Furthermore, the mean change in VA at 24 months showed a gain of 9.3 ± 14.4 letters in the study group while there was actually a loss of 1.9 ± 15.2 letters in the SOC group. This study showed a statistically significant benefit in the primary end point of resolution of retinal thickening at the center of the macula as well as improvement in VA and DR score (41).

As in the uveitis trials, adverse events with the fluocinolone acetonide implant included cataract progression and increased IOP. Seventy-seven point four percent of patients in the implant study group had "serious" cataract progression and 74.2% required cataract extraction compared with only 13.3% and 13.3%, respectively, in the SOC group; 31.7% patients in the study group had increased IOP compared with 0% in the SOC group. Increased IOP was controlled mainly with hypotensive drops; however, eight patients (study group) required trabeculectomy. These side effects were expected with the established relationship between steroids and cataract progression. No retinal detachments were reported (41). A larger Phase III study (CDS FL-005) consisting of approximately 200 patients is currently underway.

DEXAMETHASONE IMPLANT (POSURDEX®)

A biodegradable implant with a sustained-release formulation of dexamethasone was first investigated for use after cataract surgery to treat postsurgical inflammation. The device, called Surodex® (Oculex Pharmaceuticals) was placed directly into the anterior chamber during the cataract surgery. This delivery system released dexamethasone at constant, therapeutic levels (43). In studies by Tan et al. (44) and Wadood et al. (45) the Surodex device controlled intraocular inflammation as well as dexamethasone 0.1% eye drops (Maxidex®), following cataract surgery. In a later study, the dexamethasone delivery system shows a decrease in protein concentration, cell infiltrate, myeloperoxidase activity, IF-γ levels, and IL-4 levels in treated eyes compared with contralateral controls, in an animal uveitis model (46).

A Phase II randomized, multicenter, controlled trial sponsored by Oculex Pharmaceuticals, tested the Posurdex system, a biodegradable implant for extended-release dexamethasone to the posterior segment to treat macular edema that persisted longer than 90 days despite medical therapy or laser photocoagulation. A total of 165 of the 306 patients enrolled in the trial had DME. Patients received an implant that contained either 350 or 700 µg of dexamethasone. A third group was simply observed (control). After 180 days, there was a statistically significant two- and three-line improvement in VA with the 700 µg implants with a trend toward improved VA in the 350 µg implant group. There was also a statistically significant decrease in retinal thickness (measured by OCT) and fluorescein leakage in eyes treated with both the 350 and the 700 µg implants when compared with the control. Further, after 90 days, contrast sensitivity was significantly improved in patients who received the 700 µg implant compared with the control group.

Side effects were primarily related to device implantation into the vitreous base region through a small pars plana sclerostomy and included subconjunctival hemorrhage and vitreous hemorrhage. Both complications were self-limited. There was no report of cataract progression in patients receiving the dexamethasone implant. However, 17% of the patients receiving the implant had a rise in IOP ≥ 10 mmHg at some point during the study compared with only 3% in the control group (47). The Posurdex device degrades in approximately six to eight weeks as it breaks down to lactic acid and glycolic acid and then further into water and carbon dioxide. It is likely that multiple implants would be required to achieve sustained therapy, increasing chances of adverse events during implantation. A Phase III study is currently underway to evaluate the Posurdex device to treat diabetic macular edema. In this trial, a single-use 22-gauge applicator preloaded with the implant is used to insert the implant through the pars plana. This applicator may help reduce the adverse events associated with a conjunctival and scleral incision.

PROTEIN KINASE C INHIBITION

Kinases transfer adenosine triphosphate (ATP) groups to sites on target proteins (enzymes, cell membrane receptors, ion transport channels), thereby causing the activation of the protein. The PKC family is a group of enzymes that are activated by molecules such as DAG and glycation end-products; intracellular concentrations of these molecules are significantly increased in patients with diabetes. Increased DAG augments PKC affinity for calcium, causing its translocation to the cell membrane and thus activating it. Excessive PKC activation causes increased vascular permeability, basement membrane thickening, reduced Na/K ATPase activity, enhanced monocyte adhesion to the vessel wall, and impaired smooth muscle contractility (13,48). PKC inhibition enhances apoptotis and inhibits pericyte proliferation. These activities suggest that PKC inhibition may accelerate pericyte dropout during stages when these cells are still present in the retinal capillaries.

Protein kinase C enzymes are found throughout the body; therefore, widespread inhibition would likely be toxic (49). PKC-β_2, an isoenzyme of PKC, mediates the angiogenic and permeability effects of VEGF. PKC-β is present at high levels in the retina. Activation of this enzyme leads to increased VEGF expression (48,50). As described above, increased levels of VEGF are associated with ischemia and, by virtue of its permeability effect, is thought to produce accumulation of fluid within the retina that causes retinal thickening.

Several approaches to block PKC-β, and inhibit VEGF are currently undergoing testing. Multiple members of the VEGF family can be inactivated when VEGF 1 and 2 receptors and PKC are blocked. PKC412, a nonspecific kinase inhibitor, blocks VEGF 1 and 2 receptors, PDGF receptors, stem cell factor receptors, and several isoforms of PKC. PKC412 significantly suppresses VEGF-induced retinal neovascularization and VEGF-induced retinal vascular leakage (23). The effects of PKC412 on DME was studied in a recent randomized, multicenter, double-masked controlled trial; PKC412 (50, 100, or 150 mg/day) was given orally to half of the 141 enrolled patients. The other half (controls) received placebo. At three months, there was a statistically significant decrease in greatest retinal thickening area and volume and statistically significantly increased VA in patients treated with PKC412 when compared with the placebo group. It was concluded that at doses of 100 mg/day or higher, orally administered PKC412 significantly reduced macular edema (confirmed by OCT) and improved VA in diabetic subjects. There was concern, however, regarding liver toxicity with systemic therapy making local delivery of PKC inhibitors a more appealing approach (51).

Inhibitors specific for PKC-β have a more favorable toxicity profile. In fact, in a study by Aiello et al. (52), the effect of VEGF on retinal vascular permeability appeared to be mediated predominantly by the β-isoform of PKC. There was >95% inhibition of VEGF-induced permeability after administration of a PKC β-isoform-selective inhibitor (50). Studies are currently being conducted with a new PKC-β inhibitor, known as ruboxistaurin mesylate. Thus far, based on animal studies, ruboxistaurin has shown to inhibit PKC-β formation and thereby normalizing retinal vascular function (52).

The efficacy of ruboxistaurin mesylate (LY333531) to delay or stop DME progression has been evaluated in a recent trial. This trial, the PKC-DMES trial was a multicenter, double-masked, placebo-controlled trial that included 686 patients. In this study, when patients with very poor glycemic control (HbA1c > 10%) were excluded from the data, ruboxistaurin (32 mg dose) was associated with a reduction in DME progression (52).

VEGF INHIBITION

As described above, VEGF enhances vascular permeability and thereby promotes macular edema. Several studies have looked or are currently looking for ways to inhibit VEGF.

Pegaptanib sodium (Macugen®) is a synthetic oligonucleotide bound to polyethylene glycol to slow down its clearance rate, thus increasing its half-life. Intravitreal pegaptanib injections are required every six weeks to maintain adequate drug levels in the posterior pole. Pegaptanib selectively binds to VEGF165, a VEGF isoform. A Phase II randomized, placebo-controlled, double-masked, dose-finding, multicenter trial using pegaptanib in eyes with DME was performed on 169 patients. The preliminary results showed statistically significantly increased VA at 36 weeks, and a trend toward decreased retinal thickness, measured by OCT in patients receiving 0.3 mg pegaptanib compared with the control group. The odds of decreased retinal thickness of 75 μm or more at the macular center was four times larger for the group receiving 0.3 mg pegaptanib when weighed against the control group. Pegaptanib seems to be well tolerated and adverse events that included endophthalmitis, retinal detachment, and vitreous hemorrhage are mainly due to the need for repeated injections (53).

NEW AGENTS ON THE HORIZON

A variety of agents have been investigated recently. Anti-angiogenic agents also inhibit vascular permeability and, thus, may be useful drugs to treat DME. Plasminogen kringle 5 (K5), an angiogenic inhibitor, blocks retinal neovascularization in oxygen-induced retinopathy models. Zhang et al. (54) studied the effect of K5 on vascular leakage in the retina. K5 reduced vascular permeability by downregulating VEGF expression and inhibited insulin-like growth factor-1-induced hyperpermeability which is linked to the VEGF expression. Nambu et al. (55) studied the effects of Angiopoietin 1 on ocular neovascularization. Angiopoietin 1 not only significantly suppressed retinal and choroidal neovascularization in eyes with retinal ischemia or rupture of Bruch's membrane, respectively, but also significantly reduced VEGF-induced retinal vascular permeability. Angiostatin is another angiogenic inhibitor. Sima et al. (56) confirmed that intravitreal injection of angiostatin reduced retinal vascular permeability in DME models (56). Angiostatin was shown to downregulate VEGF in the retina. Saishin et al. (57) described VEGF-TRAP(R1R2), a VEGF inhibitor, as a fusion protein that combines ligand binding elements taken from the extracellular domains of VEGF receptors 1 and 2 fused to the Fc portion of IgG1. They reported strong suppression of choroidal neovascularization after

VEGF-TRAP(R1R2) intravitreal injection. VEGF-TRAP(R1R2) was shown to significantly reduce the breakdown of the BRB. The authors suggested that VEGF-TRAP(R1R2) warrants consideration as a new method to treat choroidal neovascularization and DME. Future studies will clarify the usefulness of these various agents to treat DME.

REFERENCES

1. American Diabetes Association. National Diabetes Fact Sheet, 2002. http://www.diabetes.org/about-diabetes.jsp.
2. Klein R, Klein BED. Diabetes in America. In: Harris MI, Cowie CC, Stern MP, et al., eds. 2nd ed. National Institute of Diabetes and Digestive and Kidney Diseases. Bethesda, MD: 1995:293–338. NIH Publication No. 95–1468.
3. Klein R, Moss SE, Klein BEK, et al. The Wisconsin Epidemiologic Study of Diabetic Retinopathy, XI: the incidence of macular edema. Ophthalmology 1989; 96:1501–1510.
4. Patz A, Schatz H, Berkow JW, et al. Macular edema—an overlooked complication of diabetic retinopathy. Trans Am Acad Ophthalmol Otolaryngol 1973; 77:OP34–42.
5. Aiello LM, Rand LI, Briones JC, et al. Diabetic retinopathy in Joslin Clinic patients with adult-onset diabetes. Ophthalmology 1981; 88:619–623.
6. Klein R, Klein BEK, Moss SE. Visual impairment in diabetes. Ophthalmology 1984; 91:1–9.
7. Moss SE, Klein R, Klein BEK. The incidence of vision loss in a diabetic population. Ophthalmology 1988; 95:1340–1348.
8. Early Treatment Diabetic Retinopathy Study Research Group. Photocoagulation for diabetic macular edema. Early Treatment Diabetic Retinopathy Study report number 1. Arch Ophthalmol 1985; 103:1796–1806.
9. Klein R, Klein BEK, Moss SE, et al. The Wisconsin Epidemiologic Study of Diabetic Retinopathy. IV. Diabetic macular edema. Ophthalmology 1984; 91:1464–1474.
10. Early Treatment Diabetic Retinopathy Study Research Group. Photocoagulation for diabetic macular edema. Early Treatment Diabetic Retinopathy Study report number 4. Int Ophthalmol Clin 1987; 27:265–272.
11. Khan B, Lam W. Macular edema, diabetic. http://www.emedicine.com/oph/topic399.htm
12. Flynn H, Smiddy W. Diabetes and ocular disease: past, present and future therapies. Ophthalmology monographs. The American Academy of Ophthalmology, 2000.
13. Idris I, Gray S, Donnelly R. Protein kinase C activation: isozyme-specific effects on metabolism and cardiovascular complications in diabetes. Diabetologia 2001; 44: 659–673.
14. Chew EY, Klein ML, Ferris FL III, et al. Association of elevated serum lipid levels with retinal hard exudates in diabetic retinopathy. Early Treatment Diabetic Retinopathy Study report number 22. Arch Ophthalmol 1996; 114(9):1078–1084.
15. Early Treatment Diabetic Retinopathy Study Research Group. Photocoagulation for diabetic macular edema. Early Treatment Diabetic Retinopathy Study report number 2. Ophthalmology 1987; 94:761–774.
16. Bakri SJ, Kaiser PK. Intravitreal steroid injections for macular edema: way of the future? Retin Physician 2004; 1(1):40–45.
17. Heiss JD, Papavassiliou E, Merrill MJ, et al. Mechanism of dexamethasone suppression of brain tumor-associated vascular permeability in rats. J Clin Invest 1996; 98:1400–1408.
18. Vinores SA, Youssri AI, Luna JD, et al. Upregulation of vascular endothelial growth factor in ischemic and non-ischemic human and experimental retinal disease. Histol Histopathol 1997; 12(1):99–109.

19. Hofman P, van Blijswijk BC, Gaillard PJ, et al. Endothelial cell hypertrophy induced by vascular endothelial growth factor in the retina: new insights into the pathogenesis of capillary nonperfusion. Arch Ophthalmol 2001; 119:861–866.

20. Funatsu H, Yamashita H, Noma H, et al. Increased levels of vascular endothelial growth factor and interleukin-6 in the aqueous humor of diabetics with macular edema. Am J Ophthalmol 2002; 133:70–77.

21. Seo MS, Kwak N, Ozaki H, et al. Dramatic inhibition of retinal and choroidal neovascularization by oral administration of a kinase inhibitor. Am J Pathol 1999; 154:1743–1753.

22. Robinson MR, Baffi J, Yuan P, et al. Safety and pharmacokinetics of intravitreal 2-methoxy-estradiol implants in normal rabbit and pharmacodynamics in a rat model of choroidal neovascularization. Exp Eye Res 2002; 74:309–317.

23. Penn JS, Rajaratnam VS, Collier RF, Clark AF. The effect of an angiostatic steroid on neovascularization in a rat model of retinopathy of prematurity. Invest Ophthalmol Vis Sci 2001; 42:283–290.

24. Jaffe GJ, Ben-Nun J, Guo H, et al. Fluocinolone acetonide sustained drug delivery device to treat severe uveitis. Ophthalmology 2000; 107:2024–2033.

25. Penfold PL, Wong JG, Gyory J, et al. Effects of triamcinolone acetonide on microglial morphologyand quantitative expression of MHC-II in exudative age-related macular degeneration. Clin Experiment Ophthalmol 2001; 29(3):188–192.

26. Penfold PL, Wen L, Madigan MC, et al. Triamcinolone acetonide modulates permeability and intercellular adhesion molecule-1 expression of the ECV304 cell line: implications for macular degeneration. Clin Exp Immunol 2000; 121(3):458–465.

27. Martidis A, Duker JS, Greenberg PB, et al. Intravitreal triamcinolone for refractory diabetic macular edema. Ophthalmology 2002; 109(5):920–927.

28. Micelli-Ferrari T, Sborgia L, Furino C, et al. Intravitreal triamcinolone acetonide: evaluation of retinal thickness changes measured by optical coherence tomography in diffuse diabetic macular edema. Eur J Ophthalmol 2004; 14(4):321–334.

29. Jonas JB, Kreissig I, Sofker A, et al. Intravitreal injection of triamcinolone for diffuse diabetic macular edema. Arch Ophthalmol 2003; 121(1):57–61.

30. Pollack JS. ISIS-DME: a randomized dose-escalation study of intravitreal steroid injection for diabetic macular edema. Presented at Vail Vitrectomy 2004, Vail, Colorado, March 6, 2004.

31. Massin P, Audren F, Haouchine B, et al. Intravitreal triamcinolone acetonide for diabetic diffuse macular edema: preliminary results of a prospective controlled trial. Ophthalmology 2004; 111(2):218–227.

32. Jonas JB. Intraocular availability of triamcinolone acetonide after intravitreal injection. Am J Ophthalmol 2004; 137(3):560–562.

33. Jonas JB, Degenring RF, Kamppeter BA, et al. Duration of the effect of intravitreal triamcinolone acetonide as treatment for diffuse diabetic macular edema. Am J Ophthalmol 2004; 138(1):158–160.

34. Bakri SJ, Beer PM. The effect of intravitreal triamcinolone acetonide on intraocular pressure. Ophthalmic Surg Lasers Imaging 2003; 34(5):386–390.

35. Detry-Morel M, Escarmelle A, Hermans I. Refractory ocular hypertension secondary to intravitreal injection of triamcinolone acetonide. Bull Soc Belge Ophthalmol 2004; 292:45–51.

36. Metrikin DC, Anand R. Intravitreal drug administration with depot devices. Curr Opin Ophthalmol 1994; 5:21–29.

37. Berger AS, Cheng C-K, Pearson PA, et al. Intravitreal sustained release corticosteroid-5-fluoruracil conjugate in the treatment of experimental proliferative vitreoretinopathy. Invest Ophthalmol Vis Sci 1996; 37:2318–2325.

38. Smith TJ, Pearson PA, Blandford DL, et al. Intravitreal sustained-release ganciclovir. Arch Ophthalmol 1992; 110:255–258.

39. Hainsworth DP, Pearson PA, Conklin JD, Ashton P. Sustained release intravitreal dexamethasone. J Ocul Pharmacol Ther 1996; 12:57–63.

40. Jaffe GJ, Yang CH, Guo H, Denny JP, Lima C, Ashton P. Safety and pharmacokinetics of an intraocular fluocinolone acetonide sustained delivery device. Invest Ophthalmol Vis Sci 2000; 41:3569–3575.

41. Pearson PA, Baker C, Elliott D, IP M, Morse L, Callanan D. Fluocinolone acetonide intravitreal implant for diabetic macular edema: 2 year results. Presented at the Annual Association for Research in Vision and Ophthalmology Meeting, Fort Lauderdale, FL, April 25–29, 2004.

42. Driot JY, Novack G, Rittenhouse K, Milazzo C, Pearson PA. Ocular pharmacokinetics of fluocinolone acetonide after Retisert intravitreal implantation in rabbits over a 1-year period. J Ocul Pharmacol Ther 2004; 20(3):269–270.

43. Adis R&D Profile. Dexamethasone ophthalmic—Oculex. Drugs R D 2002; 3(3):152–153.

44. Tan DT, Chee SP, Lim L, Lim AS. Randomized clinical trial of a new dexamethasone delivery system (Surodex) for treatment of post-cataract surgery inflammation. Ophthalmology 1999; 106:223–231.

45. Wadood AC, Armbrecht AN, Aspinall PA, Dhillon B. Safety and efficacy of a dexamethasone anterior segment drug delivery system in patients after phacoemulsification. J Cataract Refract Surg 2004; 30(4):761–768.

46. Kodama M, Numaga J, Yoshida A, et al. Effects of a new dexamethasone-delivery system (Surodex) on experimental intraocular inflammation models. Graefes Arch Exp Ophthalmol 2003; 241(11):927–933.

47. Haller JA. The steroid device: the Oculex study. Presented at the Retinal Subspecialty Day, American Academy of Ophthalmology Meeting, Anaheim, CA, November 15–18, 2003.

48. Pomero F, Allione A, Beltramo E, et al. Effects of protein kinase C inhibition and activation on proliferation and apoptosis of bovine retinal pericytes. Diabetologia 2003; 46:416–419.

49. Frank RN. Potential new medical therapies for diabetic retinopathy: protein kinase C inhibitors. Am J Ophthalmol 2002; 133(5):693–698.

50. Aiello LP, Bursell SE, Clermont A, et al. Vascular endothelial growth factor-induced retinal permeability is mediated by protein kinase C in vivo and suppressed by an orally effective beta-isoform-selective inhibitor. Diabetes 1997; 46(9):1473–1480.

51. Campochiaro PA, C99-PKC412-003 Study Group. Reduction of diabetic macular edema by oral administration of the kinase inhibitor PKC412. Invest Ophthalmol Vis Sci 2004; 45(3):922–931.

52. Aiello LP, Davis MD, Miton RC, Sheetz MJ, Arora V, Vignati L. Initial results of the protein kinase C beta inhibitor diabetic macular edema study (PKC-DMES). Diabetologia 2003; 46:A42.

53. Macugen™ (pegaptanib sodium injection) shows positive visual and anatomical outcomes in a Phase II trial for patients with diabetic macular edema. Eyetech Pharmaceuticals Press Release, May 3, 2004; http://www.eyetk.com/investors/press_releases.asp.

54. Zhang SX, Sima J, Shao C, et al. Plasminogen kringle 5 reduces vascular leakage in the retina in rat models of oxygen-induced retinopathy and diabetes. Diabetologia 2004; 47(1):124–131.

55. Nambu H, Nambu R, Oshima Y, et al. Angiopoietin 1 inhibits ocular neovascularization and breakdown of the blood–retinal barrier. Gene Ther 2004; 11(10):865–873.

56. Sima J, Zhang SX, Shao C, Fant J, Ma JX. The effect of angiostatin on vascular leakage and VEGF expression in rat retina. FEBS Lett 2004; 564(1–2):19–23.

57. Saishin Y, Saishin Y, Takahashi K, et al. VEGF-TRAP(R1R2) suppresses choroidal neovascularization and VEGF-induced breakdown of the blood–retinal barrier. J Cell Physiol 2003; 195(2):241–248.

20

Retinal Vein Occlusion

Michael M. Altaweel and Michael S. Ip
Department of Ophthalmology and Visual Sciences, University of Wisconsin-Madison Medical School, Madison, Wisconsin, U.S.A.

LOCAL DRUG DELIVERY APPROACH: RETINAL VEIN OCCLUSION

Central retinal vein occlusion (CRVO) and branch retinal vein occlusion (BRVO) are common retinal vascular disorders. Indeed, BRVO is second only to diabetic retinopathy in the frequency with which it produces retinal vascular disease (1). Both CRVO and BRVO have a characteristic, although sometimes variable, appearance with intraretinal hemorrhage, tortuous and dilated retinal veins, and occasionally optic disk edema. These findings are present in all quadrants of the fundus in CRVO and are segmental in BRVO (Figs. 1 and 2). Visual acuity loss in CRVO and BRVO are often the result of macular edema and neovascular complications (1–5). The majority of current local drug delivery approaches for retinal vein occlusion target macular edema because visual acuity loss in retinal vein occlusion is more often a result of macular edema than from neovascular complications.

Human Clinical Trials in CRVO: Prior Studies

The Central Vein Occlusion Study (CVOS) was conducted, in part, to evaluate the effect of grid laser photocoagulation on visual acuity and macular edema in CRVO (2). In the CVOS, 728 eyes with CRVO were studied. Of these 728 eyes, 155 (21%) had macular edema reducing visual acuity to 20/50 or worse (group M eyes, macular edema). In the largest group of eyes (group P, perfused) that included 547 eyes, 84% (460 eyes) had angiographic evidence of macular edema involving the fovea at baseline.

The CVOS found no significant difference in visual outcome between the treatment and observation groups at any follow-up point. Although there was a definite decrease in macular edema on fluorescein angiography in the treatment group when compared to the control group, this did not translate to a direct improvement in visual acuity (4). Therefore, at present, there is no proven therapy for visual acuity loss from macular edema due to CRVO.

The natural history of macular edema due to CRVO was also delineated in the CVOS (2–4). One hundred and fifty-five group M eyes (77 treated eyes and 78 control eyes) were followed over a three-year period. All eyes had macular edema for a minimum of three months prior to enrollment (4). For untreated eyes with

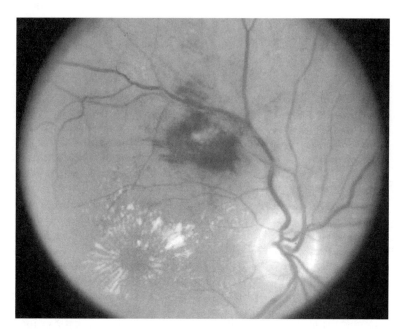

Figure 1 BRVO is characterized by intraretinal hemorrhages and a dilated retinal venule in one quadrant of the retina. This example demonstrates secondary macular edema with retinal exudates. *Abbreviation*: BRVO, branch retinal vein occlusion.

an initial visual acuity between 20/50 and 5/200 at presentation ($n = 78$ eyes), 42 eyes were available for follow-up at the three-year visit. Of these eyes, 10 (24%) gained two or more lines of visual acuity at the three-year follow-up. Twenty eyes (48%) remained within two lines of baseline visual acuity and 12 eyes (29%) lost two or

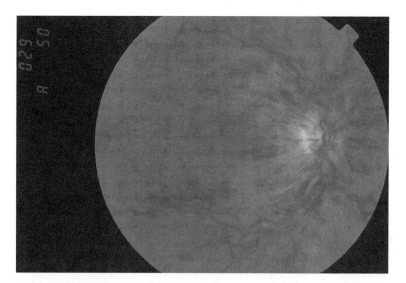

Figure 2 CRVO is characterized by retinal hemorrhages in all four quadrants, venular dilation, and frequent optic nerve edema. Nerve fiber layer infarcts and macular edema may be associated features with both types of retinal vascular occlusion. *Abbreviation*: CRVO, central retinal vein occlusion.

more lines of visual acuity at the three-year follow-up. At the three-year follow up, six eyes (14%) gained three or more lines of visual acuity. Thirty eyes (72%) remained within three lines of baseline visual acuity and six eyes (14%) lost three or more lines of visual acuity at the three-year follow-up. The final median visual acuity in untreated eyes was 20/160.

The CVOS demonstrated that the natural history of untreated macular edema is poor in many patients. Additionally, the CVOS showed that grid laser photocoagulation for macular edema does not improve visual acuity compared with the natural history of this disease. Therefore, it is important to explore other avenues for managing this common cause of vision loss.

Human Clinical Trials in BRVO: Prior Studies

The Branch Vein Occlusion Study (BVOS) was conducted, in part, to evaluate grid laser photocoagulation as a treatment for macular edema due to BRVO (Fig. 3) (1).

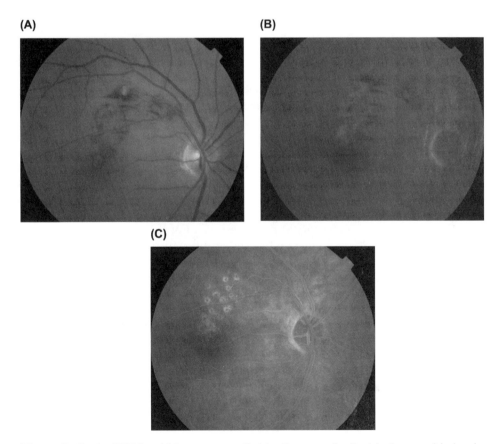

(A)

(B)

(C)

Figure 3 In the BVOS, grid laser was applied to the area of retinal leakage and ischemia. Successful treatment could lead to resolution of macular edema and improvement in visual acuity. (**A**) Baseline photo and (**B**) corresponding fluorescein angiogram. (**C**) Follow-up fluorescein angiogram with resolution of leakage; laser spots stain. *Abbreviation*: BVOS, Branch Vein Occlusion Study.

Table 1 Natural History of Macular Edema Due to Retinal Vein Occlusion in Randomized Trials

Study	Vision improved by 2 or more lines		Vision unchanged (±2 lines)		Vision worse by 2 or more lines		Number of eyes at end of study period	Follow-up period (yr)
	%	No.	%	No.	%	No.		
CVOS	19	10	59	31	22	12	53	2
BVOS	37	13	46	16	17	6	35	3

Abbreviations: CVOS, Central Vein Occlusion Study; BVOS, Branch Vein Occlusion Study.

Specifically, the group III arm of the BVOS was designed to evaluate grid photocoagulation treatment of macular edema due to BRVO that had persisted for at least three months (and less than 18 months) in eyes with visual acuity of 20/40 or worse. One hundred and thirty-nine eyes (71 treated eyes and 68 control eyes) were studied. This arm of the study did demonstrate a benefit for eyes treated with macular grid photocoagulation (1). Of 43 treated eyes available for follow-up at the three-year visit, 28 eyes (65%) had gained two or more lines of visual acuity from baseline and maintained this gain for at least eight months, as compared with the same gain in 13 of 35 (37%) untreated eyes. At the three-year visit, nearly twice as large a proportion of treated versus control eyes had visual acuity of 20/40 or better.

The BVOS also reported on the natural history of macular edema due to BRVO (1). After three years, of 35 untreated eyes available for follow-up, only 12 eyes (34%) with a presenting visual acuity of 20/40 or worse achieved a visual acuity of 20/40 or better. Furthermore, eight eyes (23%) had 20/200 or worse visual acuity at their final three-year follow-up visit.

Although the BVOS did demonstrate a visual acuity benefit for eyes treated with grid photocoagulation, the BVOS also identified a subset of patients that derive limited benefit from macular grid photocoagulation. In the BVOS, 40% of treated eyes ($n = 43$) had worse than 20/40 vision at three years and 12% of treated eyes had 20/200 or worse visual acuity at three years (1). For this subset of patients current treatment options are limited and therefore other treatment options are under investigation. Table 1 summarizes the three-year natural history data for patients with macular edema from CRVO and BRVO. These data from the CVOS and BVOS show that the natural history of such patients does not commonly include improvement in visual acuity. This is important because grid laser photocoagulation was ineffective in patients with macular edema from CRVO and effective for only some patients with macular edema from BRVO.

Pathogenesis of Retinal Vein Occlusion

Current treatment approaches to retinal vein occlusion are best explained with an understanding of the pathogenesis of this condition. Retinal venous occlusive disease results from the initial insult of thrombus formation at the lamina cribrosa or an arteriovenous crossing. Green et al. (6) in a histopathologic study of 29 eyes with CRVO, documented a fresh or recanalized thrombus of the central retinal vein in the area of the lamina cribrosa as a constant pathologic finding. Frangieh et al. (7) in a histopathologic study of nine eyes with BRVO, documented a fresh or recanalized thrombus at the site of vein occlusion in all eyes studied. Experimental work in animals has demonstrated that following venous occlusion, a hypoxic environment in the retina is produced (8). This is then followed by functional, and later structural changes, in the retinal capillaries that result in an immediate increase in retinal capillary permeability and accompanying retinal edema.

The increase in retinal capillary permeability and subsequent retinal edema may be the result of a breakdown of the blood–retina barrier mediated in part by vascular endothelial growth factor (VEGF), a 45 kDa glycoprotein (9). Aiello et al. (9) demonstrated in an in vivo model that VEGF can increase vascular permeability. Fifteen eyes of 15 albino Sprague–Dawley rats received an intravitreal injection of VEGF. The effect of intravitreal administration of VEGF on retinal vascular permeability was assessed by vitreous fluorophotometry. In all 15 eyes receiving an intravitreal injection of VEGF, a statistically significant increase in vitreous fluorescein leakage was recorded. In contrast, control eyes, which were fellow eyes injected with vehicle alone, did not demonstrate a statistically significant increase in vitreous fluorescein leakage. Vitreous fluorescein leakage in eyes injected with VEGF attained a maximum of 227% of control levels. Antonetti et al. (10) demonstrated that VEGF may regulate vessel permeability by increasing phosphorylation of tight junction proteins such as occludin and zonula occludens 1. Sprague–Dawley rats were given intravitreal injections of VEGF and changes in tight junction proteins were observed through Western blot analysis. Treatment with alkaline phosphotase revealed that these changes were caused by a change in phosphorylation of tight junction proteins. This model provides, at the molecular level, a potential mechanism for VEGF-mediated vascular permeability in the eye. Similarly, in human nonocular disease states such as ascites, VEGF has been characterized as a potent vascular permeability factor (VPF) (11).

The normal human retina contains little or no VEGF; however, hypoxia causes upregulation of VEGF production (12). Disease states characterized by hypoxia-induced VEGF upregulation include CRVO and BRVO (9,12). Vinores et al. (12) using immunohistochemical staining for VEGF, demonstrated that increased VEGF staining was found in retinal neurons and retinal pigment epithelium in human eyes with venous occlusive disease. Pe'er et al. (13) evaluated 10 human eyes enucleated for neovascular glaucoma from CRVO and used molecular localization with a VEGF-specific probe to identify cells producing VEGF messenger RNA (mRNA). All of these eyes demonstrated upregulated VEGF mRNA expression in the retina. This hypoxia-induced upregulation of VEGF may be inhibited pharmacologically. Adamis et al. (14) in a nonhuman primate model, demonstrated that anti-VEGF antibodies can inhibit VEGF-driven capillary endothelial cell proliferation. In this study, 16 eyes of nonhuman primates had retinal ischemia induced by laser retinal vein occlusion. Zero of eight eyes receiving neutralizing anti-VEGF antibodies developed iris neovascularization while five of eight control eyes eventually developed iris neovascularization.

As the preceding discussion suggests, either attenuation of the effects of VEGF or lysis/bypass of the thrombus at the site of retinal vein occlusion introduces a rationale for treatment of retinal venous occlusive disease. Thus, current treatment approaches employ pharmacologic modulation of VEGF or surgical lysis/bypass of the site of retinal vein occlusion. Other approaches discussed in this chapter include rheologic modification of systemic factors thought to be responsible for retinal vein occlusion.

Human Clinical Trials in CRVO and BRVO: Current and Future Studies

Preliminary Studies Using Intravitreal Triamcinolone Acetonide Injections

Corticosteroids, a class of substances with anti-inflammatory properties, inhibit the expression of the VEGF gene (15). In a study by Nauck et al. (16) the platelet-derived growth factor (PDGF) induced expression of the VEGF gene in cultures of human aortic vascular smooth muscle cells was abolished by corticosteroids in a dose-dependent manner. A separate study by Nauck et al. (16) demonstrated that corticosteroids abolished the induction of VEGF by the proinflammatory mediators PDGF and platelet-activating factor (PAF) in a time- and dose-dependent manner. This study was performed using primary cultures of human pulmonary fibroblasts and pulmonary vascular smooth muscle cells.

Triamcinolone acetonide is a corticosteroid that is commercially available and inexpensive. Intravitreal injection of triamcinolone acetonide is nontoxic in animal studies (17–19). McCuen et al. (17) injected 1 mg of triamcinolone acetonide into the vitreous cavity of 21 rabbit eyes. Throughout the three-month course of follow-up ophthalmoscopy, intraocular pressure, electroretinography (scotopic and photopic responses), and light and electron microscopy all remained normal. Schindler et al. (18) studied the clearance of intravitreally injected triamcinolone acetonide (0.5 mg) in 30 rabbit eyes. In nonvitrectomized eyes, the average clearance rate was 41 days. In eyes having undergone vitrectomy or combination vitrectomy and lensectomy, the average clearance rate was 17 and 7 days, respectively (18). It was found that the ophthalmoscopic disappearance of injected triamcinolone acetonide correlated well with a spectrophotometric analysis for clearance of the drug. Scholes et al. (19) also studied the clearance of intravitreally injected triamcinolone acetonide (0.4 mg) in 24 rabbit eyes. Using high-performance liquid chromatography complete clearance of the drug was noted by 21 days. Nondetectable drug levels were present before ophthalmoscopic disappearance.

As discussed previously, corticosteroids downregulate VEGF production in experimental models and possibly reduce breakdown of the blood–retinal barrier (15,16). Similarly, corticosteroids have antiangiogenic properties possibly due to attenuation of the effects of VEGF (20,21). These properties of steroids are commonly used. Clinically, triamcinolone acetonide is used locally as a periocular injection to treat cystoid macular edema secondary to uveitis or as a result of intraocular surgery (22,23). In animal studies, intravitreal triamcinolone acetonide has been used to prevent proliferative vitreoretinopathy and retinal neovascularization (24–27). Intravitreal triamcinolone acetonide has been used clinically to treat proliferative vitreoretinopathy and choroidal neovascularization (28–31).

Recently, intravitreal triamcinolone acetonide has been used clinically to treat retinal vascular disease (Fig. 4). A case report by Jonas and Sofker (32) described a patient with nonproliferative diabetic retinopathy and a six-month history of persistent, diffuse macular edema despite grid photocoagulation. Following one intravitreal injection of triamcinolone acetonide, the visual acuity of this patient improved from 20/200 to 20/50 over a five-month follow-up period. It was also noted that there was marked regression of macular edema on clinical examination. Martidis et al. (33,34) reported on the use of intravitreal triamcinolone for refractory diabetic macular edema. Sixteen eyes with a macular thickness of at least 300 μm despite prior photocoagulation were treated with 4 mg injections of triamcinolone. At three-month follow-up the mean decrease in central retinal thickness was 57.5%, with a visual acuity increase of 2.4 Snellen lines. Those with six-month follow-up demonstrated some recurrence of edema and visual acuity improvement was reduced to 1.3 lines.

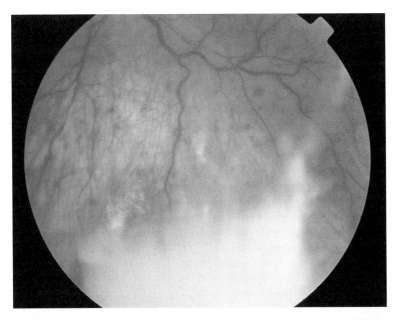

Figure 4 Intravitreal kenalog is suspended in the vitreous following injection.

Greenberg and Martidis (35) studied both eyes of one patient with bilateral diffuse macular edema secondary to CRVO. The right eye of this 80-year-old patient had macular edema from a CRVO of nine-month duration when the patient presented with a two-week history of visual acuity loss due to macular edema from a CRVO in the left eye. Because of the poor natural history of untreated macular edema in the right eye of this patient, the left eye received an intravitreal injection of triamcinolone acetonide. It did well both anatomically and functionally, with visual acuity improvement from 20/400 to 20/30 after three months of follow-up. Central foveal thickness as measured by optical coherence tomography decreased from 589 to 160 μm with restoration of a normal foveal contour following treatment. Six months following injection, visual acuity decreased to 20/400 because of recurrence of retinal thickening that measured 834 μm by optical coherence tomography. A second injection was performed and one month later visual acuity returned to 20/50 with a decrease in central foveal thickness to 158 μm and a normal foveal contour. This patient has maintained this level of visual acuity for over six months following the second injection. Given the response to treatment in the left eye, the right eye (now with 16 months of untreated macular edema) was treated with an intravitreal injection of triamcinolone acetonide. There was a prompt reduction in central foveal thickness as measured by optical coherence tomography from 735 to 195 μm. However, possibly as a result of the duration of macular edema, no visual benefit was noted. No significant elevation of intraocular pressure was noted in either eye. Other clinical case reports by Ip et al. (36) and Jonas et al. (37) have demonstrated similar results in the treatment of macular edema due to CRVO with intravitreal injections of triamcinolone acetonide (Fig. 5). Although macular edema can improve markedly with intravitreal triamcinolone administration in both nonischemic and ischemic CRVO, individuals with nonischemic CRVO are more likely to acheive visual acuity improvement as a result.

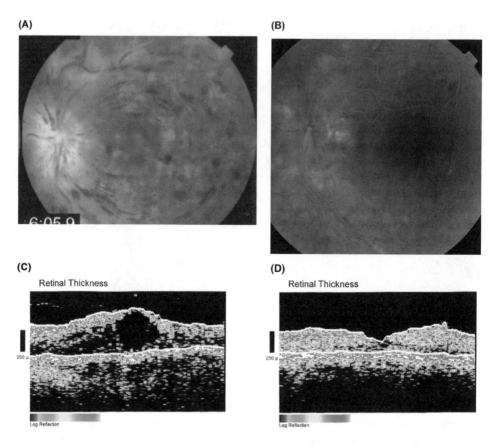

Figure 5 Patient with macular edema secondary to CRVO had an intravitreal injection of $0.1\,cm^3$ of triamcinolone acetonide (4 mg) with rapid resolution of retinal thickening. (**A**) Late-frame fluorescein angiogram before injection showing significant fluorescein leakage in macular area. (**B**) Late-frame angiogram following steroid injection shows resolution of fluorescein leakage. (**C**) (*See color insert*) OCT before injection demonstrates pronounced macular thickening. (**D**) (*See color insert*) OCT following injection shows macular edema resolution. *Abbreviations*: CRVO, central retinal vein occlusion; OCT, optical coherence tomography.

Park et al. (38) reported the first case series of triamcinolone acetonide to treat macular edema in eyes with perfused CRVO and visual acuity of 20/50 or worse. Following a single 4 mg intravitreal triamcinolone injection, the mean visual acuity improved from 58 ETDRS letters to 78 letters at last follow-up. Before treatment, all ten treated eyes had both angiographic and optical coherence tomographic evidence of macular edema. Following treatment, macular edema, as demonstrated by optical coherence tomography, improved in all eyes. The mean macular volume decreased from 4.2 to $2.6\,mm^3$ at last follow-up. One eye required reinjection five months following initial therapy because of cystoid macular edema recurrence. Four of ten eyes required either initiation or escalation of glaucoma therapy secondary to increased intraocular pressure during the follow-up period. There were no other significant complications. The rate of secondary cataract formation, glaucoma, and endophtalmitis are expected to rise with repeated administration of triamcinolone. Other studies have reported similar beneficial results with

treatment of macular edema due to BRVO and CRVO with intravitreal injection
of triamcinolone acetonide (39,40).

OTHER TREATMENTS

The previous section described treatment of retinal vein occlusion with one of the
most recently proposed treatments, that of intravitreal corticosteroids. However, a
variety of other treatments have either been proposed or are currently being used
to treat patients with retinal vein occlusions. This section describes these treatments
in greater detail.

Systemic Thrombolytic Therapy

Most cases of CRVO result from the formation of a thrombus at or just posterior to
the lamina cribrosa (6). Systemic thrombolytic therapy was first attempted with
anticoagulants in 1938. Costen et al. (41) described one case treated within seven
hours of onset who had visual acuity improve from 20/200 to 20/30 over a period
of 30 minutes due to presumed lysis of the thrombus, with venous filling time
improving from 34.6 to 23.3 seconds on repeated fluorescein angiography.

Kohner et al. (42,43) published the first controlled randomized trial using
streptokinase for the treatment of central retinal vein occlusion. Of 40 patients,
20 were treated with an intravenous streptokinase bolus followed by continuous
infusion for 72 hours and then use of oral warfarin for six months. Twenty patients
served as controls. At one-year follow-up, the treatment group had an average gain
of 1.3 lines versus the control group who lost an average of 1.5 lines ($p < 0.03$).

It was noted, however, that 3 of the 20 treated patients developed sudden
severe vitreous hemorrhage, which led to functional blindness and therefore the
authors advised caution with use of this treatment. Vitreoretinal surgical techniques
have improved significantly since this report and it is possible that such adverse
events could be more easily managed at this time.

Elman reported on the use of systemic tissue plasminogen activator (tPA) rather
than streptokinase for the management of CRVO (44). Tissue plasminogen activator
has a better safety profile, is less antigenic, has a shorter half life, and induces less
risk of causing systemic hemorrhage. Of 96 patients, 55 had systemic tPA plus
aspirin, with treatment administered an average of 21 days after the onset of central
retinal vein occlusion. Forty-two percent improved greater than or equal to three
lines of visual acuity at six-month follow-up (average of 5.1 lines). Thirty-seven percent remained stable and 21% lost three or more lines of visual acuity. When comparing the 44 patients in this trial who had baseline visual acuity between 20/50 and 20/
200, it is noted that 48% gained at least three lines of visual acuity compared to 6% of
patients in the treatment arm of the CVOS which had a similar entry visual criteria.
However, 30% lost at least three lines of visual acuity in this trial versus 13% in the
CVOS (4). Three patients sustained intraocular bleeding and one died from hemorrhagic stroke. Despite the favorable visual results, the potential for fatal adverse events
has prevented this form of management from becoming more commonly utilized.

More selective application of thrombolysis has been attempted with injection
of urokinase into the ophthalmic artery. Paques et al. (45) reported retrospectively
on 26 eyes treated in this method, nine of which were combined central retinal artery
occlusion (CRAO) and CRVO. The visual acuity improved significantly in only six

eyes and four of these were patients with combined CRVO/CRAO. The results in CRVO did not warrant the use of this approach.

Intravitreal Thrombolytic Therapy

To bypass the potential fatal side effects of systemically applied thrombolytic therapy, thrombolytic agents have been delivered locally into the vitreous. Elman et al. (46) reported on a retrospective series of nine eyes treated with 100 μg (0.2 mL) of intravitreal tPA followed by paracentesis. At six months of follow-up there were no adverse events, and four of the nine eyes had improved by at least three lines of visual acuity. However, two eyes sustained a loss of six or more lines. None of the four cases deemed ischemic at baseline improved beyond 20/200. In a similar series, Glacet-Bernard et al. (47) reported on 15 patients with CRVO of 1 to 21-day duration (mean of eight days) treated with intravitreal tPA. Eight of 15 patients had a baseline visual acuity of 20/50 or better. Of those patients available for the eight-month follow-up, visual acuity increased in five, remained unchanged in five, and decreased in four. Six of 15 patients developed an increased amount of intraretinal hemorrhage. The results were deemed to be no better than the natural history.

Lahey et al. (48) reported on 26 eyes (23 with CRVO, 3 with BRVO) treated with 65–110 μg of intravitreal tPA within 21 days of the onset of CRVO. There was no control group. This trial had a short follow-up of six weeks. At six weeks, the visual acuity was stable or improved in 16 of 23 eyes (69.6%). One eye developed vitreous hemorrhage and two others were found to have an increase in macular edema and a subsequent decrease in visual acuity. The short follow-up precludes making substantive conclusions.

Controversy exists over whether intravitreally injected tPA can diffuse across the retina to act on the site of thrombosis. Some authors also question whether thrombolytic therapy can have any effect in older cases of central vein occlusion where the thrombus has formed fibrin cross linking and become organized. It has been suggested that such treatment would be more effective if applied closer to the time of onset of the retinal vein occlusion.

Injection of tPA into Venules

Weiss and Bynoe (49) developed the technique of treating CRVO with pars plana vitrectomy, cannulation of a branch retinal vein and injection of tPA toward the thrombus. In a prospective study without controls, 28 eyes with CRVO of average duration of 4.9 months were treated. Baseline visual acuity was 20/63. The procedure involves a pars plana vitrectomy, elevation of the posterior hyaloid if no prior posterior vitreous detachment is noted, lowering of the intraocular pressure for the canulation of the branch vein and injection of 0.6–7.5 mL of tPA (200 μg/mL).

Complications included the development of vitreous hemorrhage in 25%, hemolytic glaucoma in one patient and retinal detachment in one. The authors report a 79% rate (22/28) of visual acuity improvement of one line or more. With an average follow-up of 11.8 months (3–24 months range), 14 of 28 eyes (50%) gained at least three lines. This procedure requires specialized equipment and the study has been critiqued for including eyes (29%) that had a prior procedure for treatment of central retinal vein occlusion. With the lack of a control group it is unclear whether this treatment is indeed superior to the natural history for these cases. This procedure is undergoing further investigation.

Rheologic Therapy

Pharmacologic Agents

Rheological abnormalities, including increased red blood cell (RBC) aggregation, increased plasma viscosity, and increased hematocrit and fibrinogen have been described in CRVO (50). Blood viscosity is determined by plasma viscosity, RBC deformability, RBC concentration, and RBC aggregation. Rheological drugs such troxerutin and pentoxyfylline can inhibit platelet and RBC aggregation and increase RBC deformability and therefore decrease blood viscosity and increase flow (51,52). A recent prospective double-masked placebo-controlled randomized clinical trial evaluated treatment of CRVO (27 patients) and BRVO (26 patients) with troxerutin versus placebo. The treatment group received troxerutin for a period of four months after which both the treatment and placebo groups received troxerutin for the remainder of a two-year period. The mean follow-up was two years. In the treatment group, retinal circulation time improved and a smaller percentage progressed to ischemic CRVO and required panretinal photocoagulation than the placebo group. The RBC aggregation index and fibrinogen level had decreased significantly at four months when compared to controls. As well, although both groups had a similar baseline visual acuity, at four months the mean visual acuity in the treatment group was 20/63 versus 20/100 in the controls.

Hemodilution

Hemodilution has been proposed as a method to reduce abnormal blood viscosity to treat retinal vein occlusion. In this procedure, whole blood is withdrawn and the same volume of a plasma expander such as hydroxyethyl hemadon or hydroxyethyl starch is reinjected until the hematocrit is decreased to approximately 35%.

Glacet-Bernard et al. (53) reported on a prospective study of isovolemic hemodilution to treat 142 eyes with central retinal vein or hemiretinal vein occlusion. They had a target hematocrit of 35% and noted an average decrease in hematocrit with treatment from 42% to 32% without major side effects. They distinguished patients who were treated within two weeks of symptom onset and those treated later. With a mean follow-up of 10 months, 41% of the early treatment group had improved visual acuity to 20/40 or better. In the late treatment group, 23% achieved this visual acuity ($p < 0.01$). Overall, the early treatment group lost 0.2 ETDRS lines while the late treatment group lost 1.9 lines on average. Retinal arteriole–venous transit time was found to decrease from 8.8 to 6.2 seconds on average. Although the visual acuity results appear promising in the early treatment group, it should be noted that there was a significant increase in retinal ischemia, with individuals having greater than 100 disk diameters of ischemia on fluorescein angiography rising from 3% at baseline to 57% at final follow-up. Fifty-seven of 142 eyes required panretinal laser or cryopexy for treatment of ischemia-related changes.

Hattenbach et al. (54) studied 22 patients with CRVO and distinguished those treated within 11 days of onset versus those treated later. Visual acuity improved in two of the nine patients treated early and one of the 13 patients in the late treatment group. Hansen et al. (55) studied 83 eyes, 35 of whom were deemed ischemic at baseline. Hemodilution was conducted over a period of six weeks with side effects including fainting spells in 5% of patients and weakness in 16%. Maximum venous filling time was measured on fluorescein angiography and found to decrease from 18.4 to 13.1 seconds in nonischemic CRVO and from 24.5 to 14.8 seconds in ischemic

CRVO at the three-month follow-up. At baseline, 50% of the patients with nonischemic CRVO started with visual acuity of at least 20/50 and at one year 59% of these patients maintained this vision. In patients with ischemic CRVO 48.5% improved by two or more lines at one year with 25% achieving 20/50 or better visual acuity (one of 35 started with this level of visual acuity). In this study, all ischemic patients received xenon panretinal photocoagulation (PRP) prophylactically. In a separate study, Hansen et al. (56) demonstrated that hemodilution plus PRP led to superior visual acuity outcome when compared to administration of PRP alone. Five of 19 in the treatment group improved their visual acuity by at least two lines at one year follow-up compared with none of the control group (PRP alone).

There is controversy regarding hemodilution. Wiek et al. (57) have found no evidence in their studies that there is a significantly higher plasma viscosity or hematocrit in patients with central retinal vein occlusion when compared with controls. They found no change in viscosity with hemodilution. In a prospective randomized trial of 59 patients who developed CRVO less than three months prior to entering the study, Luckie et al. (58) demonstrated no significant difference in the rate of visual acuity improvement or in the development of neovascularization of the iris at six-month follow-up. However, they found the incidence of vision loss to be five times greater in the treatment group.

Hemodilution has been attempted to manage BRVO. In one series, 34 patients (18 treated and 16 controls) were managed within three months of symptom onset and were treated for six weeks with a target hematocrit of 35%. Visual acuity improved in the treatment group from a mean of 20/100 – 2 units to 20/40 at one-year follow-up. In the control group, baseline visual acuity was similar (20/100 – 2) and improved to a mean of 20/80. However, in this study the results are confounded by the application of macular grid laser in 28% of the hemodiluted patients and 44% of the control group (59).

Wolfe et al. (60) added treatment with pentoxifylline for six months following hemodilution. At one-year follow-up the mean visual acuity improved by 1.5 lines in the 19 treated patients versus a decline of 1.5 lines in the 21 control patients. The mean arteriole–venous transit time normalized more quickly in the treatment group but was equal at one year in both groups. Plasma viscosity decreased with treatment. It is difficult in this study to determine whether the treatment was more beneficial for nonischemic or ischemic CRVO. In this study, CRVOs were considered ischemic when they met two or more of the following conditions: two disk areas of nonperfusion, visual acuity less than or equal to 20/200, greater than 10 cotton wool spots. The central vein occlusion study has demonstrated that a better benchmark for ischemic CRVO may be the presence of more than 10 disk areas of nonperfusion.

Plasmapheresis has been added to hemodilution as a further method to decrease viscosity and improve blood flow. Plasmapheresis has been performed in a study of five eyes with central retinal vein occlusion where all patients improved visual acuity by at least one line at final follow-up. These authors recommended use of plasmapheresis when hemodilution has failed.

In summary, there has been controversy over whether central and branch retinal vein occlusions are associated with increased plasma viscosity or elevated hematocrit and whether the results for nonischemic CRVO are better than the natural history. There have been further contradictory findings regarding the efficacy of this treatment for ischemic CRVO. Although promising results have been demonstrated, the technique requires evaluation with larger randomized controlled clinical trials.

Surgical Decompression Procedures

For both CRVO and BRVO the pathophysiology includes the formation of thrombosis within the venule, typically at the crossing point of an arteriole and venule in the case of BRVO, and at or just posterior to the lamina cribrosa in the case of CRVO. In both instances the compartment syndrome model has been proposed whereby the vein is partially compressed by the adjacent arteriole artery within a restricted space. This can lead to turbulence of blood flow, and secondary development of thrombosis. Surgical decompression of the vein or the surrounding structures has been proposed as a therapeutic procedure for both disorders.

Radial Optic Neurotomy

A compartment syndrome model has been proposed to explain the development of CRVO. According to this model, an unyielding scleral ring at the optic disk forms the boundary of the compartment. Thrombosis of the central retinal vein may occur when vascular thickening or collagen changes in the scleral ring allow increased pressure on the central retinal vein within a confined space.

Surgical decompression of the scleral ring from a posterior approach was first advocated by Vasco Posada in 1972 (61). Twenty-two patients with CRVO had the scleral ring and dural sheath of the optic nerve cut posterior to the globe. The baseline visual acuity was count fingers to 20/100 and improved to 20/80–20/20. The complication rate was high. In 1984, Arciniegas (62) used the same procedure on eyes of 44 patients with CRVO. Forty-eight percent had no change in visual acuity, 38.6% had improved visual acuity (that averaged 4.1 lines), and 13.6% had decreased visual acuity. This study lacked a control group.

Opremcak (63) reported an anterior approach to decompress the central retinal vein. In this procedure, a pars plana vitrectomy is followed by a radial incision at the nasal junction of the optic nerve and retina extended to the depth of the lamina cribrosa and adjacent sclera. In the initial report, 11 patients with baseline visual acuity of 20/400 or worse were treated with radial optic neurotomy. With an average follow-up of nine months, 7 of the 11 (73%) achieved 20/200 or better, and 5 of the 11 (45%) achieved 20/70 or better. Overall, 73% of the patients had improved visual acuity. All patients at baseline had macular edema. Five were deemed nonischemic and six were indeterminate. The author compared the results with the CVOS and concluded that this trial had favorable results as only 20% of patients in the CVOS who had a baseline visual acuity of less than or equal to 20/200 ended with a final visual acuity of greater than 20/200. In a larger series of 110 patients with initial visual acuity of 20/200 or worse, radial optic neurotomy was performed (64). Of 84 patients who had at least six-month data, 63% had improved visual acuity (one to nine lines) and 7% developed decreased visual acuity. Complications included vitreous hemorrhage in 6% and subretinal hemorrhage in 7%. The potential complication of laceration of one of the central vessels was not reported, nor was retinal detachment or globe perforation. In a cadaver histological study, Altaweel et al. found that the potential for such perforation was reduced by approaching the radial optic neurotomy incision with the blade entered from the nasal sclerotomy (Fig. 6) (65).

Hayreh (66) has argued that since the lamina cribrosa comprises a firm rather than elastic band of collagen tissue, an incision in one aspect should not relax the tissue and allow decompression of the central retinal vein. He has also commented that as the thrombosis is often posterior to the lamina cribrosa, decompression at

(A)

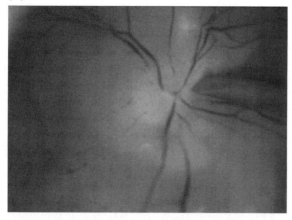

(B)

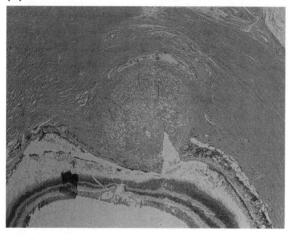

Figure 6 (**A**) Radial optic neurotomy involves incision of the optic nerve sheath with an MVR blade at the junction of the optic disk and retina. (**B**) (*See color insert*) Cross section of the optic nerve demonstrates sectioning of the scleral ring in a cadaver eye. *Abbreviation*: MVR, microvitreoretinal.

the lamina cribrosa may not affect the thrombus. As the circle of Zinn–Haller is cut with this procedure, it is possible that the procedure could adversely affect the circulation of the optic nerve head. Further evaluation of this procedure would benefit from a randomized controlled clinical trial.

An alternate method of decompressing the central retinal vein has been described (67). In an animal study using enucleated pig eyes and rabbit eyes in vivo, a blade with one sharp side and one blunt side was directed posteriorly adjacent to the vessels rather than at the edge of the optic nerve as performed in radial optic neurotomy. In "lamina puncture" the blunt side is adjacent to the central retinal vessels and a penetration is conducted to strip connective tissue from around the vessel wall. It is proposed that this procedure would release constriction of the central retinal vein by surrounding connective tissue. The vein lumen would then increase, which may allow improved blood flow around the thrombus or passage of the thrombus.

Arteriovenous Sheathotomy for BRVO

BRVO occurs at the crossing site of an arteriole and venule. The adventia of the two vessels are fused at the crossing site. The arteriole may compress the venule and alter the blood flow in the venule, leading to secondary thrombosis (68). In a histological study, medial layer hypertrophy in the arteriole wall was observed in 90% of cases (7). The first reported case of attempted separation of the arteriole and venule to allow decompression of the venule and improved blood flow or passage of the thrombosis was from Osterloh and Charles (69). They performed the procedure on 12 porcine eyes, 2 eye-bank eyes, and 1 human eye. In this technique, a pars plana vitrectomy was performed and the adventitia was separated from the vein with an intraocular horizontal scissors. The patient's visual acuity improved from 20/200 to 20/25 over a period of eight months. In a prospective series, Opremcak and Bruce (70) reported on 15 patients who had pars plana vitrectomy, elevation of the intraocular pressure, and separation of the arteriole from the venule with a bent microvitreoretinal (MVR) blade. An inner retinal incision was created along the vessel walls proximal to the arteriovenous crossing and the blade was slid along the vessel towards the common adventitia, which was incised, allowing separation of the vessels (Fig. 7). In this series, the average BRVO duration was five months, and the average patient age was 69 years. The preoperative visual acuity was less than 20/70 in all patients and less than 20/200 in many. The average follow-up duration was 3.3 months (range, 1–12 months). Twelve eyes (80%) had stabilized or improved visual acuity. Sixty-seven percent gained visual acuity, and, in this group, the mean gain was four lines. Twenty percent had decreased visual acuity, (that averaged two lines). Many of the patients included in this study were not eligible for standard treatment with grid laser as managed in the BRVO study, as 46% of the patients in this trial had macular nonperfusion. All patients improved clinically with decreased macular edema, decreased intraretinal hemorrhage, and improved perfusion on fluorescein angiography. In a later unpublished prospective nonrandomized series of 50 eyes, visual acuity improved in 76% (a mean of 4.5 lines in this group) and outcome was not related to initial perfusion status.

One small report of three eyes undergoing this procedure contradicts these findings, with visual acuity remaining unchanged in two patients and decreasing in one. In these three cases, increased areas of nonperfusion were noted. It is possible that the vessels were damaged during the procedure (70).

Shah (71) reported 6.5-years follow-up in an uncontrolled retrospective study of five eyes with BRVO and initial visual acuity of 20/200 or worse that were treated with arteriovenous sheathotomy. Four of the five (80%) obtained a visual acuity of 20/70 or better at the extended follow-up.

The only study to include a control group is that of Mester and Dillinger (72). Forty-three study eyes and 25 controls with entry visual acuity of less than 20/40 were enrolled. This study is not completely comparable with the other studies as all patients had isovolemic hemodilution for 10 days and the treatment group had the surgical procedure as well. The follow-up period in this study was extremely short (six weeks) and the mean duration of venous obstruction was only 4.6 weeks. In the treatment group, the mean visual acuity improved from 20/40−2 to 20/30+1 and in the control group it decreased from 20/30−1 to 20/30−2. One-third of the control group experienced a decrease in visual acuity over the six-week period. However, in the treatment group, 60% had improved visual acuity by at least two lines and 28% improved by four lines or more. Only one out of 43 patients experienced a moderate decrease in visual acuity (more than or equal to two lines). Macular

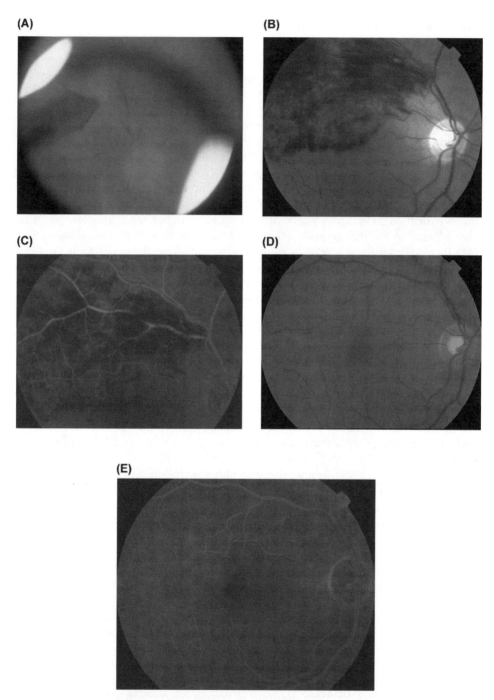

Figure 7   (**A**) (*See color insert*) In branch vein desheathing surgery, a bent MVR blade or scissors are used to incise the common adventitia at the arteriovenous crossing allowing separation of the vessels. (**B**) Eye with BRVO has extensive intraretinal hemorrhage, (**C**) and macular edema corresponding to quadrantic area of fluorescein leakage, in part obscured by intraretinal hemorrhage. (**D**) Following treatment, there is resolution of intraretinal hemorrhages, and (**E**) decreased venous caliber, improved flow, and resolution of leakage on a fluorescein angiogram. *Abbreviations*: BRVO, branch retinal vein occlusion; MVR, microvitreoretinal.

edema and intraretinal hemorrhage resolved more in the treatment group than in the control group (72).

The potential complications with this procedure include vitreous hemorrhage, nerve fiber layer defect, retinal detachment, accelerated cataract formation, and increased retinal nonperfusion. The rates of these complications have been low in the reported studies.

Laser-Induced Chorioretinal Venous Anastomosis for Treatment of CRVO

Macular edema secondary to CRVO occurs largely because of venous outflow obstruction. McAllister and Constable (73) devised a method to create new venous outflow routes. In this technique, Bruch's membrane is ruptured with a high-energy laser directed adjacent to the wall of a venule distal to the obstruction to create an anastomosis with the choroidal circulation. They initially demonstrated the feasibility of the procedure in dog and rat models. Subsequently, they reported a case series of 24 eyes with nonischemic CRVO. An argon green laser with a 50 μm spot size, a pulse duration of 0.1 seconds, and 1.5–2.5 W of energy was used to disrupt Bruch's membrane and the venule wall. One to five laser attempts were required. A successful anastamosis was created in 8 of the 24 eyes. Flow was found in the anastomotic region at 3–7 weeks postprocedure. Six of the eight patients had an initial visual acuity of 20/200 or worse, and three of these six improved more than two lines. Two patients with an initial visual acuity of 20/120 improved to 20/30. The eight patients with successful anastomosis remained nonischemic, whereas five of the remaining 16 (31%) converted to ischemic CRVO. In subsequent reports, other authors have used a YAG laser for the same purpose.

Significant adverse events were reported with laser chorioretinal anastamosis. Complications included hemorrhage from the retinal vein in 40%, subretinal hemorrhage in 7%, and choroidal hemorrhage in 6%. These hemorrhages generally resolved spontaneously. Late consequences included branch retinal vein occlusion in 11%, preretinal (13%), and subretinal fibrosis (5%) at the site of laser application, vitreous hemorrhage, and neovascularization associated with conversion to ischemic central retinal vein occlusion.

The rate of anastomosis formation was similarly low in a series of 24 eyes collected by Fekrat et al. (74). Nine eyes (38%) had successful anastomosis formation. Of these nine, visual acuity improved in four and the remainder had no change. Forty-two percent of eyes developed a transient vitreous hemorrhage and 21% developed localized choroidal neovascularization at the site of the laser treatment. Overall, visual acuity decreased by at least one line in 63%.

Browning and Antoszyk (75) created an anastomosis in only two of eight eyes. These two eyes did not have improved visual acuity while two others without successful anastomosis did have improved visual acuity. Two of the eight patients developed tractional retinal detachment, three developed iris neovascularization, and one developed neovascular glaucoma.

Overall, chorioretinal anastamosis is a procedure that may be best performed for nonischemic CRVO as the major complication of pre- and subretinal fibrosis is more likely in ischemic CRVO. These uncontrolled trials have not proven that this procedure is superior in outcome to the natural history of non-ischemic CRVO. Additionally, the risks as demonstrated in these trials are significant. As a result, this procedure has not gained wide acceptance.

Quiroz-Mercado et al. (76) have described the combination of pars plana vitrectomy with posterior hyaloid detachment and formation of a chorioretinal venous anastomosis with YAG laser. It was postulated that the addition of the vitrectomy may prevent formation of preretinal fibrosis as described in previous trials. Although the initial results indicated moderate benefit in two patients, this modality requires more study. In one case study described as having a pars plana vitrectomy followed by the use of an MVR blade to create an anastomosis for the treatment of ischemic CRVO, preretinal fibrosis still developed and visual acuity improvement was minimal.

Laser-Induced Chorioretinal Venous Anastomosis for BRVO

As in the management for CRVO, the purpose of such a procedure is to reestablish venous outflow by creating a chorioretinal anastomosis distal to the site of obstruction, to attempt to improve visual acuity by decreasing macular edema, and to decrease the conversion from nonischemic to ischemic vein occlusion. Fekrat and de Juan (77) reported on six eyes with branch vein occlusion of whom, three had successful anastomoses. Of the six eyes, the visual acuity improved one to three lines in two, remained unchanged in one, and decreased in three. This result does not appear to be significantly better than the natural history of visual acuity loss due to BRVO.

Pars Plana Vitrectomy for Macular Edema Secondary to Retinal Vein Occlusion

Vitrectomy has been advocated by some authors to manage retinal vein occlusion with persistent macular edema. It has been postulated that removal of the vitreous and elevation of the hyaloid may decrease breakdown of the blood–retina barrier and secondary leakage from blood vessels, and as well may allow better maintenance of preretinal oxygen tension. Stefansson et al. (78) created branch retinal vein occlusion in 10 nonvitrectomized and 5 vitrectomized cat eyes. In the vitrectomized eyes, there was no significant alteration in oxygen tension at the retinal surface whereas in the nonvitrectomized eyes there was a significant reduction, from 20 to 6 mmHg. They postulated that vitrectomy may, therefore, prevent the development of neovascularization (78).

Saika et al. (79) reported on a group of 19 eyes that had pars plana vitrectomy, elevation of the posterior hyaloid, fluid gas exchange, and cataract extraction with intraocular lens implant. Ten of the 19 eyes had decreased in macular edema as identified on optical coherence tomography; the mean thickness decreased from 383 to 208 μm. However, there was no statistically significant improvement in visual acuity. If only patients with more recent onset of branch retinal vein occlusion were included in the analysis, a visual improvement was noted.

Another study evaluated 29 eyes with BRVO (average duration of 9.5 months) and 14 eyes with CRVO (average duration of 2.8 months), all with macular edema. Treatment included pars plana vitrectomy, elevation of the posterior hyaloid, cataract extraction, and intraocular lens implantation. Most patients had received prior unsuccessful grid laser. At one year, the branch vein occlusion group had improved from 20/50−2 to 20/30 and the CRVO group had improved from 20/40 to 20/30 + 1. Adverse events included the development of retinal tears in four of 43 eyes and a macular hole in one. Tachi et al. (80) concluded that this procedure was beneficial in reducing macular edema and improving visual acuity. However, this form of treatment carries with it the risk of vitrectomy and, to date, has not been compared with a control group.

SUMMARY

There have been many therapeutic approaches to manage CRVO and BRVO which attempt to address the primary pathology (thrombosis at or just posterior to the lamina cribrosa in CRVO, thrombosis at the arteriovenous crossing in BRVO) or to address the secondary consequences of the occlusion (macular edema, ischemia, neovascularization). Many of these studies suffer from lack of a randomized controlled prospective design and others have small control groups. Differing entry criteria may limit the applicability of the results of one study to the general population. The current plethora of varying treatments for retinal vein occlusions points to the lack of a definitive therapy.

Currently, a local drug delivery approach has been most favored because it is possible to avoid systemic toxicity with this method. As discussed earlier, corticosteroids given by intravitreal injection, or in a sustained drug delivery system may be an effective method to treat macular edema secondary to CRVO and BRVO. However, the case series discussed previously have described the potential adverse side effects of intravitreal kenalog administration which include secondary cataract formation, glaucoma, and injection-related side effects such as endophthalmitis and retinal detachment. This stresses the importance of randomized controlled trials of a longer duration. The SCORE (Standard Care vs. Corticosteroid for Retinal Vein Occlusion) study is a National Eye Institute–funded Phase III controlled, randomized clinical trial that will compare a preservative-free triamcinolone preparation versus grid laser photocoagulation to treat patients with BRVO and secondary macular edema and versus observation in individuals with CRVO and macular edema. Other steroid applications that will be tested for these conditions include an injectable biodegradable dexamethasone implant and a sustained release, nonbiodegradable fluocinolone acetonide intravitreal implant.

In addition to local drug delivery of corticosteroids to the eye to treat retinal vein occlusion, many other pharmaceutical agents are either in development or are in clinical testing. These agents have anti-permeability and anti-angiogenic properties that may address the common complications of retinal vascular occlusions. For example, the anti-VEGF aptamer, pegaptanib sodium (macugen), is currently being evaluated in clinical trials as a treatment for macular edema associated with CRVO. This compound will be delivered to the eye by intravitreal injection. Other compounds that may eventually enter clinical trials for retinal vein occlusion include antibodies directed against VEGF (ranibizumab and bevacizumab), modified steroid compounds and small interfering RNA (siRNA) technology among others.

Within the next five years there will likely be further data available to help guide the management of CRVO and BRVO with its associated macular edema. The future for patients with these conditions looks more promising than ever because of new pharmacologic agents that can be effectively and efficiently delivered to the posterior segment of the eye.

Side Bar: Authors' intravitreal injection procedure

- Patient is reclined at 60° or is placed in supine position.
- Preoperative application of fourth-generation fluoroquinolone, one drop q. 5 min × 3.
- Anesthetic: proparacaine drops plus apply a proparacaine-soaked cotton-tipped pledget to the site of injection with moderate pressure for 30 seconds.

- Prep: povidone iodine solution 5%, one drop 5 minutes before procedure and repeat just before procedure; cleanse lids with iodine-soaked cotton-tipped pledget.
- Place lid speculum.
- Drape—optional.
- Site marking: utilize calipers or the hub of a TB syringe to mark 3.5–4 mm from the limbus in the inferior or inferotemporal location.
- Injection: $0.1\,cm^3$ (4 mg) of triamcinolone acetonide through a 27 gauge needle on a $1\,cm^3$ syringe; insert needle to one-half depth directed towards optic nerve. Inject slowly. Withdraw needle and rub site with sterile q-tip. Apply one drop of antibiotic and remove lid speculum.
- Examination with indirect ophthalmoscopy to ensure proper location of steroid, perfusion at optic nerve, and lack of complications.
- If visual acuity has decreased to no light perception and very high intraocular pressure is found, consider paracentesis.
- Recheck intraocular pressure in 5–10 minutes.
- Use antibiotic four times a day for 3–4 days.
- Follow-up examination day three to seven.
- Symptoms of decreasing visual acuity, ocular pain, or conjunctival erythema should lead to early reassessment.

REFERENCES

1. The Branch Vein Occlusion Study Group. Argon laser photocoagulation for macular edema in branch vein occlusion. Am J Ophthalmol 1984; 98:271–282.
2. Central Vein Occlusion Study Group. Baseline and early natural history report. The Central Vein Occlusion Study. Arch Ophthalmol 1993; 111:1087–1095.
3. The Central Vein Occlusion Study Group. Natural history and clinical management of central retinal vein occlusion. Arch Ophthalmol 1997; 115:486–491.
4. The Central Vein Occlusion Study Group. Evaluation of grid pattern photocoagulation for macular edema in central vein occlusion. The Central Vein Occlusion Study Group M Report. Ophthalmology 1995; 102:1425–1433.
5. Central Vein Occlusion Study Group. A randomized clinical trial of early panretinal photocoagulatioin for the treatment of macular edema associated with central retinal vein occlusion. Central Retinal Vein Occlusion Study Group N. Arch Ophthalmol 1995; 102:1434–1444.
6. Green WR, Chan CC, Hutchins GM, Terry JM. Central retinal vein occlusion: a prospective histopathologic study of 29 eyes in 28 cases. Trans Am Ophthalmol Soc 1981; 79:371–422.
7. Frangieh GT, Green WR, Barraquer-Somers E, Finkelstein D. Histopathologic study of nine branch retinal vein occlusions. Arch Ophthalmol 1982; 100:1132–1140.
8. Hockley DJ, Tripathi RC, Ashton N. Experimental branch vein occlusion in rhesus monkeys. III. Histopathological and electron microscopical studies. Br J Ophthalmol 1979; 63:393–411.
9. Aiello LP, Bursell SE, Clermont A, et al. Vascular endothelial growth factor-induced retinal permeability is mediated by protein kinase C in vivo and suppressed by an orally effective beta-isoform-selective inhibitor. Diabetes 1997; 46:1473–1480.
10. Antonetti DA, Barber AJ, Hollinger LA, Wolpert EB, Gardner TW. Vascular endothelial growth factor induces rapid phosphorylation of tight junction proteins occludin and zonula occluden 1. J Biol Chem 1999; 274:23,463–23,467.

11. Senger DR, Galli SJ, Dvorak AM, Peruzzi CA, Harvey VS, Dvorak HF. Tumor cells secrete a vascular permeability factor (VPF) that promotes accumulation of ascites fluid. Science 1983; 219:983–985.
12. Vinores SA, Youssri AI, Luna JD, et al. Upregulation of vascular endothelial growth factor in ischemic and non-ischemic human and experimental retinal disease. Histol Histopathol 1997; 12:99–109.
13. Pe'er J, Folberg R, Itin A, Gnessin H, Hemo I, Keshet E. Vascular endothelial growth factor upregulation in human central retinal vein occlusion. Ophthalmology 1998; 105: 412–416.
14. Adamis AP, Shima DT, Tolentino MJ, et al. Inhibition of vascular endothelial growth factor prevents retinal ischemia-associated iris neovascularization in a nonhuman primate. Arch Ophthalmol 1996; 114:66–71.
15. Nauck M, Karakiulakis G, Perruchoud AP, Papakonstantinou E, Roth M. Corticosteroids inhibit the expression of the vascular endothelial growth factor gene in human vascular smooth muscle cells. Eur J Pharmacol 1998; 341:309–315.
16. Nauck M, Roth M, Tamm M, et al. Induction of vascular endothelial growth factor by platelet-activating factor and platelet-derived growth-factor is downregulated by corticosteroids. Am J Resp Cell Mol Biol 1997; 16:398–406.
17. McCuen B, Bressler N, Tano Y, Chandler D, Machemer R. The lack of toxicity of intravitreally administered triamcinolone acetonide. Am J Ophthalmol 1981; 91:785–788.
18. Schindler RH, Chandler DB, Thresher R, Machemer R. The clearance of intravitreal trimacinolone acetonide. Am J Ophthalmol 1982; 93:415–417.
19. Scholes GN, O'Brien WJ, Abrams GW, Kubicek MF. Clearance of triamcinolone from vitreous. Arch Ophthalmol 1985; 103:1567–1569.
20. Folkman J, Ingber DE. Angiostatic steroids. Ann Surg 1987; 206:374–383.
21. Diaz-Flores L, Gutierrez R, Varela H. Angiogenesis: an update. Histol Histopathol 1994; 9:807–843.
22. Yoshikawa K, Kotake S, Ichiishi A, Sasamoto Y, Kosaka S, Matsuda H. Posterior sub-Tenon injections of repository corticosteroids in uveitis patients with cystoid macular edema. Jpn J Ophthalmol 1995; 39:71–76.
23. Thach AB, Dugel PU, Flindall RJ, Sipperley JO, Sneed SR. A comparison of retrobulbar versus sub-Tenon's corticosteroid therapy for cystoid macular edema refractory to topical medications. Ophthalmology 1997; 104:2003–2008.
24. Tano Y, Chandler D, Machemer R. Treatment of intraocular proliferation with intravitreal injection of triamcinolone acetonide. Am J Ophthalmol 1980; 90:810–816.
25. Machemer R, Sugita G, Tano Y. Treatment of intraocular proliferations with intravitreal steroids. Trans Am Ophthalmol Soc 1979; 77:171–178.
26. Antoszyk AN, Gottlieb JL, Machemer R, Hatchell DL. The effects of intravitreal triamcinolone acetonide on experimental pre-retinal neovascularization. Graefes Arch Clin Exp Ophthalmol 1993; 231:34–40.
27. Danis RP, Bingaman DP, Yang Y, Ladd B. Inhibition of preretinal and optic nerve head neovascularization in pigs by intravitreal triamcinolone acetonide. Ophthalmology 1996; 103:2099–2104.
28. Jonas JB, Hayler JK, Panda-Jonas S. Intravitreal injection of crystalline cortisone as adjunctive treatment of proliferative vitreoretinopathy. Br J Ophthalmol 2000; 84:1064–1067.
29. Penfold PL, Gyory JF, Hunyor AB, Billson FA. Exudative macular degeneration and intravitreal triamcinolone. A pilot study. Aust N Z J Ophthalmol 1995; 23:293–298.
30. Challa JK, Gillies MC, Penfold PL, Gyory JF, Hunyor ABL, Billson FA. Exudative macular degeneration and intravitreal triamcinolone: 18 month follow up. Aust N Z J Ophthalmol 1998; 26:277–281.
31. Danis RP, Ciulla TA, Pratt LM, Anliker W. Intravitreal triamcinolone acetonide in exudative age-related macular degeneration. Retina 2000; 20:244–250.
32. Jonas JB, Sofker A. Intraocular injection of crystalline cortisone as adjunctive treatment of diabetic macular edema. Am J Ophthalmol 2001; 132:425–427.

33. Martidis A, Duker JS, Greenberg PB, et al. Intravitreal triamcinolone for refractory diabetic macular edema. Ophthalmology 2002; 109:920–927.
34. Martidis A, Rogers AH, Greenberg PB, et al. Intravitreal triaminoclone acetonide for refractory diabetic macular edema. Invest Ophthalmol Vis Sci (suppl) 2001; 42:S741.
35. Greenberg PB, Martidis A, Rogers AH, Duker JS, Reichel E. Intravitreal triamcinolone acetonide for macular oedema due to central retinal vein occlusion. Br J Ophthalmol 2002; 86:247–248.
36. Ip MS, Gottlieb JL, Kahana A, et al. Intravitreal triamcinolone for the treatment of macular edema associated with central retinal vein occlusion. Arch Ophthalmol 2004; 122:1131–1136.
37. Jonas JB, Kreissig I, Degenring RF. Intravitreal triamcinolone acetonide as treatment of macular edema in central retinal vein occlusion. Graefes Arch Clin Exp Ophthalmol 2002; 240:782–783.
38. Park CH, Jaffe GJ, Fekrat S. Intravitreal triamcinolone acetonide in eyes with cystoid macular edema associated with central retinal vein occlusion. Am J Ophthalmol 2003; 136:419–425.
39. Bashshur ZF, Ma'luf RN, Allam S, Jurdi FA, Haddad RS, Noureddin BN. Intravitreal triamcinolone for the management of macular edema due to nonischemic central retinal vein occlusion. Arch Ophthalmol 2004; 122:1137–1140.
40. Krepler K, Ergun E, Sacu S, et al.Intravitreal triamcinolone acetonide in patients with macular oedema due to central retinal vein occlusion. Acta Ophthalmol Scand 2005; 83:71–75.
41. Costen MT, Donaldson WB, Olson JA. Acute central retinal vein occlusion successfully treated with intravenous thrombolysis. Br J Ophthalmol 1999; 83:1196–1197.
42. Kohner EM, Hamilton AM, Bulpitt CJ, Dollery CT. Streptokinase in the treatment of central retinal vein occlusion. Trans Ophthalmol Soc UK 1974; 94:599–603.
43. Kohner EM, Pettit JE, Hamilton AM, Bulpitt CJ, Dollery CT. Streptokinase in central retinal vein occlusion: a controlled clinical trial. Br Med J 1976; 1:550–553.
44. Elman MJ. Thrombolytic therapy for central retinal vein occlusion: results of a pilot study. Trans Am Ophthalmol Soc 1996; 94:471–504.
45. Paques M, Vallee JN, Herbreteau D, et al. Superselective ophthalmic artery fibrinolytic therapy for the treatment of central retinal vein occlusion. Br J Ophthalmol 2000; 84: 1387–1391.
46. Elman MJ, Raden RZ, Carrigan A. Intravitreal injection of tissue plasminogen activator for central retinal vein occlusion. Trans Am Ophthalmol Soc 2001; 99:219–221; discussion 222–223.
47. Glacet-Bernard A, Kuhn D, Vine AK, Oubraham H, Coscas G, Soubrane G. Treatment of recent onset central retinal vein occlusion with intravitreal tissue plasminogen activator: a pilot study. Br J Ophthalmol 2000; 84:609–613.
48. Lahey JM, Fong DS, Kearney J. Intravitreal tissue plasminogen activator for acute central retinal vein occlusion. Ophthalmic Surg Lasers 1999; 30:427–434.
49. Weiss JN, Bynoe LA. Injection of tissue plasminogen activator into branch retinal vein in eyes with central retinal vein occlusion. Ophthalmology 2001; 108:2249–2257.
50. Glacet-Bernard A, Zourdani A, Milhoub M, Maraqua N, Coscas G, Soubrane G. Effect of isovolemic hemodilution in central retinal vein occlusion. Graefes Arch Clin Exp Ophthalmol 2001; 239:909–914.
51. Sonkin PL, Sinclair SH, Hatchell D. The effect of pentoxyfylline on retinal capillary blood flow velocity and whole blood velocity. Am J Ophthalmol 1993; 115:775–780.
52. Boisseau MR, Freyberger G, Busquet M, Beylot C. Pharmacological aspects of erythrocyte aggregation. Effect of high doses of troxerutin. Clin Haemorheol 1989; 9:871–876.
53. Glacet-Bernard A, Coscas G, Chabanel A, Zourdani A, Lelong F, Samama MM. A randomized, double-masked study on the treatment of retinal vein occlusion with troxerutin. Am J Ophthalmol 1994; 118:421–429.

54. Hattenbach LO, Wellermann G, Steinkamp GW, Sharrer I, Koch FH, Ohrloff C. Visual outcome after treatment with low dose recombinant tissue plasminogen activator or hemodilution in ischemic central retinal vein occlusion. Ophthalmologica 1999; 213:360–366.

55. Hansen LL, Weik J, Schade M, Muller-Stolzenburg N, Wiederholt M. Effect and compatibility of isovolaemic haemodilution in the treatment of ischaemic and nonischaemic central retinal vein occlusion. Ophthalmologica 1989; 199:90–99.

56. Hansen LL, Daniesevskis P, Arntz HR, Hovener G, Wiederholt M. A randomized prospective study on treatment of central retinal vein occlusion by isovolaemic haemodilution and photocoagulation. Br J Ophthalmol 1985; 69:108–116.

57. Weik J, Schade M, Widerholt M, Arntz HR, Hansen LL. Haemorheological changes in patients with retinal vein occlusion after isovolaemic haemodilution. Br J Ophthalmol 1990; 74:665–669.

58. Luckie AP, Wroblewski JJ, Hamilton P, et al. A randomised prospective study of outpatient haemodilution for central retinal vein obstruction. Aust N Z J Ophthalmol 1996; 24:223–232.

59. Chen HC, Wiek J, Gupta A, Luckie A, Kohner EM. Effect of isovolaemic haemodilution on visual outcome in branch retinal vein occlusion. Br J Ophthalmol 1998; 82:162–7.

60. Wolfe S, Arend O, Bertram B, et al. Hemodilution therapy in central retinal vein occlusion. One-year results of a prospective randomized study. Graefes Arch Clin Exp Ophthalmol 1994; 232:33–39.

61. Vasco Posada J. Modification of the circulation in the posterior pole of the eye. Ann Ophthalmol 1972; 4:48–59.

62. Arciniegas A. Treatment of the occlusion of the central retinal vein by section of the posterior ring. Ann Ophthalmol 1984; 16:1081–1086.

63. Opremcak EM, Bruce RA, Lomeo MD, Ridenour CD, Letson AD, Rehmar AJ. Radial optic neurotomy for central retinal vein occlusion: a retrospective pilot study of 11 consecutive cases. Retina 2001; 21:408–415.

64. Opremcak EM. Radial optic neurotomy for central retinal vein occlusion. 20th Annual Vitreous Society Meeting, 35th Annual Retina Society Meeting (abstract), 2002, 135.

65. Altaweel MM, Freisberg L, Dawson D, Gleiser J, Ryan E, Albert D. Radial optic neurotomy for central retinal vein occlusion: a histologic perspective. 20th Annual Vitreous Society Meeting, 35th Annual Retina Society Meeting (abstract), 2002, 136.

66. Hayreh SS. Radial optic neurotomy for central retinal vein occlusion. Retina 2002; 22:827.

67. Lit ES, Tsilimbaris M, Gotzaridis E, D'Amico DJ. Lamina puncture: pars plana optic disc surgery for central retinal vein occlusion. Arch Ophthalmol 2002; 120:495–499.

68. Cahill MT, Fekrat S. Arteriovenous sheathotomy for branch retinal vein occlusion. Ophthalmol Clin North Am 2002; 15:417–442.

69. Osterloh MD, Charles S. Surgical decompression of branch retinal vein occlusions. Arch Ophthalmol 1998; 106:1469–1471.

70. Opremcak EM, Bruce RA. Surgical decompression of branch retinal vein occlusion via arteriovenous crossing sheathotomy: a prospective review of 15 cases. Retina 1999; 19:1–5.

71. Shah GK. Adventitial sheathotomy for treatment of macular edema associated with branch retinal vein occlusion. Invest Ophthalmol Vis Sci 2000; 41:877–879.

72. Mester U, Dillinger P. Vitrectomy with arteriovenous decompression and internal limiting membrane dissection in branch retinal vein occlusion. Retina 2002; 22:740–746.

73. McAlister IL, Constable IJ. Laser-induced chorioretinal venous anastomosis for treatment of nonischemic central retinal vein occlusion. Arch Ophthalmol 1995; 113:456–462.

74. Fekrat S, Goldberg MF, Finkelstein D. Laser-induced chorioretional venous anastomosis for nonischemic central or branch retinal vein occlusion. Arch Ophthmol 1998; 116: 43–52.

75. Browning DJ, Antosyzk AM. Laser chorioretinal anastamosis for nonischemic central retinal vein occlusion. Ophthalmologica 1998; 212:389–393.

76. Quiroz-Mercado H, Sanchez-Buenfil E, Guerro-Naranjo JL, et al. Successful erbium: YAG laser-induced chorioretinal venous anastomosis for the management of ischemic central retinal vein occlusion. A report of two cases. Graefes Arch Clin Exp Ophthalmol 2001; 239:872–875.
77. Fekrat S, de Juan E. Chorioretinal venous anastomosis for central retinal vein occlusion: transvitreal venipuncture. Ophthalmic Surg Lasers 1999; 30:52–55.
78. Stefansson E, Novack RL, Hatchell DL. Vitrectomy prevents retinal hypoxia in branch retinal vein occlusion. Invest Ophthalmol Vis Sci 1990; 31:284–289.
79. Saika S, Tanaka T, Miyamoto T, Ohnishi Y. Surgical posterior vitreous detachment combined with gas/air tamponade for treating macular edema associated with branch retinal vein occlusion: retinal tomography and visual outcome. Graefes Arch Clin Exp Ophthalmol 2001; 239:729–732.
80. Tachi N, Hashimoto Y, Ogini N. Vitrectomy for macular edema combined with retinal vein occlusion. Doc Ophthalmol 1999; 97:465–469.

21

Cytomegalovirus Retinitis

Caroline R. Baumal
*Department of Ophthalmology, Vitreoretinal Service, New England Eye Center,
Tufts University School of Medicine, Boston, Massachusetts, U.S.A.*

INTRODUCTION

Cytomegalovirus (CMV) retinitis typically develops in severely immunocompromised individuals. CMV retinal infection was extremely rare until the early 1980s when its incidence rose rapidly due to its occurrence in immunosuppressed patients with acquired immunodeficiency syndrome (AIDS) (1). It may also develop in individuals with reduced systemic immunity secondary to organ transplantation, congenital, or acquired immunosuppressive disorders, medications, malignancy, or congenital CMV infection (2). Improvements in organ transplantation and systemic immunosuppressive medications have drastically increased survival of immunocompromised individuals, who are at greater overall risk of developing opportunistic infections such as CMV retinitis.

CMV produces progressive retinal destruction leading to blindness unless anti-CMV treatment is commenced and/or the underlying cause of systemic immunosuppression is reversed (3). Multiple new therapies and different routes of medication delivery have been investigated to treat this previously devastating infection in an effort to prevent or limit visual loss.

Data regarding the clinical presentation of CMV retinitis, pattern of infection, disease course, and complications have been obtained in the last two decades because of its association with AIDS. Therapy for AIDS has been greatly improved since the mid-1990s with the development and use of highly active antiretroviral therapy (HAART). HAART therapy improves systemic immune function in many patients with AIDS, and thus has drastically altered the incidence and clinical features of CMV retinitis and other opportunistic infections (4).

VIROLOGY AND EPIDEMIOLOGY OF CMV INFECTION

Cytomegalovirus is a double-stranded DNA virus belonging to the human herpesvirus family (1,5). The other members include herpes simplex virus (HSV), varicella zoster virus (VZV), and Epstein–Barr virus (EBV). While the herpesviruses are indistinguishable by electron microscopy, CMV is specifically diagnosed by its characteristic appearance on histopathology and culture, as well as antigenic features (5).

Systemic CMV infection was initially described in 1905 (6). It had been known as salivary gland disease, and then as cytomegalic inclusion disease based on the characteristic cellular inclusions produced by CMV on light microscopy. Transmission of CMV virus can occur by body secretions, infected blood products, or via the placenta (1). The incidence of CMV seropositivity increases with age and exposure to CMV has usually occurred by early adulthood. By age 35, nearly all persons living in North America have been exposed to CMV and over 80% demonstrate detectable circulating viral antibodies (7,8). The prevalence of CMV may be particularly high in AIDS due to the potential for sexual transmission of CMV virus. The incidence of newborn CMV infection ranges from 0.5% to 2.5%, and fortunately, only 10% of serologically positive newborns will show clinical manifestations of the virus (2,7).

CMV infection rarely produces clinical disease in immunocompetent persons, but it can produce considerable morbidity and mortality in immunocompromised individuals. Severe CMV infection is typically associated with abnormal cell-mediated immunity. It affects two major categories of individuals; immunosuppressed patients and newborns. After exposure and systemic infection with CMV, the virus typically attains a dormant state within specific host cells in a similar fashion to other members of the herpesvirus family (9). CMV may infect various sites in immunosuppressed individuals including the reticuloendothelial system, liver, kidneys, lungs, gastrointestinal system, and the central nervous system. One of the most common target sites for CMV infection is the retina. The features of CMV retinitis were described in the 1950s (10–12). In 1964, this virus was identified as the causative agent of CMV retinitis (12,13). During periods of severe immunosuppression, CMV virus appears to reactivate and infect the retina by hematogenous spread (14). The development of new onset, active CMV retinitis may indicate the presence of systemic CMV infection, although this can be subclinical or asymptomatic. The presence and site(s) of active or symptomatic CMV infection is important when considering the mode of delivery of antiviral medication directed against CMV.

Infection with human immunodeficiency virus (HIV) is the most common cause of immunosuppression leading to reactivation of CMV and symptomatic infection. Severe CMV infection has also been associated with congenital immunodeficiency syndromes, pharmacologic immunosuppression, organ transplantation, malignancy, and autoimmune disorders (2,6). When CMV disease occurs in AIDS, it typically manifests as a retinal infection in well over 70% of patients although it can affect other sites. CMV was the major cause of ocular morbidity in the late stages of AIDS (9,15). Prior to the introduction of effective highly active anti-retroviral therapy (HAART) for individuals with HIV infection, estimates for the prevalence of CMV retinitis varied between 10% and 40% (15–18). CMV retinitis was bilateral at presentation in approximately one-third of patients and it was the AIDS-defining diagnosis in about 5% of HIV infections (16,18,19). Certain AIDS subpopulations have a lower incidence of CMV retinitis such as pediatric patients, and this may result from lower CMV seropositivity rates in this population (16,20–22). The onset of CMV retinitis appears to be highly dependent on the CD4+ T-lymphocyte cell count in adults with AIDS. In children, an age-adjusted CD4+ count should be considered (2). The risk for CMV retinitis increases as the number of CD4+ cells or their function is diminished. The mean CD4+ cell count at the time of diagnosis of CMV retinitis in AIDS is typically less than 50 cells/mm^3.

It is rare for patients with CD4+ counts greater than 200 cells/mm^3 to develop this disease, although it could potentially occur in an individual who previously had a very low CD4+ count less than 50 cells/mm^3 who is initially experiencing a favorable response to HAART with rapid elevation of the CD4+ count (4).

CMV RETINITIS IN THE ERA OF HAART

Highly active antiretroviral therapy, also known as HAART, refers to a combination of medications to treat HIV infection including protease inhibitors and/or nucleoside analogues. It has been defined as an antiretroviral regimen that can be expected to reduce the viral load to less than 50 copies per milliliter in treatment of naive patients (23). The use of HAART in developed countries commenced in 1995–1996 and it has markedly improved survival in AIDS. The increased CD4+ lymphocyte cell count and the decreased HIV viral load in the peripheral blood secondary to HAART in patients who respond to this therapy, results in a reduction of all opportunistic infections including CMV retinitis (4). Since the introduction of HAART, there has been reduction in the incidence of primary and relapsing CMV disease, CMV viral load and antigenemia (24). The North American incidence of CMV retinitis in the post-HAART era has decreased to approximately one-fourth that of the pre-HAART era (4,24–27).

The CD4+ lymphocyte cell count continues to be a reliable indicator of the immune status while undergoing HAART therapy in most instances. Prior to HAART, immunocompromised individuals with CMV retinitis required anti-CMV medications for the duration of their immune suppression, which was typically chronic for AIDS. This treatment initially required daily prolonged intravenous therapy with systemically toxic agents. Local treatment with the sustained release ganciclovir implant and intravenous therapy with longer acting agents was subsequently developed for CMV retinitis. This led to an enormous improvement in the quality of life for these patients as well as improved control of the CMV retinal disease. Discontinuation of anti-CMV therapy in AIDS patients with improved immune function due to HAART is now an option for this disorder that previously required lifelong anti-CMV therapy (28). Spontaneous healing of CMV retinal lesions with HAART alone has also been described, although HAART therapy is not typically recommended as the sole therapy for acute CMV retinitis as it may take a prolonged period to produce elevation of functional CD4+ cells (29). It is advised to wait several months to ensure that the immune recovery induced by HAART is stable before discontinuing prophylaxis or treatment of opportunistic infections such as CMV. It has been recommended that discontinuation of anti-CMV therapy without risk of developing recurrent retinal infection may be considered when the CD4+ T-cell count rises above 100–150 cells/μL for 3–6 months (30). A new entity known as "immune recovery uveitis," or IRU, has been described in HIV-infected patients typically with inactive CMV retinitis who are undergoing HAART treatment (31,32). It has been hypothesized that the recovery of immunity produces a renewed inflammatory reaction against the infectious antigen. This inflammatory reaction had not previously been generated in the individual with AIDS due to the coexisting immunosuppression. Features of IRU include marked inflammation localized to the posterior segment and a more normal CD4+ count compared to the immunosuppressed AIDS patient. The CD4+ count averaged 300 cells/μL in one study of IRU (33). The complications of IRU can be visually disabling and chronic. It remains to be seen whether any other ocular complications will arise related to chronic HAART use. It is notable that the prevalence of AIDS overall has increased due to the improved survival of HIV-infected individuals secondary to HAART combined with a relatively steady HIV infection rate in the United States.

CMV disease may still develop or relapse when HAART is not effective or the CD4+ count remains low. As well, incomplete immune recovery may not fully protect against CMV retinitis. Furthermore, HAART medications may not be easily obtained, especially in some underdeveloped countries. For these reasons, CMV retinitis continues to be a prevalent and serious opportunistic infection in AIDS. It is important to be aware of the features and treatment options for this potentially multisystemic infection in AIDS and other immunocompromised individuals. The treatment of CMV retinal infection remains challenging, especially due to the multiple side effects of anti-CMV medications.

CLINICAL FEATURES OF CMV RETINITIS

The presence of symptoms may be related to the location of affected retina, and posterior pole CMV retinitis often produces more symptoms than peripherally located disease. However, it is not unusual to diagnose active retinitis in an asymptomatic individual on routine screening examination. When symptoms occur, these may include floaters, scotomata, decreased peripheral or central vision, and metamorphopsia. A significant proportion of patients may present to the ophthalmologist with sight-threatening macular retinitis. In a large series of 648 patients with AIDS, the prevalence of visual impairment at the time of diagnosis of CMV retinitis was high and varied based on patient demographics (34). The prevalence of visual acuity of 20/50 or worse or 20/200 or worse at the time of CMV retinitis diagnosis was 33% and 17%, respectively. White race and injection drug use were associated with a lower and a higher prevalence of visual impairment, respectively. The incidence of visual impairment at one year was also high. Individuals who received HAART had a 75% lower risk of visual impairment, and the greatest benefit occurred in those who experienced immune recovery. Pain, external ocular injection, and severe uveitis are not typical features of CMV retinitis.

CMV produces a retinitis with full-thickness retinal cell necrosis. This is in contrast to some of the other herpesvirus retinal infections that can preferentially involve the outer retinal layers. The retinal tissue adjacent to major retinal blood vessels and/or the optic disk are often affected by CMV retinitis, which may be secondary to hematogenous viral spread. CMV retinitis has a characteristic appearance, allowing very reliable clinical diagnosis in most cases. Active CMV retinitis has either a yellow-white fluffy or granular appearance with adjacent intraretinal hemorrhages. Areas of burned-out necrosis show absence of any retinal tissue, and the underlying retinal pigment epithelium has a mottled or "salt and pepper" appearance. Areas of burnt-out and active retinitis may be adjacent to each other within the eye. The CMV retinal infection can be either a large area of hemorrhagic retinal necrosis or small, focal areas of retinal whitening (1,14,15). Over an interval that usually spans weeks, untreated CMV retinitis tends to assume one of two different patterns (9,14). The first pattern is called hemorrhagic. It is characterized by broad geographic zones of retinal whitening with adjacent retinal hemorrhages, leading to the description as either "pizza-pie" or "cottage cheese and ketchup." The border between necrotic and unaffected retina is sharply demarcated and jagged. The retinal blood vessels, both arteries and veins, within the areas of necrosis may appear sheathed due to vasculitis. Secondary branch retinal vascular occlusion and vasculitis resembling "frosted branch angiitis" have been reported (14,35). The second pattern of CMV retinitis is called "granular" or "brushfire border," where focal granular infiltrates enlarge slowly leaving behind areas of destroyed retina and atrophic retinal pigment epithelium. Hemorrhages and vitreous cells are less prominent. This pattern of infection may result from to direct cell-to-cell virion transfer. The brushfire border may be seen with CMV retinitis anterior to the equator. In some eyes, both patterns of disease can occur simultaneously or sequentially. In most clinical trials, progression of CMV retinitis has been defined as one of the following: movement of a lesion border at least 750 μm along a front that is 750 μm or greater in length, development of a new CMV lesion in a previously involved eye or in the uninvolved fellow eye of an individual with baseline unilateral disease (36). CMV retinitis in zone 1 is defined as retinitis located within 3000 μm from the fovea or within 1500 μm from the optic nerve. Infection in this zone is considered visually threatening compared to more peripherally located disease in zone 2 or zone 3.

Mild vitreous cells are almost always present in CMV retinitis. Hypopyon or severe vitritis is rare (6). The new entity of immune recovery uveitis (IRU) is a chronic, inflammatory, sight-threatening syndrome associated with immune reconstitution due to HAART in AIDS patients who typically have inactive CMV retinitis (37–39). This disorder is differentiated from CMV retinitis by the history of HAART therapy, as well as by the CD4+ T-lymphocyte count and clinical examination. The mean CD4+ T-lymphocyte count in one study was 393 cells/mm^3 when IRU was diagnosed (37). The main feature is significant vitritis that is more pronounced than the mild vitreous cells in primary CMV retinitis. Complications may include macular or optic disk edema, proliferative vitreoretinopathy, epiretinal membrane formation, synechia and posterior subcapsular cataracts which can produce loss of vision from IRU (40). The IRU prevalence varies with HAART immune recovery but may be as high as 23%. Risk factors for IRU are immune recovery with HAART and the presence of a large area of CMV retinitis. Treatment includes subtenons and topical corticosteroids, which do not appear to reactivate the CMV retinitis.

Without treatment or improvement in the host's immune system, CMV retinitis is a relentless, progressive infection that can produce blindness from one of the following: retinal necrosis, retinal detachment, and/or optic nerve involvement. CMV infection can affect the optic nerve either directly or by extension from adjacent retinitis (41–43). Exudative retinal detachment can occur, typically with inferiorly located shifting fluid (6,14,41). It may be difficult to assess whether there is a rhegmatogenous component with a full-thickness retinal break located within the thin necrotic retina. Exudative retinal detachment related to CMV is usually nonprogressive and may resolve with anti-CMV viral therapy. Rhegmatogenous retinal detachment was reported in 20–30% of eyes with CMV retinitis in AIDS prior to HAART (44–47). Risk factors include a large retinal lesion and CMV affecting the anterior retina. The risk of detachment increases if greater than 25% of peripheral retina is involved (48,49). The retinal breaks in eyes with CMV retinitis may be located within or at the border of the necrotic, atrophic retina (50). The retinal breaks may also be posterior and multiple, and thus different from rhegmatogenous retinal detachment that occurs secondary to posterior vitreous detachment. It may be difficult to visualize all of the breaks in necrotic translucent retina and there may be primary proliferative vitreoretinopathy. For these reasons, CMV-related retinal detachments are difficult to repair with a scleral buckle alone, although buckling may be considered in small peripheral detachments when the entire lesion can be completely placed on the element. Laser photocoagulation demarcation has been described to delimit macula-sparing CMV-related retinal detachment (51). In many cases, pars plana vitrectomy with silicone oil retinal tamponade (or less commonly long-acting intraocular gas) is indicated (44,52,53). The success rate of macular reattachment with vitrectomy and silicone oil is high, although the visual results may be limited by the underlying disease (54). The visual potential, status of the fellow eye and the individual's systemic status should be considered when evaluating the method for surgical repair. HAART therapy has reduced the rates of retinal detachment, progression of retinitis, and visual loss; the overall risk of vision loss is reduced by approximately 75% in patients who respond to HAART.

DIAGNOSIS OF CMV RETINITIS

The diagnosis of CMV retinitis is primarily clinical (3). In most cases, ophthalmoscopic examination combined with clinical history is all that is required to confirm the diagnosis. As the disease usually spreads slowly, close observation and fundus photography may be considered when the diagnosis is uncertain (9). Although most patients with CMV retinitis have concurrent diffuse systemic CMV infection, they are usually asymptomatic from a systemic point of view. As the majority of immunosuppressed patients at risk for CMV retinitis will show serologic or culture evidence of CMV in body fluids, documented CMV viremia and/or viruria does not confirm the diagnosis of CMV retinitis (9). CMV retinitis can be diagnosed from infected retinal tissue, but an invasive procedure to obtain tissue carries potential risk. A retinal biopsy during retinal detachment repair may be performed when the diagnosis is unclear (44,46). Retina biopsy may show cytomegalic cells, although severe tissue necrosis may preclude this finding. Standard culture of an aqueous or vitreous sample is typically of little assistance in diagnosing CMV retinitis because of the limited involvement of these tissues. Newer techniques such as the polymerase chain reaction (PCR) allow detection of CMV-DNA from small amounts of intraocular fluid leading to diagnosis (55,56).

It has been recommended that patients with AIDS should be screened on the basis of their CD4+ lymphocyte count although the efficacy of screening asymptomatic patients at risk has not been clearly evaluated. Some patients may be asymptomatic with active retinitis and earlier diagnosis and treatment should result in a reduced area of retina involvement and less extension of CMV into the macula.

TREATMENT OF CMV RETINITIS

Treatment for CMV retinitis and systemic CMV infection has markedly improved over the last decade (57). There are two treatment principles for CMV infection. The first principle is to reverse or improve the underlying cause of immunosuppression. This may be possible in some cases by decreasing immunosuppressive medications, for example, after organ transplantation. This is now possible in AIDS by commencing or altering HAART therapy to improve the immune status (4,24).

Prior to the development of any specific anti-CMV medications, the major approach to CMV retinitis therapy was to alter immune function and this was not particularly successful. In most cases, it was not possible to improve systemic immune function or it took a prolonged time interval for immune recovery. Thus, CMV retinitis often progressed to involve the fovea with irreversible visual loss. Then in the 1980s, two systemic drugs became available for intravenous treatment of CMV retinitis: ganciclovir and foscarnet. Before their availability, some of the medical therapies that were attempted but were not effective included corticosteroids, gamma globulin, antifungal agents, vidarabine, human leukocyte interferon, interferon alpha, and acyclovir (58,59). This led to the second principle of therapy, which is to treat the CMV infection with a medication that has specific anti-CMV activity. Multicenter studies have shown that both ganciclovir and foscarnet are effective initially to halt progression of CMV retinitis and to induce regression of retinal infection (3). However, these drugs are virostatic and thus require administration for the entire time that a patient is immunocompromised, which can be prolonged in patients with an irreversible cause of immunosuppression. These agents were initially given intravenously, requiring chronic venous access with daily prolonged administration. Oral ganciclovir subsequently became available but it is limited by poor gastrointestinal absorption. Systemic administration of either ganciclovir or foscarnet is associated with significant systemic toxicities to the bone marrow or kidneys, respectively, that may limit or even prohibit their use.

SYSTEMIC ANTI-CMV THERAPY

Ganciclovir is an acyclic nucleoside that is a cogener of acyclovir, but it is between 10 and 100 times more effective against CMV (9,60). Most isolates of CMV are inhibited at dosages of 0.1–0.3 µg/mL (60). Virostatic levels can be achieved with intravenous doses of 2.3–5 mg/kg (61). Ganciclovir is also active against the other members of the herpesvirus family. The recommended intravenous dosage of ganciclovir is 5–7.5 mg/kg (3,9). Initially, during the induction phase of treatment, the drug is given twice a day for two to three weeks. Clearing of active retinitis usually takes several weeks. The initial response rate to intravenous ganciclovir induction varies between 80% and 100%. Maintenance intravenous ganciclovir consists of 5 mg/kg once daily, five or seven times per week (3,9,62). If ganciclovir therapy is stopped and the patient remains immunocompromised, reactivation usually occurs within four weeks (9). Reactivation of CMV retinitis despite maintenance intravenous ganciclovir therapy is also common in individuals who remain immunocompromised (2,14,62,63). Reactivation has been attributed to low ocular drug bioavailability, progressive decline in immune function, and development of CMV resistance to ganciclovir. Resistance to ganciclovir has been associated with a mutation in both the UL97 and UL54 genes (64,65). Evidence indicates that ganciclovir-treated AIDS patients with CMV retinitis live longer than those who receive no treatment (66). Despite ganciclovir treatment, contralateral CMV retinitis may develop in up to 15% of previously unaffected eyes (18). Ganciclovir may cause bone marrow toxicity; up to 70% of treated patients may develop some bone marrow suppression (3,62). Severe thrombocytopenia or neutropenia can develop, leading to cessation of intravenous therapy. The availability of hematopoietic stimulating factors has improved tolerance of ganciclovir in the face of bone marrow suppression.

The second intravenous agent approved for the treatment of CMV retinitis was foscarnet. Foscarnet (trisodium phosphoformate) is a synthetic, water-soluble pyrophosphate analogue that inhibits replication of herpesviruses in vitro (3,14). It noncompetitively binds to the exchange site of viral DNA polymerase, thereby rendering it inactive. Foscarnet lacks the bone marrow toxicity of ganciclovir, allow-ing for its concurrent use with zidovudine (AZT®). It requires extensive hydration during intravenous therapy, and it may induce nephrotoxicity and seizures (3,67). The foscarnet induction regimen is 60 mg/kg every 8 hours or 90 mg/kg twice daily for two to three weeks (3,67). Induction is followed by maintenance therapy at 90–120 mg/kg daily. The efficacy of foscarnet to induce regression of retinitis and maintain CMV inactivity closely parallels that of intravenous ganciclovir (3,68). The Studies of the Ocular Complications of AIDS (SOCA) Research Group showed that the two drugs appear equivalent in controlling CMV retinitis and preserving vision (68). In general, intravenous ganciclovir appears to be better tolerated than foscarnet for long-term therapy, although foscarnet may be associated with longer survival than ganciclovir.

Besides their side effects, both ganciclovir and foscarnet have other disadvan-tages. Neither is viricidal, so their use must be maintained for as long as the affected individual remains immunosuppressed (3,62). The relapse rate during maintenance anti-CMV therapy is high. Relapses may be controlled by reinduction with either agent or a combination of both intravenous ganciclovir and foscarnet or switching to a new therapeutic modality. The improved efficacy of combining intra-venous ganciclovir and foscarnet to control relapsing retinitis is counteracted by its potential for serious side effects and the negative impact on quality of life measures (69). For this reason, combination therapy is reserved for severely resistant cases. An alternative to combination intravenous therapy that is available combines place-ment of a ganciclovir intraocular implant combined with intravenous or intravitreal foscarnet. Both foscarnet and ganciclovir are costly and inconvenient to administer intravenously.

Cidofovir is an antiviral nucleotide analogue with significant activity against CMV and other herpesviruses. Cidofovir has a long intracellular half-life which allows for a prolonged interval (2 weeks) between intravenous maintenance doses, in contrast to daily administered intravenous ganciclovir and foscarnet (70). The efficacy of intra-venous cidofovir has been demonstrated in AIDS patients with untreated CMV retini-tis and with previously treated, relapsing CMV retinitis (71,72). Indirect comparisons of clinical trial data suggest that intravenous cidofovir appears to have similar efficacy to intravenous ganciclovir or foscarnet in delaying progression of CMV retinitis. Intra-venous cidofovir is less invasive, more convenient due to its prolonged dosage interval and an indwelling catheter is not required. The major treatment-limiting side effect is potentially irreversible nephrotoxicity; thus, renal function tests, hydration, and simul-taneous administration of probenecid are required. A relatively high rate of anterior uveitis, up to 40%, has been reported and a small number of patients have developed hypotony with intravenous cidofovir (73). These ocular complications were even more prevalent with trials of intravitreal cidofovir administration and have percluded its administration by this route (74).

Oral ganciclovir became available in the mid-1990s. It can be used as daily maintenance therapy in patients who respond well to the initial intravenous ganciclo-vir induction. It has poor ocular bioavailability and thus large doses in the range of 3–6 g daily are required. The median interval to progression of retinitis is less with oral ganciclovir than intravenous form (29 vs. 49 days, respectively) (75). The risk of developing CMV retinitis in the contralateral eye is greater with oral than with intravenous ganciclovir. Oral ganciclovir does play a role in decreasing the risk of fellow eye retinitis in patients with unilateral CMV retinitis who are treated with local therapy such as the ganciclovir implant (76).

Development of newer oral anti-CMV medications has been slowed by the decreasing incidence of the disease. Valganciclovir (also known as Valcyte®) is a prodrug of ganciclovir. It has excellent oral bioavailability and is the most recently approved oral anti-CMV medication for therapy of CMV retinitis. Valganciclovir is rapidly metabolized into ganciclovir and thus high ganciclovir blood levels are achieved without the complications associated with chronic intravenous access and administration. Orally administered valganciclovir appears to be as effective as intravenous ganciclovir or induction treatment of newly diagnosed CMV retinitis and it is more convenient to administer. In a study comparing induction therapy with either intravenous ganciclovir or oral valganciclovir, the median time to progression of CMV retinitis was 125 and 160 days in the intravenous ganciclovir and the oral valganciclovir groups, respectively (77). Approximately 10% of patients in either treatment group progressed photographically within the first four weeks of therapy and the frequency and severity of adverse events were similar. While there are no comparative trials of oral valganciclovir as a maintenance treatment, pharmacokinetic data suggests that it is about as effective as intravenous ganciclovir. The adverse effects of oral valganciclovir are similar to those of intravenous ganciclovir, except that the oral route avoids the risk of local complications at the infusion site and inconveniences of injection. Valganciclovir may produce more frequent diarrhea and oral candidiasis than intravenous ganciclovir.

All of the systemic and intravenous treatment options for CMV retinitis are associated with potentially serious adverse events. Selection of pharmacotherapy should be individualized and depends on a number of factors including the CMV lesion characteristics, patient quality of life issues and efficacy and tolerability profiles of available therapies. Two to three weeks of systemic therapy are required to stabilize, and then achieve regression of CMV retinitis. While on maintenance systemic therapy, patients are examined approximately every four weeks for evidence of reactivation of CMV retinitis. Reactivation is common when the individual remains systemically immunocompromised. If CMV retinitis reactivation occurs, the options for further treatment include re-induction with a more frequent, higher dose of the same agent, changing to a different intravenous anti-CMV agent, using more than one anti-CMV intravenous agent or adding local adjunctive therapy (78).

LOCAL MODES OF INTRAOCULAR DRUG DELIVERY

Because of the potential for severe toxicity associated with systemic anti-CMV agents and inconvenient intravenous administration that carries a risk of sepsis, new local treatment options have been developed. These treatments represent major therapeutic advances, and include the sustained release ganciclovir intraocular implant and intravitreal injection of ganciclovir, foscarnet, or fomivirsen. These newer therapies avoid the negative impact on quality of life associated with prolonged intravenous therapy. While there may be associated systemic CMV infection, most patients lack extraocular symptoms. Thus, local therapy treats the infection that requires immediate attention. Local therapy is an attractive option as it delivers a specific concentration of anti-CMV medication directly to the infection site. While two to three weeks are required for stabilization and regression of CMV retinitis with intravenous therapy, local therapy with the implant, or intra vitreal injections has an immediate effect by depositing a therapeutic drug level directly to the infected retina.

INTRAVITREAL DRUG INJECTION

Intravitreal injections of either ganciclovir or foscarnet have been used successfully to control CMV retinitis in some patients, especially those with recurrent or refractory disease. Initial animal studies demonstrated that intravitreal injections in excess of 400 μg of ganciclovir and 1200 μg of foscarnet were nontoxic to the retina (79,80). The published dosages of intravitreal ganciclovir range from 200 to 2000 μg in 0.1 mL sterile normal saline. Intravitreal ganciclovir appears to control active CMV retinitis initially in over 80% of eyes. During the induction phase, injections are given three times per week, followed by injections given once or twice per week during long-term maintenance therapy. Multiple injections are required due to the short intraocular drug half-life. About 30% of immunocompromised patients will develop recurrent disease if only treated with this regimen of intravitreal injection (81,82). As an alternative for CMV retinitis resistant to or poorly responsive to ganciclovir, intravitreal foscarnet at a dose of up to 2400 μg in 0.1 mL may be used to control CMV retinitis (83). No retinal toxicity has been associated with the above doses. Rare but potentially vision-threatening complications associated with the technique of intravitreal injection include exogenous endophthalmitis, vitreous hemorrhage, and retinal detachment. However, chronic intravitreal injections are often not well tolerated due to the inconvenience and risk of complications with frequent injections (61,81,82). While cidofovir for intravitreal injection may have had hypothetical advantages due to its longer intracellular half-life requiring less frequent injections, it has been associated with multiple serious side effects including uveitis and chronic hypotony with permanent visual loss (73,84). For these reasons, cidofovir is not presently used for intravitreal anti-CMV therapy. A recent observational study showed that the use of cidofovir either by an intravitreal or intravenous route was a major risk factor for the development of IRU (85). As well, continued therapy of healed CMV retinitis after immune recovery did not appear to protect against the development of IRU. These authors recommended that the association of IRU and cidofovir may preclude use of this agent.

Fomivirsen (or Vitravene®, Isis Pharmaceuticals, Inc.) is an antisense oligonucleotide which inhibits replication of human CMV by binding to complementary sequences on messenger RNA transcribed from a major transcriptional unit of the virus (86,87). Fomivirsen is the first in a class of novel therapeutics based on the antisense mechanism approved for marketing in the United States. Fomivirsen utilizes a mechanism of action different than that of ganciclovir, foscarnet, and cidofovir. Thus, it may be useful for treatment of CMV retinitis that is resistant to ganciclovir or foscarnet. It is an intravitreal injection administered as an alternative therapy for CMV retinitis in AIDS. It is specifically recommended for individuals who are intolerant of or have a contraindication to other treatments for CMV retinitis or who were insufficiently responsive to previous CMV retinitis treatments. It has a longer intravitreal elimination half-life than either ganciclovir or foscarnet. Induction and maintenance schedules are required as with the other anti-CMV therapies and different regimens have been studied (88). Fomivirsen at a dose of 165 μg was injected weekly for three doses as induction therapy, followed by injection every other week as maintenance therapy in a randomized clinical trial for treatment of newly diagnosed peripheral CMV retinitis (89). The median time to first progression of disease for the formivirsen-treated group was 71 days compared to 13 days in the deferral of anti-CMV treatment group. Eventual progression occurred in 44% of patients in the formivirsen treatment group during the study. The study concluded that formivirsen was an effective treatment for CMV retinitis in patients with AIDS. Common side effects include increased intraocular pressure and mild to moderate uveitis, which occurs in up to one-quarter of treated individuals and is generally transient and reversible with topical steroid drops (90). Pigmentary bull's eye maculopathy has been reported that was reversible with discontinuation of therapy (91,92). Electrophysiological abnormalities have also been reported (93). It is recommended that intravitreal fomivirsen should not be administered within two to four weeks of cidofovir treatment due to an increased risk of intraocular inflammation.

THE GANCICLOVIR INTRAOCULAR IMPLANT

The ganciclovir intraocular implant (commercially available as Vitrasert® from Bausch & Lomb, Rochester, New York, U.S.A.) is a nonerodible device that delivers this drug in a sustained release fashion directly to the posterior segment (94,95). Studies have demonstrated that it is extremely well tolerated and effective to halt activity and delay further progression of CMV retinitis (96–98). In fact, it has the highest efficacy of all of the approved agents used to treat CMV retinitis. The implant consists of a compressed ganciclovir pellet covered by a semipermeable membrane (Fig. 1). The membrane is comprised of polyvinyl alcohol and ethylene vinyl acetate, which are permeable and impermeable, respectively, to ganciclovir. This device permits slow continuous diffusion of ganciclovir from the implant into the vitreous cavity. The intraocular ganciclovir level produced by the implant (mean 4.1 µg/mL) is higher than can be attained with intravenous administration (mean 0.93 µg/mL). The implant contains approximately 5 mg of drug that is released at a rate of 1.5 µg/hr. The ganciclovir intraocular implant typically has a therapeutic effect releasing drug of approximately six to eight months (99). It is placed surgically into the vitreous cavity through a pars plana incision where it releases ganciclovir linearly in a time-release fashion.

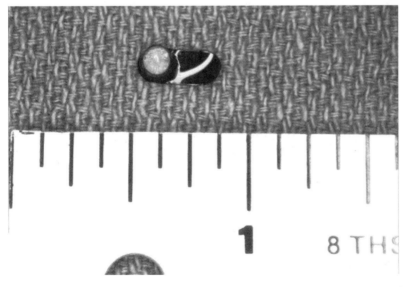

Figure 1 (*See color insert*) The ganciclovir implant. Note the yellow pellet of ganciclovir on the left and the strut which is secured to the sclera to the right.

The surgical technique has been described in detail elsewhere but there are some key points to be considered (100). First, it is prudent to prepare the implant before an ocular incision is started, to ensure that the implant is not damaged before implantation. The implant is usually placed inferiorly or inferotemporally through a 5.5–6 mm pars plana incision (Fig. 2). This location also produces a better cosmetic effect as the lower lid hides the incision site. The inferior location also maximizes the ability of the implant to release the drug into an inferior aqueous humor meniscus if silicone oil tamponade for CMV-related retinal detachment repair is required in the future in these eyes predisposed to develop complicated retinal detachments (Fig. 3). The sclerotomy should be inspected to ensure that the pars plana has been completely incised before implant insertion to avoid accidental suprachoroidal placement. Removal of prolapsed vitreous with a cutter at the incision should be performed. The implant should only be grasped at the strut and not by the drug pellet and its position should be confirmed in the vitreous cavity before securing its position to the sclera. It is anchored in the incision with a double armed 8.0 nylon suture and the strut should not protrude in the incision. Careful watertight closure of the sclera incision with crossed sutures and re-establishment of the intraocular pressure are performed at the end of surgery.

The effectiveness and safety of the ganciclovir implant device have been verified by multiple clinical trials. In a randomized clinical trial comparing the ganciclovir implant to intravenous ganciclovir, the median time to progression of retinitis was 221 days versus 71 days in implant and intravenous treated patients, respectively (98). The median time to retinitis progression exceeded one year when the implant was combined with oral ganciclovir, and this greatly exceeds the efficacy of any other

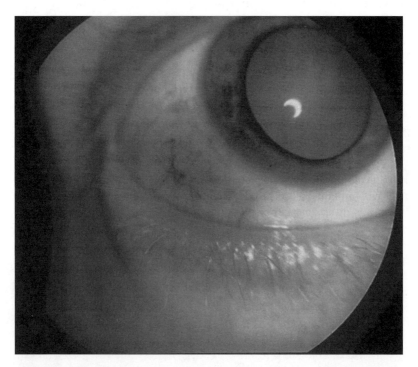

Figure 2   (*See color insert*) The scleral incision to place the implant into the vitreous is at the pars plana. Note the crossed suture securing the incision below the conjunctiva.

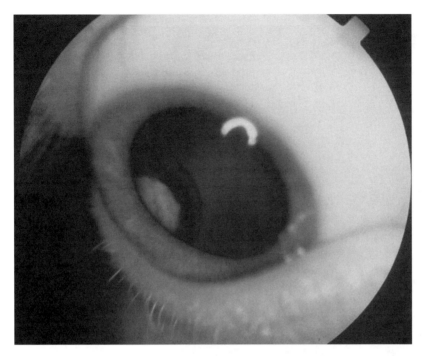

Figure 3 The ganciclovir intravitreal implant is noted behind the lens and iris in the infer-otemporal quadrant.

therapeutic modalities for CMV retinitis (76). Overall, the implant offers the longest median time of control of retinitis reported to date when compared with other therapies. While the time to progression of CMV retinitis was significantly longer after treatment with the ganciclovir implant than with intravenous ganciclovir therapy, there may be a higher rate of contralateral eye retinitis and systemic CMV disease in immunocompromised individuals when the implant alone is used (98). For this reason, the implant is usually combined with oral ganciclovir to reduce the risk of contralateral CMV retinitis and extraocular CMV infection especially if immune compromise persists or until the immune status recovers with HAART. This combination of ganciclovir in oral and implant forms has also been shown to prolong the time to progression of retinitis and reduce the risk of developing Kaposi sarcoma (101).

The ganciclovir implant has released patients from the inconvenience of daily prolonged intravenous infusions. Other advantages of the implant include less frequent ocular manipulation than with intravitreal injection and a constant steady rate of drug release into the vitreous. This intraocular ganciclovir implant, as seen in Figure 1, was approved by the Food and Drug Administration (FDA) in 1996, and it has subsequently become widely used for treatment of CMV retinitis (94,95). The ganciclovir implant has been used to treat primary, recurrent, and bilateral CMV retinitis (bilateral implants are placed for bilateral disease). It has also been effective in eyes with silicone oil tamponade for CMV-related retinal detachment (102–104). Clinical management of CMV retinitis with an associated retinal detachment often involves concurrent use of silicone oil and a ganciclovir implant (Fig. 4A and B). Ganciclovir is water-soluble and would not be expected to partition within the silicone oil. In most retinal reattachment procedures only 80–90% of the posterior

(A)

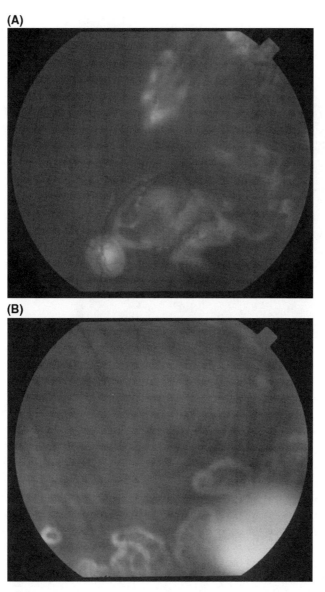

(B)

Figure 4 (**A**) The reflection of silicone oil is noted in the posterior pole in. (**A, B**) Silicone oil tamponade was used for CMV-related retinal detachment repair and a ganciclovir implant was used to treat active CMV retinitis. The object noted inferotemporally is the ganciclovir implant as visualized when the camera is focused posteriorly on the retina. *Abbreviation*: CMV, cytomegalovirus.

segment is filled with silicone oil and there is an inferior layer of aqueous humor where the inferiorly placed implant may release ganciclovir (Fig. 5). Effective ganciclovir levels are maintained in the aqueous phase of the vitreous cavity of silicone oil-filled eyes. In fact, ganciclovir levels may be maintained longer in silicone-filled eyes than in those without, supporting combined use of ganciclovir implants with silicone oil tamponade. Additional silicone oil re-infusion may be required if some of the silicone oil volume is lost externally during implant exchange.

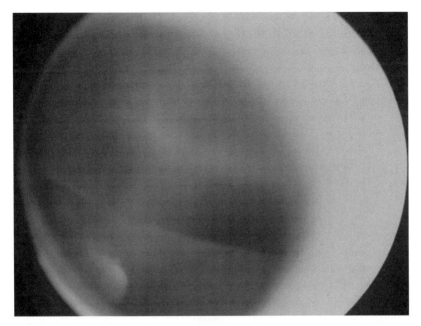

Figure 5 (*See color insert*) The ganciclovir implant is noted in the inferior aqueous layer in a silicone oil–filled eye. The silicone oil meniscus is noted above the implant.

INDICATIONS FOR THE GANCICLOVIR IMPLANT

The implant releases the drug over six to eight months, after which time the device is depleted of medication and replacement may be considered. The implant was developed at a time when median survival after diagnosis of CMV retinitis was approximately 12 months, lifelong anti-CMV therapy was required and HAART was not readily available. With the introduction of HAART, which can improve immune function in AIDS, long-term anti-CMV therapy may not be necessary if immune function recovers. As well, the beneficial effect of HAART on immunologic status and survival in HIV has altered the incidence of primary and relapsing CMV retinitis. Thus, the indications for the ganciclovir implant have been modified as the clinical course and management of CMV retinitis has changed with HAART therapy. Davis and colleagues reported that the use of HAART was associated with improved outcomes in individuals with AIDS and recurrent CMV retinitis who had been treated with the ganciclovir implant (105). In 1999, a panel of physicians with expertise in management of CMV retinitis and use of the ganciclovir implant was convened by the International AIDS Society—U.S.A. to clarify the risks, benefits and provide recommendations for utilization and replacement of the ganciclovir implant for treatment of AIDS-related CMV retinitis in the HAART era (100). The panel recommended that selection of therapy for CMV retinitis should be individualized depending on multiple factors including the patients' antiretroviral history, the potential for immune improvement, CD4+ count, plasma HIV RNA level, lifestyle choices, compliance, living conditions, and the location of CMV retinitis. The implant can be used for CMV retinitis in any zone but it is particularly useful in zone 1 disease to provide the most rapid effective intraocular drug dose to infected retina.

There are multiple advantages offered by the implant including the longest length of effective control of CMV retinitis compared to the other therapies, lack of systemic toxicity, improved quality of life issues, and lack of necessity for intravenous access. Disadvantages include the surgical discomfort, potential transient decrease in visual acuity, and risk of postoperative complications. In patients who remain chronically immunocompromised and are treated with local implant therapy, systemic anti-CMV therapy in the form of oral ganciclovir or Valganciclovir is recommended to avoid CMV infection in the fellow eye as well as symptomatic extraocular CMV disease (76).

Use of the ganciclovir implant and eventual discontinuation of systemic therapy may be possible in selected patients with improved immunity due to HAART and this approach may result in better long-term visual outcomes (100,106,107). In patients who have not yet received HAART when the CMV retinitis is treated with an implant, systemic therapy for CMV is indicated until the CD4+ count has stabilized appropriately at an elevated level. In patients who develop their initial episode of CMV retinitis while undergoing HAART therapy, this is a sign of progressive immune dysfunction and reassessment of the antiretroviral therapy is indicated. Use of the implant may be especially useful in these patients who are undergoing readjustment of HAART or who have failed HAART as longer term CMV control can be provided by the implant.

Relapsed CMV retinitis while undergoing HAART may be a sign of progressive immune dysfunction and requires reassessment of HAART and other medications. Individuals treated with HAART may experience reactivation of CMV retinitis when their CD4 count decreases (108). The threshold CD4 count below which reactivation of CMV retinitis occurred in patients for whom HAART was not successful is approximately $50\,\text{cells}/\text{mm}^3$. Thus, despite an initial response to HAART, there is a risk for CMV retinitis reactivation when the CD4 count decreases below $50\,\text{cells}/\text{mm}^3$. The HIV viral load does not appear to predict CMV reactivation.

Other causes of CMV retinitis relapse include development of antiviral medication resistance and inadequate intraocular drug levels. The implant is effective for relapsed retinitis and should be considered especially if immune improvement is unlikely. However previous exposure to intravenous or oral ganciclovir reduces the probability that the implant will be as effective as it is in individuals who have never had systemic ganciclovir exposure (109). If the implant appears ineffective, intravenous or intravitreal foscarnet may be added. A therapeutic trial of intravitreal ganciclovir may be considered before implant placement to assess for the possibility of ganciclovir viral resistance, although the dose and frequency of intravitreal ganciclovir injection to stimulate the concentration and drug release of the implant are not known. Other options for relapsed CMV retinitis include reinduction with the same or another systemic intravenous agent or combination therapy. When relapse has occurred, the interval between subsequent relapses continues to shorten with continued intravenous therapy.

Each eye should be considered independently in individuals with bilateral disease. Simultaneous implant surgery is not usually indicated although only a few days are required between operating on each eye.

REPLACEMENT OF THE GANCICLOVIR IMPLANT

The primary reason for CMV relapse in an eye treated initially with a ganciclovir implant is drug depletion from the device, which typically occurs by six to eight months after placement (110,111). The device can be replaced with a second implant when it is depleted of drug for continued CMV control. Some clinicians prefer to observe for early signs recurrent CMV activity before replacement while in other situations, preemptive replacement of the implant may be preferable (99). The clinical situation including the immune status and access and response to HAART should be considered when evaluating whether to replace the implant when it is depleted or to observe for reactivation (100). In general, preemptive replacement should be considered in patients with persistent immunocompromise especially if the CMV retinitis is located in zone 1. The empty implant may be removed and exchanged for a new one through the same incision site or it may be placed in a new incision site leaving the initial implant in situ. There have been reported cases with multiple implants remaining in situ in a single eye.

COMPLICATIONS OF THE GANCICLOVIR IMPLANT

Complications associated with the ganciclovir intraocular implant are uncommon, but these should be discussed with a surgical candidate preoperatively. Complications may be related to the surgical procedure, the implant device or the medication contained within the implant. Patients with AIDS and CMV retinitis may be at increased risk for various intraocular complications independent of the mode of therapy. In some instances, it may be difficult to discern whether the negative event is related to the implant, the underlying CMV retinitis disease process or to other systemic therapy. For example, retinal detachment has been described after ganciclovir implantation surgery but this is a common event to occur in these predisposed eyes with necrotic retina secondary to CMV retinitis irrespective of the implantation surgery (112). Cystoid macular edema has been described after ganciclovir implant but is also a feature of IRU secondary to HAART and has been reported secondary to systemic cidofovir use. Thus, the complication rates should be considered for other types of systemic therapy and for the natural course of the disease as well as for the ganciclovir implant.

Endophthalmitis is fortunately a rare but potentially visually devastating complication. The largest series from 30 clinical practices identified 24 cases from 5185 ganciclovir implant procedures (0.46%) (113). This rate is higher than that reported after cataract surgery, 0.072%, and after vitrectomy, 0.01%, although the procedures are completely unrelated and not comparable (114). This rate of implant-associated endophthalmitis is much less than early hypothesized rates of infection. The initial concerns about increased rates of infection related to placement of a foreign device intravitreally in immunocompromised individuals have not been substantiated. However, the rate of simultaneous or subsequent retinal detachment after implant-related endophthalmitis was over 50% in these already compromised eyes. Late onset endophthalmitis occurring more than 30 days after implant surgery has been associated with wound problems such as implant strut or suture exposure. In the cases with a wound abnormality or with non-clearing intraocular infection despite intravitreal antibiotic therapy, implant removal may be warranted.

Some other uncommon complications include macular edema, vitreous hemorrhage, hypotony, cataract (Fig. 6), temporary reduced vision secondary to astigmatism, implant malposition, and retinal detachment, which is more likely if the CMV infection involves over 25% of the retina (100). Lim and colleagues evaluated a series of 110 ganciclovir implant procedures and noted posterior segment complications in 12% (111). Some of these eyes had undergone multiple prior implant procedures and

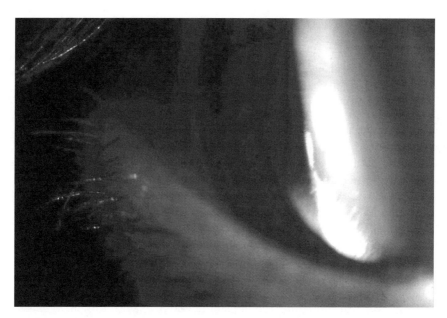

Figure 6   A small nonvisually significant lens opacity is noted in the region of the ganciclovir implant.

this series was compiled before the use of HAART, which may have affected the overall complication rate. A rare complication is separation of the ganciclovir medication pellet from the suture strut upon its removal (115). To minimize the chance of this complication, care should be taken to open adequately the scleral incision, to grasp the anterior scleral lip, and to handle the suture strut rather than the pellet during implant removal.

There are some relative contraindications to use of the ganciclovir implant. The risk of using the implant in an individual with limited life expectancy should be carefully considered, although the immune status can be markedly improved with HAART. Other relative contraindications include implantation when there is an ocular surface infection or systemic coagulopathy. Extraocular ganciclovir resistant viral strains may be a relative contraindication, although more than one strain of CMV can infect an individual and the resistant isolate from the blood or extraocular tissue may not reflect the CMV strain in the eye (116).

SUMMARY

The introduction of HAART has notably changed the incidence, features, and course of CMV retinitis in AIDS. This therapy has altered the treatment of CMV retinitis as well as the indications for the ganciclovir implant. Factors that play a role in therapeutic management and utilization of the ganciclovir implant include the potential for immune improvement, location, severity, and laterality of CMV retinitis, coexisting retinal detachment, risks of implantation, and the costs of ganciclovir implant with oral ganciclovir. Local intraocular treatments as alternatives to systemic medications have been a significant advance in management of CMV retinitis. Local ocular therapy has proven effective and provides significant advantages over systemic therapy with regards to control of retinitis, prevention of progression, and avoidance of systemic bone marrow and renal toxicity. Disadvantages of local therapy include the risk of CMV retinitis in the contralateral eye and the lack of protection against systemic extraocular CMV infection, although this appears to be less of an issue with concomitant oral anti-CMV therapy. The ganciclovir intraocular implant is extremely effective therapy for primary and relapsed CMV retinitis and should be considered especially for zone 1 disease and also if immune improvement is unlikely. Prior to HAART, lifelong anti-CMV therapy was required to prevent progression of CMV retinal disease and subsequent loss of vision in AIDS. The beneficial effect of HAART on the immunologic status of some patients has made the role of maintenance anti-CMV treatment less clear. Maintenance anti-CMV medications have been safely stopped in some small series of AIDS patients with stable inactive CMV retinitis and elevated CD4+ cell counts (greater than 100 cells/μL) without reactivation of CMV retinitis (26,90,91). Thus immune recovery following potent HAART may be effective to control the major opportunistic infection of CMV retinitis, even in patients with a history of previous severe immunosuppression. Patients may be at risk for reactivation of CMV retinitis despite initial favorable response to HAART if their CD4 lymphocyte count falls below 50 cells/mm^3 (92). The use of HAART has led to a new entity, IRU. Lifelong continued ophthalmologic evaluation is required to ensure that this infection remains quiescent and does not produce further visual loss.

REFERENCES

1. Hennis HL, Scott AA, Apple DJ. Cytomegalovirus retinitis. Surv Ophthalmol 1989; 34:193.
2. Baumal CR, Levin AV, Read S. Cytomegalovirus retinitis in immunosuppressed children. Am J Ophthalmol 1999; 106:301–305.
3. Studies of the ocular complications of AIDS research group, in collaboration with the AIDS clinical trials group. Mortality in patients with the acquired immune deficiency syndrome treated with either foscarnet or ganciclovir for cytomegalovirus retinitis. N Engl J Med 1992; 326 (4):213–220. [Erratum in: N Engl J Med 1992; 326(17):1172.]
4. Deayton JR, Wilson P, Sabin CA, et al. Changes in the natural history of cytomegalovirus retinitis following the introduction of highly active antiretroviral therapy. AIDS 2000; 14:1163–1170.
5. Freidman AH, Orellana J, Freeman WR, et al. Cytomegalovirus retinitis: a manifestation of the acquired immune deficiency syndrome (AIDS). Br J Ophthalmol 1983; 67:372.
6. Chumbley LC, Robertson DM, Smith TF, Campbell RJ. Adult cytomegalovirus inclusion retino-uveitis. Am J Ophthalmol 1975; 80:807.
7. Nicholson DH. Cytomegalovirus infections of the retina. Int Ophthalmol Clin 1975; 15(4):151.
8. Meredith TA, Aaberg TM, Reeser FH. Rhegmatogenous retinal detachment complicating cytomegalovirus retinitis. Am J Ophthalmol 1979; 87:793.
9. Bloom JN, Palestine AG. The diagnosis of cytomegalovirus retinitis. Ann Intern Med 1988; 109:963.
10. Murray HW, Knox DL, Green WR, Susel RM. Cytomegalovirus retinitis in adults: a manifestation of disseminated viral infection. Am J Med 1977; 63:574.
11. Christensen L, Beeman HW, Allen A. Cytomegalic inclusion disease. Arch Ophthalmol 1957; 57:90.
12. Foester HW. Uveitis symposium. Surv Ophthalmol 1959; 4:296.
13. Smith ME. Retinal involvement in adult cytomegalic inclusion disease. Arch Ophthalmol 1964; 72:44.

14. Palestine AG. Clinical aspects of cytomegalovirus retinitis. Rev Infect Dis 1988; 10:515.

15. Schuman JS, Orellana J, Friedman AH, Teich SA. Acquired immunodeficiency syndrome (AIDS). Surv Ophthalmol 1987; 31:384.

16. Jabs DA, Green WR, Fox R, et al. Ocular manifestations of acquired immune deficiency syndrome. Ophthalmology 1989; 96:1092.

17. Freeman SR, Lerner CW, Mines JA, et al. A prospective study of the ophthalmologic findings in the acquired immune deficiency syndrome. Am J Ophthalmol 1984; 97:133.

18. Gross JS, Bozzette SA, Mathews WC, et al. Longitudinal study of cytomegalovirus retinitis in acquired immune deficiency syndrome. Ophthalmology 1990; 97:681.

19. Sison RF, Holland GN, MacArthur LJ, et al. Cytomegalovirus retinopathy as the initial manifestation of the acquired immunodeficiency syndrome. Am J Ophthalmol 1991; 112:243.

20. Kestelyn P, Van de Perre P, Rouvoy D, et al. A prospective study of the ophthalmologic findings in the acquired immune deficiency in Africa. Am J Ophthalmol 1985; 100:230.

21. Levin AV, Zeichner S, Duker JS, et al. Cytomegalovirus retinitis in an infant with acquired immune deficiency syndrome (AIDS). Pediatrics 1989; 84:683.

22. Baumal CR, Levin AV, Kavalec C, Petric M, Khan H, Read S. Screening for cytomegalovirus retinitis in children. Arch Pediatr Adolesc Med 1996; 150:1186–1192.

23. Bartlett JG, Gallant JE. Medical management of HIV infection. Baltimore, MD: Division of Infectious Diseases, Johns Hopkins University, 2000.

24. Varani S, Spezzacatena P, Manfredi R, et al. The incidence of cytomegalovirus (CMV) antigenemia and CMV disease is reduced by highly active antiretroviral therapy. Eur J Epidemiol 2000; 16:433–437.

25. Jacobson MA, Stanley H, Holtzer C, et al. Natural history and outcome of new AIDS-related cytomegalovirus diagnosed in the era of highly active antiretroviral therapy. Clin Infect Dis 2000; 30:231.

26. Baril I, Jouan M, Agher R, et al. Impact of highly active antiretroviral therapy on onset of mycobacterium avium complex infection and cytomegalovirus disease in patients with AIDS. AIDS 2000; 14:2593.

27. Jabs DA, Bartlett JG. AIDS and ophthalmology: a period of transition. Am J Ophthalmol 1997; 124:227–233.

28. Vrabec TR, Baldassano VF, Whitcup SM. Discontinuation of maintenance therapy in patients with quiescent cytomegalovirus retinitis and elevated CD4+ counts. Ophthalmol 1998; 105:1259–1264.

29. Reed JB, Schwab IR, Gordon J, Morse LS. Regression of CMV retinitis associated with protease inhibitor treatment of AIDS. Am J Ophthalmol 1997; 124:199–205.

30. Masur H, Kaplan JE, Holmes KK. Guidelines for preventing opportunistic infections among HIV-infected persons B 2002. Recommendations of the U.S. Public Health Service and the Infectious Diseases Society of America. Ann Intern Med 2002; 137:435.

31. Karavellas MP, Lowder CY, MacDonald JC, Avila CP, Freeman WR. Immune recovery vitritis associated with inactive cytomegalovirus retinitis: a new syndrome. Arch Ophthalmol 1998; 116:169–175.

32. Zegans ME, Walton RC, Holland GN, O'Donnell JR, Jacobson MA, Margolis TP. Transient vitreous inflammatory reactions associated with combination antiretroviral therapy in patients with AIDS and cytomegalovirus retinitis. Am J Ophthalmol 1998; 125:292–300.

33. Karavellas MP, Azen SP, Macdonald JC, et al. Immune recovery vitritis and uveitis in AIDS: clinical predictors, sequelae, and treatment outcomes. Retina 2001; 21:1.

34. Kempen JH, Jabs DA, Wilson LA, Dunn JP, West SK, Tonascia JA. Risk of vision loss in patients with cytomegalovirus retinitis and the acquired immunodeficiency syndrome. Arch Ophthalmol 2003; 121:466–476.

35. Mansour AM, Li HK. Frosted retinal periphlebitis in the acquired immunodeficiency syndrome. Ophthalmologica 1993; 207:182–186.

36. Holbrook JT, Davis MD, Hubbard LD, Martin BK, Holland GN, Jabs DA. Risk factors for advancement of cytomegalovirus retinitis in patients with acquired immuno-deficiency syndrome. Studies of the ocular complications of AIDS research group. Arch Ophthalmol 2000; 11:1196–1204.

37. Robinson MR, Reed G, Csaky KG, Polis MA, Whitcup SM. Immune-recovery uveitis in patients with cytomegalovirus retinitis taking highly active antiretroviral therapy. Am J Ophthalmol 2000; 130:49–56.

38. Kuppermann BD, Holland GN. Immune recovery uveitis. Am J Ophthalmol 2000; 130:103–106.

39. MacDonald JC, Karavellas MP, Torriani FJ, et al. Highly active antiretroviral therapy-related immune recovery in AIDS patients with cytomegalovirus retinitis. Ophthalmol 2000; 107:877–881.

40. Karavellas MP, Song M, Macdonald JC, Freeman WR. Long-term posterior and anterior segment complications of immune recovery uveitis associated with cytomegalovirus retinitis. Am J Ophthalmol 2000; 130:57–64.

41. Gross JG, Sadun AA, Wiley CA, Freeman WR. Severe visual loss related to isolated peripapillary retinal and optic nerve cytomegalovirus infection. Am J Ophthalmol 1989; 108:691.

42. Grossniklaus HE, Frank KE, Tomsak RL. Cytomegalovirus retinitis and optic neuritis in acquired immune deficiency syndrome. Ophthalmology 1987; 94:1601.

43. Berger BB, Weinberg RS, Tessler HH, et al. Bilateral cytomegalovirus panuveitis after high-dose corticosteroid therapy. Am J Ophthalmol 1979; 88:1020.

44. Freeman WR, Henderly DE, Wan WL, et al. Prevalence, pathophysiology, and treatment of rhegmatogenous retinal detachment in treated cytomegalovirus retinitis. Am J Ophthalmol 1987; 103:527.

45. Freeman WR, Schneiderman TE. Invasive posterior segment diagnostic procedures in immunosuppressed patients. Int Ophthalmol Clin 1989; 29:119.

46. Holland GN, Sidikaro Y, Kreiger AE, et al. Treatment of cytomegalovirus retinopathy with ganciclovir. Ophthalmology 1987; 94:815.

47. Broughton WL, Cupples HP, Parver LM. Bilateral retinal detachment following cyto-megalovirus retinitis. Arch Ophthalmol 1978; 96:618.

48. The Studies of Ocular Complications of AIDS (SOCA) Research Group in collabora-tion with the AIDS clinical trials group (ACTG). Rhegmatogenous retinal detachment in patients with cytomegalovirus retinitis: the Foscarnet–Ganciclovir cytomegalovirus retinitis trial. Am J Ophthalmol 1997; 124 (1):61–70.

49. Freeman WR, Friedberg DN, Berry C, et al. Risk factors for development of rhegmato-genous retinal detachment in patients with cytomegalovirus retinitis. Am J Ophthalmol 1993; 116:713–720.

50. Baumal CR, Reichel E. Management of cytomegalovirus-related rhegmatogenous retinal detachment. Ophthalmic Surg Lasers 1998; 29:916–925.

51. Vrabec TR. Laser photocoagulation repair of macula-sparing cytomegalovirus-related retinal detachment. Ophthalmology 1997; 104:2062–2067.

52. Regillo CD, Vander JF, Duker JS, et al. Repair of retinitis-related retinal detach-ments with silicone oil in patients with acquired immune deficiency syndrome. Am J Ophthalmol 1992; 113:21..

53. Dugel PU, Liggett PE, Lee MB, et al. Repair of retinal detachment caused by cytome-galovirus retinitis in patients with the acquired immune deficiency syndrome. Am J Ophthalmol 1991; 112:235.

54. Azen SP, Scott IU, Flynn HW Jr, et al. Silicone oil in the repair of complex retinal detachments. A prospective observational multicenter study. Ophthalmology 1998; 105:1587–1597.

55. Dodt KK, Jacobsen PH, Hoffman B, et al. Development of cytomegalovirus (CMV) disease may be predicted in HIV-infected patients by CMV polymerase chain reaction and the antigenemia test. AIDS 1997; 11:F21–F28.

56. Smith IL, Macdonald JC, Freeman WR, Shapiro AM, Spector SA. Cytomegalovirus (CMV) retinitis activity is accurately reflected by the presence and level of CMV DNA in aqueous humor and vitreous. J Infect Dis 1999; 179:1249–1253.

57. Hoffman VF, Skiest DJ. Therapeutic developments in cytomegalovirus retinitis. Expert Opin Investig Drugs 2000; 9:207–220.

58. Guyer DR, Jabs DA, Brant AM, et al. Regression of cytomegalovirus retinitis with zidovudine. Arch Ophthalmol 1989; 107:868.

59. Pollard RB, Egbert PR, Gallaher JG, Merigan T. Cytomegalovirus retinitis in immunosuppressed hosts. Ann Intern Med 1980; 93:655.

60. Buhles WC, Mastre BJ, Tinker AJ, et al. Ganciclovir treatment of life- or sight-threatening cytomegalovirus infection: experience with 314 immunocompromised patients. Rev Infect Dis 1988; 10:495.

61. Ussery FM, Gibson SR, Conklin RH, et al. Intravitreal ganciclovir in the treatment of AIDS-associated cytomegalovirus retinitis. Ophthalmology 1988; 95:640.

62. Jabs DA, Newman C, De Bustros S, Polk BF. Treatment of cytomegalovirus retinitis with ganciclovir. Ophthalmology 1987; 94:824.

63. Holland GN, Buhles WC, Mastre B, et al. A controlled retrospective study of ganciclovir treatment for cytomegalovirus retinopathy. Arch Ophthalmol 1989; 107:1759.

64. Jabs DA, Martin BK, Forman MS, et al. Mutations conferring ganciclovir resistance in a cohort of patients with acquired immunodeficiency syndrome and cytomegalovirus retinitis. Infect Dis 2001; 183:333–337.

65. Jabs DA, Martin BK, Forman MS, et al. Cytomegalovirus retinitis and viral resistance study group. Cytomegalovirus resistance to ganciclovir and clinical outcomes of patients with cytomegalovirus retinitis. Am J Ophthalmol 2003; 135(1):26–34.

66. Holland GN, Sison RF, Jatulis DE, et al. Survival of patients with the acquired immune deficiency syndrome after development of cytomegalovirus retinopathy. Ophthalmology 1990; 97:204.

67. Lehoang P, Girard B, Robinet M, et al. Foscarnet in the treatment of cytomegalovirus retinitis in acquired immune deficiency syndrome. Ophthalmology 1989; 96:865.

68. Studies of the Ocular Complications of AIDS Research Group, in collaboration with the AIDS Clinical Trials group. Foscarnet–ganciclovir cytomegalovirus retinitis trial. 4. Visual outcomes. Ophthalmology 1994; 101(7):1250–1261.

69. The Studies of Ocular Complications of AIDS Research Group in Collaboration with the AIDS Clinical Trials Group. Combination foscarnet and ganciclovir therapy vs monotherapy for the treatment of relapsed cytomegalovirus retinitis in patients with AIDS. The cytomegalovirus retreatment trial. Arch Ophthalmol 1996; 114(1):23–33.

70. Plosker GL, Noble S. Cidofovir: a review of its use in cytomegalovirus retinitis in patients with AIDS. Drugs 1999; 58:325–345.

71. The Studies of Ocular Complications of AIDS Research Group in collaboration with the AIDS Clinical Trials Group. Long-term follow-up of patients with AIDS treated with parenteral cidofovir for cytomegalovirus retinitis: the HPMPC Peripheral Cytomegalovirus Retinitis Trial. AIDS 2000; 14:1571.

72. Studies of Ocular Complications of AIDS Research Group in Collaboration with the AIDS Clinical Trials Group. Parenteral cidofovir for cytomegalovirus retinitis in patients with AIDS: the HPMPC peripheral cytomegalovirus retinitis trial. A randomized, controlled trial. Ann Intern Med 1997; 126(4):264–274.

73. Davis JL, Taskintuna I, Freeman WR, et al. Iritis and hypotony after treatment with intravenous cidofovir for cytomegalovirus retinitis. Arch Ophthalmol 1997; 115:733–740.

74. Taskintuna I, Rahhal FM, Rao NA, et al. Adverse events and autopsy findings after intravitreous cidofovir (HPMPC) therapy in patients with acquired immune deficiency syndrome (AIDS). Ophthalmology 1997; 104:1827–1836.

75. Drew WL, Ives D, Lalezari JP, et al. Oral ganciclovir as maintenance treatment for cytomegalovirus retinitis in patients with AIDS. N Engl J Med 1995; 333:615–620.

76. Martin DF, Kuppermann BD, Wolitz RA, et al. Oral ganciclovir for patients with cytomegalovirus retinitis treated with a ganciclovir implant. N Engl J Med 1999; 340:1063–1070.

77. Martin DF, Sierra-Madero J, Walmsley S, et al. A controlled trial of valganciclovir as induction therapy for cytomegalovirus retinitis. N Engl J Med 2002; 346:1119–1126.

78. Jacobson MA, Wilson S, Stanley H, Holtzer C, Cherrington J, Safrin S. Phase I study of combination therapy with intravenous cidofovir and oral ganciclovir for cytomegalovirus retinitis in patients with AIDS. Clin Infect Dis 1999; 28:528–533.

79. Pulido J, Palacio M, Peyman GA, et al. Toxicity of intravitreal antiviral drugs. Ophthalmic Surg 1984; 15:666.

80. She SC, Peyman GA, Schulman JA. Toxicity of intravitreal injection of foscarnet in the rabbit eye. Int Ophthalmol 1988; 12:151.

81. Heinemann MH. Long-term intravitreal ganciclovir therapy for cytomegalovirus retinopathy. Arch Ophthalmol 1989; 107:1767.

82. Cantrill HL, Henry K, Melroe NH, et al. Treatment of cytomegalovirus retinitis with intravitreal ganciclovir. Ophthalmology 1989; 96:367.

83. Diaz-Llopis M, Chipont E, Sanchez S, et al. Intravitreal foscarnet for cytomegalovirus retinitis in a patient with acquired immune deficiency syndrome. Am J Ophthalmol 1992; 114:742.

84. Taskintuna I, Rahhal FM, Arevalo JF, et al. Low-dose intravitreal cidofovir (HPMPC) therapy of cytomegalovirus retinitis in patients with acquired immune deficiency syndrome. Ophthalmology 1997; 104:1049–1057.

85. Song M, Azen SP, Buley A, et al. Effect of anti-cytomegalovirus therapy on the incidence of immune recovery uveitis in AIDS patients with healed cytomegalovirus retinitis. Am J Ophthalmol 2003; 136;696–702.

86. Perry CM, Balfour J. Fomivirsen. Drugs 1999; 57:375–380.

87. Amin HI, Ai E, McDonald HR, Johnson RN. Retinal toxic effects associated with intravitreal fomivirsen. Arch Ophthalmol 2000; 118(3):426–427.

88. The Vitravene Study Group. Randomized dose-comparison studies of intravitreous fomivirsen for treatment of cytomegalovirus retinitis that has reactivated or is persistently active despite other therapies in patients with AIDS. Am J Ophthalmol 2002; 133(4):475–483.

89. The Vitravene Study Group. A randomized controlled clinical trial of intravitreous fomivirsen for treatment of newly diagnosed peripheral cytomegalovirus retinitis in patients with AIDS. Am J Ophthalmol 2002; 133(4):467–474.

90. Vitravene Study Group. Safety of intravitreous fomivirsen for treatment of cytomegalovirus retinitis in patients with AIDS. Am J Ophthalmol 2002; 133(4):484–498.

91. Stone TW, Jaffe GJ. Reversible bull's-eye maculopathy associated with intravitreal fomivirsen therapy for cytomegalovirus retinitis. Am J Ophthalmol 2000; 130:242–243.

92. Uwaydat SH, Li HK. Pigmentary retinopathy associated with intravitreal fomivirsen. Arch Ophthalmol 2002; 120:854–857.

93. Zambarakji HJ, Mitchell SM, Lightman S, Holder GE. Electrophysiological abnormalities following intravitreal Vitravene (ISIS 2922) in two patients with CMV retinitis. Br J Ophthalmol 2001; 85:1142.

94. Sanborn GE, Anand R, Torti RE, et al. Sustained-release ganciclovir therapy for treatment of cytomegalovirus retinitis. Arch Ophthalmol 1992; 110:188.

95. Anand R, Nightingale SD, Fish RH, et al. Control of cytomegalovirus retinitis using sustained release of intraocular ganciclovir. Arch Ophthalmol 1993; 111:223.

96. Duker JS, Robinson M, Anand R, Ashton P. Initial experience with an eight month sustained-release intravitreal ganciclovir implant for the treatment of CMV retinitis associated with AIDS. Ophthalmic Surg Lasers 1995; 26:442–448.

97. Duker JS, Ashton P, Davis JL, Keller R, Chuang E. Long-term successful maintenance of bilateral cytomegalovirus retinitis using exclusively local therapy. Arch Ophthalmol 1996; 114:881–882.

98.   Musch DC, Martin DF, Gordon JF, et al. Treatment of CMV retinitis with a sustained-release ganciclovir implant. The ganciclovir implant study group. N Engl J Med 1997; 115:733–740.

99.   Morley MG, Duker JS, Ashton P, Robinson MR. Replacing ganciclovir implants. Ophthalmology 1995; 102:388–392.

100. Martin DF, Dunn JP, Davis JL, et al. Use of the ganciclovir implant for the treatment of cytomegalovirus retinitis in the era of potent antiretroviral therapy: recommendations of the International AIDS Society—USA panel. Am J Ophthalmol 1999; 127:329–339.

101. Martin D, Kupperman B, Wolitz R, Palestine C, Robinson C. Combined oral ganciclovir and intravitreal ganciclovir implant for treatment of patients with cytomegalovirus (CMV) retinitis; a randomized controlled study. In: 37th Interscience Conference on Antimicrobial Agents and Chemotherapy, Toronto, Ontario, Canada, 1997. Abstract LB-9.

102. Hatton MP, Duker JS, Reichel E, Morley MG, Puliafito CA. Treatment of relapsed CMV retinitis with the sustained-release ganciclovir implant. Retina 1998; 18:50–55.

103. Roth DB, Feuer WJ, Blenke AJ, Davis JL. Treatment of recurrent cytomegalovirus retinitis with the ganciclovir implant. Am J Ophthalmol 1999; 127:276–282.

104. McGuire DE, McAulife P, Heinemann MH, Rahhal FM. Efficacy of the ganciclovir implant in the setting of silicone oil vitreous substitute. Retina 2000; 20:520–523.

105. Davis JL, Tabandeh H, Feuer WJ, Kumbhat S, Roth DB, Chaudhry NA. Effect of potent antiretroviral therapy on recurrent cytomegalovirus retinitis treated with the ganciclovir implant. Am J Ophthalmol 1999; 127:283–287.

106. Jabs DA, Bolton SG, Dunn JP, Palestine AG. Discontinuing anticytomegalovirus therapy in patients with immune reconstitution after combination antiretroviral therapy. Am J Ophthalmol 1998; 126:817–822.

107. Whitcup SM, Fortin E, Lindblad AS, et al. Discontinuation of anticytomegalovirus therapy in patients with HIV infection and cytomegalovirus retinitis. J Am Med Assoc 1999; 282:1633–1637.

108. Song MK, Karavellas MP, MacDonald JC, Plummer DJ, Freeman WR. Characterization of reactivation of cytomegalovirus retinitis in patients healed after treatment with highly active antiretroviral therapy. Retina 2000; 20:151–155.

109. Roth DB, Feuer WJ, Blenke AJ, Davis JL. Treatment of recurrent cytomegalovirus retinitis with the ganciclovir implant. Am J Ophthalmol 1999; 127:276–282.

110. Martin DF, Ferris FL, Parks DJ, et al. Ganciclovir implant exchange; timing, surgical procedure, and complicatins. Arch Ophthalmol 1997; 115:1389–1394.

111. Martin DF, Parks DJ, Mellow SD, et al. Treatment of cytomegalovirus retinitis with an intraocular sustained release ganciclovir implant. Arch Ophthalmol 1994; 112:1531–1539.

112. Lim JI, Wolitz RA, Dowling AH, et al. Visual and anatomic outcomes associated with posterior segment complications after ganciclovir implant procedures in patients with AIDS and cytomegalovirus retinitis. Am J Ophthalmol 1999; 127:288–293.

113. Shane TS, Martin DF. for the Endophthalmitis-Ganciclovir Implant Study Group. Am J Ophthalmol 2003; 136:649–654.

114. Kattan HM, Flynn HW Jr, Pflugfelder SC, Robertson C, Forster RK. Nonsocomial endophthalmitis survey: current incidence of infection after intraocular surgery. Ophthalmology 1991; 98:227–238.

115. Boyer DS, Posalski J. Potential complication associated with removal of ganciclovir implants. Am J Ophthalmol 1999; 127:349–350.

116. Chern KC, Margolis TP, Chandler DB, et al. Glycoprotein B subtyping of cytomegalovirus strains in the vitreous of patients with AIDS and cytomegalovirus retinitis. J Infect Dis 1998; 178:1149–1153.

22
Endophthalmitis

Travis A. Meredith
*Department of Ophthalmology, University of North Carolina, Chapel Hill,
North Carolina, U.S.A.*

OVERVIEW

Endophthalmitis is an uncommon, but perhaps the most feared complication of ocular surgery. Endophthalmitis is defined as a microbial infection involving the vitreous cavity: organisms are often isolated from anterior chamber as well. Retinal, choroidal, and scleral invasion can also occur. Most cases of endophthalmitis occur after elective ocular surgery. In the first six weeks after operation, endophthalmitis is caused by microbes introduced into the eye during the time of the surgery or in the immediate postoperative period before the wound is securely sealed. Delayed onset or chronic endophthalmitis can occur as the result of slow-growing bacteria such as *Propionibacterium acnes*. Occasionally, bacterial endophthalmitis has a late onset and some delayed cases are caused by fungi. The second most common cause of endophthalmitis is penetrating ocular trauma, while infected filtering blebs constitute a less common but clinically significant presentation. The least common form of endophthalmitis is the endogenous type. These infections may be either bacterial or fungal, originating from an infection elsewhere in the body transmitted to the eye by hematogenous route.

PRESENTATION

The typical presenting complaints of endophthalmitis are pain, decreased vision, and conjunctival hyperemia (1). The earliest findings are periphlebitis accompanied by anterior chamber reaction and vitreous cellular debris (2). In the typical case, the vitreous is too opaque to view retinal detail. Hypopyon is considered the hallmark of an infection although severe sterile inflammation may cause hypopyon in some instances. Wound abnormalities are commonly noted in endophthalmitis cases. Lid edema is a frequent finding.

PREDISPOSING FACTORS

Predisposing factors in postoperative cases are related both to preoperative patient factors and operative complications. External infections including blepharitis and lacrimal sac infection predispose patients to postoperative endophthalmitis. Diabetics and immunocompromised patients are more likely to develop postoperative infections (3). Capsular rupture during phacoemulsification, vitreous loss, and wound complications are all thought to predispose patients to postoperative infection (3).

In traumatic globe injury, the incidence of infection depends on the type of trauma. City dwellers rarely develop endophthalmitis because of the nature of their injuries in this group, infections occur in only 1% to 2% of eyes following trauma. Intraocular foreign bodies become infected in 11% to 17% of the reported cases (4,5). Endophthalmitis following trauma that occurs in a rural setting may complicate as many as 30% of penetrating ocular injuries (6).

BACTERIAL SPECTRUM

The bacterial spectrum of the infections depends of the origin of the infection. Postoperative endophthalmitis results almost uniformly from the patient's flora inadvertently introduced into the eye (7). The incidence of positive anterior chamber cultures obtained at the end of cataract surgery has been reported to be as high as 22% to 43% but only a small minority of patients actually become infected (8). Gram-positive coagulase negative-micrococci account for about 70% of the culture-positive cases of postoperative endophthalmitis. *Staphylococcus* and *streptococcus* are less common organisms while gram-negative infections occur infrequently; for example, gram-negative organisms accounted for only 6% of cases in the Endophthalmitis Vitrectomy Study (EVS) patients (1). Approximately one-third of patients with postoperative endophthalmitis have negative cultures. In some of these cases, polymerase chain reaction (PCR) indicates that bacteria were present but cannot be cultured from the material obtained at vitreous biopsy (1,9).

In cases of ocular trauma, *Bacillus* species are the most common cause of postoperative infection, particularly in a rural setting (6,10–12). Staphylococcal organisms are the next most common while gram-negative infections have an increased likelihood by comparison to postoperative cases. Mixed infections involving multiple bacteria are also more common than after elective surgery (5,6,13).

Bleb-related endophthalmitis is caused by streptococcal infection approximately half of the time and *Hemophilus influenza* is seen more commonly than after postcataract infections (14,15). The cause of endogenous endophthalmitis depends on the population studied. *Staphylococcus* and *Streptococcus* are frequent bacterial causes. Candida septicemia is frequently found in hospitalized patients.

ANTIMICROBIAL THERAPY

History

Leopold and Scheie (16,17) first introduced intravitreal antimicrobial therapy for treatment of endophthalmitis shortly after the development of antibiotics, but intraocular antibiotics did not gain favor for treatment of endophthalmitis for several decades. Up until the 1970s intravenous antimicrobials, frequent topical drops, and subconjunctival injections were the favored routes of antimicrobial delivery. Intravenous antimicrobial treatment was demonstrated to be effective in some cases of *Staphylococcus epidermidis* endophthalmitis, but effectiveness of intravenous therapy has not been demonstrated in randomized clinical trials (18).

In the late 1970s and early 1980s, Peyman and Baum et al. (19,20) popularized the use of intraocular antibiotics and this route of administration became the mainstay of therapy for intraocular infection. Intraocular antimicrobials are now given in essentially all cases of endophthalmitis; some patients are also treated with vitrectomy. In the EVS, all patients received intraocular antibiotics but only half the patients received systemic therapy. Systemic antimicrobials did not improve prognosis in the EVS, but amikacin, which has poor intraocular penetration, was used for gram-positive coverage in the study (1).

Choice of Intravitreal Antimicrobial

The choice of antimicrobials for intravitreal injection is guided by several factors: (i) Since the organism causing infection is unknown at the time when the antimicrobial injection is done, both gram-positive and gram-negative coverage is necessary. Thus, it is generally necessary to give injections of two different antibiotics to assure coverage of the potential pathogens. (ii) The therapeutic window is considered particularly with regard to the potential for ocular toxicity (see Chapter 6). (iii) The duration of antimicrobial effect plays a role in choice of antibiotic since some agents such as ciprofloxacin have a very short half-life.

Intravenous Antimicrobials

The role of intravenous or oral antimicrobials in the prophylaxis and treatment of endophthalmitis remains controversial. In the EVS, amikacin and ceftazidime were administered in half of the cases but did not result in better visual outcomes compared to cases treated without intravenous antimicrobials (1). Since the aminoglycosides have very poor penetration into the vitreous after intravenous administration and ceftazidime has poor gram-positive coverage, this result is not surprising (21–26). Better penetration into the vitreous cavity is a characteristic of some of the fluoroquinolones, especially gatifloxacin and some of the beta-lactam antimicrobials (27–34). Inflammation, trauma, or prior surgery all increase the likelihood of penetration to levels above the minimum inhibitory concentration (MIC) of the relevant organisms, so that even vancomycin administrated intravenously can reach potentially therapeutic levels in infected eyes (35,36). Clinical studies have demonstrated that intravenously administered ceftazidime can modulate experimental *Pseudomonas* infections, and that intravenous cefazolin is effective in treatment of experimental traumatic endophthalmitis (29,37). Intravenous antimicrobials may be effective in endogenous endophthalmitis in which there is a presumed site of infection or entry into the eye in the uvea or immediately adjacent to it.

Sterilization of the Vitreous Cavity

The goal of antimicrobial therapy is sterilization of the vitreous cavity and cure of the intraocular infection. Clinical results show good treatment responses for some species of bacteria. In the EVS and in studies by Omerod et al. (38), gram-positive, coagulase-negative micrococci had a good clinical outcome from intravitreal therapy (39,40). Gram-positive coagulase-negative staphylococci are easy to eradicate in experimental models, and in fact, some strains have been demonstrated to die spontaneously after injection into the vitreous cavity (41,42). In other species, the visual results were not as good, and when reculture was done in some eyes persistent infection was documented.

In the EVS, a second culture was positive in eyes infected initially with gram-positive, coagulase-negative micrococci (20%), *Staphylococcus aureus* and *Streptococcus* (47%), and gram-negative organisms (62%) (1,43). Shaarawy et al. (44) also demonstrated persistent infection from a variety of organisms after appropriate intravitreal antimicrobials were injected.

Single injections of antimicrobials do not completely sterilize the vitreous cavity in a variety of animal models. For example, in several studies in which *S. aureus* was the infecting organism, single injections were not effective in sterilizing the vitreous cavity (45–47). In a second study, a significant percentage of infected eyes had persistent positive cultures after injection of appropriate antimicrobials (48). *Streptococci* and *Enterococcus* persist in the vitreous cavity despite intravitreal antimicrobial injection (45). Failure of ciprofloxacin therapy when bacteria were injected into the eye 18 or 24 hours prior to treatment was demonstrated by Kim et al. (49) in models of *Pseudomonas* infection. Perhaps most surprising was a study to show that antimicrobials affected the growth of *Pseudomonas aeruginosa* in the vitreous cavity as a time-dependent phenomenon (50). In this study, early treatment successfully eradicated the bacteria but treatment was ineffective when instituted 48 hours or later after the bacterial infection was established.

Reinjection Strategies

Because single injections of antimicrobials may not be effective, toxicity studies after reinjection have been investigated. In one study, injection of vancomycin and ceftazadime was followed 48 hours later with injection of one or the other drug, reasoning that by that time culture results would be available and it would be not necessary to cover for only gram-positive or gram-negative organisms in most cases. Significant toxicity was not noted after a second injection although a third injection another 48 hours later did create mild signs of toxicity (51). In a separate study, when eyes were reinjected with both antibiotics most demonstrated retinal toxic reactions that worsened with a third injection was performed (52).

INJECTION OF INTRAOCULAR CORTICOSTEROIDS

Injection of corticosteroids into the eye at the time of endophthalmitis therapy simultaneously with the injection of antimicrobials has been advocated. Corticosteroids administered by other routes have been demonstrated in numerous studies to improve the outcome of antimicrobial treatment of intraocular infection. Each patient in the EVS received prednisone 30 mg twice a day (1). Animal models consistently demonstrated improved inflammatory scores without interfering with the ability of the antimicrobial to sterilize the vitreous cavity. However, in an animal model negative as well as positive effects have been noted in studies of intraocular injection (48,51,53).

Several human studies have attempted to analyze the outcome of intraocular dexamethasone injection along with intraocular antimicrobials. In a small randomized study by Das et al. (54), an early beneficial effect on inflammatory scores was noted when patients were treated with vitrectomy and intraocular antibiotics and intraocular corticosteroids. No significant influence could be demonstrated on visual outcome 12 weeks after therapy (54). Others have attempted to analyze the effect of intraocular corticosteroids on the outcome in retrospective reviews. While animal studies have suggested a benefit in one human study those eyes treated with the intraocular corticosteroids actually had a worse outcome than those without (55).

PROGNOSIS

The outcome of infectious endophthalmitis treatment depends on multiple factors, most notably on the infecting organism. In the EVS, slightly more than half of the eyes achieved 20/40 vision and three quarters were 20/100 or better (1). Infections with gram-positive, coagulase-negative micrococci have a relatively good prognosis; 50% of eyes have 20/40 vision after treatment (1). In the EVS, treatment of infection caused by other organisms produced good visual acuity less commonly. While 84% of eyes with gram-positive, coagulase-negative micrococci achieved 20/100 vision or better, only 30% of eyes with streptococcal infection and 14% of eyes with entero-coccal infections attained this level. Approximately one half of the eyes with *S. aureus* infections and gram-negative infections were 20/100 or better (1,38). Eyes with streptococcal infections have also been shown to have a poor outcome in other series (56). These data help explain the poor treatment results in bleb related infections which are most often caused by *Streptococcus* (15,57).

In traumatic endophthalmitis, both the effects of the injury and the damage added by the infection contribute to the visual outcome. Generally, few patients recover useful vision and most see less than 20/400 (4,5). Many of these infections are caused by *Bacillus* for which successful treatment is rare (6,10–12). The onset of *Bacillus* infection is explosive, often destroying the eye within 12 hours. Endogenous endophthalmitis may have a good outcome when it is mild and linked to successfully treated systemic septicemia. Diagnosis is often delayed, however, and the resulting infections can be severe and cause significant visual loss.

SUMMARY

Injection of intraocular antibiotics is the mainstay of endophthalmitis therapy. High concentrations of the drug are achieved by intraocular injection and therapeutic outcomes are favorable following infection by less virulent bacteria such as gram-positive, coagulase-negative micrococci. Despite the ability to sterilize the vitreous cavity in many infections, bacteria persist in others even after treatment. Outcomes are often poor in cases such as those caused by *Streptococcus* and especially those attributable to *Bacillus* infections. Intravenous antibiotics are capable of eradicating mild intraocular infections but their additive effect to intraocular injections has not been demonstrated. Corticosteroids are thought to be beneficial in the treatment of endophthalmitis, although the superiority of intraocular corticosteroid injections over other forms of therapy has not been demonstrated, and potential harm has been noted in clinical and animal studies.

REFERENCES

1. Endophthalmitis Vitrectomy Study Group. Results of the endophthalmitis vitrectomy study. A randomized trial of immediate vitrectomy and of intravenous antibiotics for the treatment of postoperative bacterial endophthalmitis. Arch Ophthalmol 1995; 113(12):1479–1496.

2. Packer AJ, Weingeist TA, Abrams GW. Retinal periphlebitis as an early sign of bacterial endophthalmitis. Am J Ophthalmol 1983; 96:66–71.

3. Speaker MG, Menikoff JA. Postoperative endophthalmitis: pathogenesis, prophylaxis, and management. Int Ophthalmol Clin 1993; 33:51–70.

4. Affeldt JC, Flynn HW Jr, Forster RK, Mandelbaum S, Clarkson JG, Jarus GD. Microbial endophthalmitis resulting from ocular trauma. Ophthalmology 1987; 94:407–413.

5. Brinton GS, Topping TM, Hyndiuk RA, Aaberg TM, Reeser FH, Abrams GW. Posttraumatic endophthalmitis. Arch Ophthalmol 1984; 102:547–550.

6. Boldt HC, Pulido JS, Blodi CS, Folk JC, Weingeist TA. Rural endophthalmitis. Ophthalmology 1989; 96:1722–1726.

7. Speaker MG, Milch FA, Shah MK. Role of external bacterial flora in the pathogenesis of acute postoperative endophthalmitis. Ophthalmology 1991; 98:639.

8. Ariyasu RG, Nakamura T, Trousdale MD, Smith RE. Intraoperative bacterial contamination of the aqueous humor. Ophthalmic Surg 1993; 24(6):367–373.

9. Okhravi N, Adamson P, Carroll N, et al. PCR_based evidence of bacterial involvement in eyes with suspected introcular infection. Invest Ophthalmol Vis Sci 2000; 41: 3474–3479.

10. Das T, Choudhury K, Sharma S, Jalali S, Nuthethi R. Clinical profile and outcome in Bacillus endophthalmitis. Ophthalmology 2001; 108(10):1819–1825.

11. Davey RT, Tauber WB. Posttraumatic endophthalmitis: the emerging role of Bacillus cereus infection. Rev Infect Dis 1987; 9:110–123.

12. O'Day DM, Smith RS, Gregg CR, et al. The problem of Bacillus species infection with special emphasis on the virulence of *Bacillus cereus*. Ophthalmology 1981; 88:833–838.

13. Kunimoto DY, Das T, Sharma S, et al. Microbiologic spectrum and susceptibility of isolates: part II. Posttraumatic endophthalmitis. Endophthalmitis Research Group. Am J Ophthalmol 1999; 128(2):242–244.

14. Mandelbaum S, Forster RK. Endophthalmitis associated with filtering blebs. Int Ophthalmol Clin 1987; 27:107–111.

15. Song A, Scott IU, Flynn HW Jr, Budenz DL. Delayed-onset bleb-associated endophthalmitis: clinical features and visual acuity outcomes. Ophthalmology 2002; 109(5):985–991.

16. Leopold IH, Scheie HG. Studies with microcrystalline sulfathiazole. Arch Ophthalmol 1943; 29:811.

17. Leopold IH. Intravitreal penetration of penicillin and penicillin therapy of infections of the vitreous. Arch Ophthalmol 1945; 33:211–216.

18. O'Day DM, Jones DB, Patrinely J, Elliott JH. Staphylococcus epidermidis endophthalmitis: visual outcome following noninvasive therapy. Ophthalmology 1982; 89:354–360.

19. Peyman GA. Antibiotic administration in the treatment of bacterial endophthalmitis. II. Intravitreal injections. Surv Ophthalmol 1977; 21:332–346.

20. Baum J, Peyman GA, Barza M. Intravitreal administration of antibiotic in the treatment of bacterial endophthalmitis. III. Consensus. Surv Ophthalmol 1982; 26:204–206.

21. Alfaro DV III, Hudson SJ, Kasowski EJ, et al. Experimental pseudomonal posttraumatic endophthalmitis in a swine model. Treatment with ceftazidime, amikacin, and imipenem. Retina 1997; 17(2):139–145.

22. Fiscella RG, Gieser J, Phillpotts B, et al. Intraocular penetration of gentamicin after once-daily aminoglycoside dosing. Retina 1998; 18(4):339–342.

23. Mandell BA, Meredith TA, Aguilar E, el Massry A, Sawant A, Gardner S. Effects of inflammation and surgery on amikacin levels in the vitreous cavity. Am J Ophthalmol 1993; 115(6):770–774.

24. Ng EW, D'Amico DT. Therapeutic efficacy of intravitreal ceftazidime, imipenem, and amikacin in a swine model of posttraumatic *Pseudomonas aeruginosa* endophthalmitis. Retina 1997; 17(5):464–465.

25. Verbraeken H, Verstraete A, Van de Velde E, Verschraegen G. Penetration of gentamicin and ofloxacin in human vitreous after systemic administration. Graefes Arch Clin Exp Ophthalmol 1996; 234(suppl 1):S59–S65.

26. Yoshizumi MO, Leinwand MJ, Kim J. Topical and intravenous gentamicin in traumatically lacerated eyes. Graefes Arch Clin Exp Ophthalmol 1992; 230:175–177.
27. Hariprasad SM, Mieler WF, Holz ER. Vitreous penetration of orally administered gatifloxacin in humans. Trans Am Ophthalmol Soc 2002; 100:153–159.
28. Hariprasad SM, Mieler WF, Holz ER. Vitreous and aqueous penetration of orally administered gatifloxacin in humans. Arch Ophthalmol 2003; 121(3):345–350.
29. Alfaro DV, Runyan T, Kirkman E, Tran VT, Liggett PE. Intravenous cefazolin in penetrating eye injuries. II. Treatment of experimental posttraumatic endophthalmitis. Retina 1993; 13:331–334.
30. Alfaro DV, Liggett PE. Intravenous cefazolin in penetrating eye injuries. Effects of trauma and multiple doses on intraocular delivery. Graefes Arch Clin Exp Ophthalmol 1994; 232:238–241.
31. Barza M, Kane A, Baum J. Ocular penetration of subconjunctival oxacillin, methicillin, and cefazolin in rabbits with staphylococcal endophthalmitis. J Infect Dis 1982; 145:899–903.
32. Barza M, Kane A, Baum J. Pharmacokinetics of intravitreal carbenicillin, cefazolin, and gentamicin in rhesus monkeys. Invest Ophthalmol Vis Sci 1983; 24:1602–1606.
33. Martin DF, Ficker LA, Aguilar HA, Gardner SK, Wilson LA, Meredith TA. Vitreous cefazolin levels after intravenous injection. Effects of inflammation, repeated antibiotic doses, and surgery. Arch Ophthalmol 1990; 108(3):411–414.
34. Schech JM, Alfaro DV, Laughlin RM, Sanford EG, Briggs J, Dalgetty M. Intravenous gentamicin and ceftazidime in penetrating ocular trauma: a swine model. Retina 1997; 17:28–32.
35. Meredith TA. Antimicrobial pharmacokinetics in endophthalmitis treatment: studies of ceftazidime. Trans Am Ophthalmol Soc 1993; 91:653–699.
36. Meredith TA, Aguilar HE, Shaarawy A, Kincaid M, Dick J, Niesman MR. Vancomycin levels in the vitreous cavity after intravenous administration. Am J Ophthalmol 1995; 119:774–778.
37. Liang C, Meredith TA, Aguilar HE. Prophylaxis of experimental *Pseudomonas* endophthalmits with intravenous ceftazidime. Invest Ophthalmol Vis Sci 1994; 35:156.
38. Ormerod LD, Ho DD, Becker LE, et al. Endophthalmitis caused by the coagulase-negative staphylococci. 1. Disease spectrum and outcome. Ophthalmology 1993; 100:715–723.
39. Han DP, Wisniewski SR, Wilson LA, et al. Spectrum and susceptibilities of microbiologic isolates in the endophthalmitis vitrectomy study. Am J Ophthalmol 1996; 122:1–17.
40. Endophthalmitis Vitrectomy Study Group. Microbiologic factors and visual outcome in the endophthalmitis vitrectomy study. Am J Ophthalmol 1996; 122(6):830–846.
41. Meredith TA, Trabelsi A, Miller MJ, Aguilar E, Wilson LA. Spontaneous sterilization of experimental *Staphylococcus epidermidis* endophthalmitis. Invest Ophthalmol Vis Sci 1990; 31:181–186.
42. Smith MA, Sorenson JA, Lowy FD. Treatment of experimental methicillin-resistant *Staphylococcus epidermidis* endophthalmitis with intravitreal vancomycin. Ophthalmology 1986; 93:1328–1335.
43. Doft BH, Kelsey SF, Wisniewski SR. Additional procedures after the initial vitrectomy or tap-biopsy in the Endophthalmitis Vitrectomy Study. Ophthalmology 1998; 105(4):707–716.
44. Shaarawy A, Grand MG, Meredith TA, Ibanez H. Persistent infection after intravitreal antibimicrobial therapy. Ophthalmology 1995; 102:382–387.
45. Forster RK. Experimental postoperative endophthalmitis. Trans Am Ophthalmol Soc 1992; 90:505–559.
46. Aguilar HE, Meredith TA, Drews CD, Sawant A, Gardner S, Wilson LA. Treatment of experimental *S. aureus* endophthalmitis with vancomycin, cefazolin and corticosteroids. Invest Ophthalmol Vis Sci 1990; 31(suppl):308.

47. Yoshizumi MO, Lee GC, Equi RA, et al. Timing of dexamethasone treatment in experi-
     mental *Staphylococcus aureus* endophthalmitis. Retina 1998; 18(2):130–135.
48. Aguilar HE, Meredith TA, Drews C, Grossniklaus H, Sawant AD, Gardner S. Compara-
     tive treatment of experimental *Staphylococcus aureus* endophthalmitis. Am J Ophthalmol
     1996; 121(3):310–317.
49. Kim IT, Chung KH, Koo BS. Efficacy of ciprofloxacin and dexamethasone in experi-
     mental *Pseudomonas* endophthalmitis. Korean J Ophthalmol 1996; 10(1):8–17.
50. Davey PG, Barza M, Stuart M. Dose response of experimental *Pseudomonas* endophtha-
     lmitis to ciprofloxacin, gentamicin, and imipenem: evidence of resistance to "late" treatment
     of infections. J Infect Dis 1987; 155:518–523.
51. Meredith TA, Abdala C, Aguilar HE, Chanping L, Hageman GS. Toxicity of intravitreal
     ceftazidime and vancomycin. Invest Ophthalmol Vis Sci 1995; 36:3647.
52. Oum BS, D'Amico DJ, Wong KW. Intravitreal antibiotic therapy with vancomycin and
     aminoglycoside. An experimental study of combination and repetitive injections. Arch
     Ophthalmol 1989; 107:1055–1060.
53. Maxwell DP Jr, Brent BD, Diamond JG, Wu L. Effect of intravitreal dexamethasone on
     ocular histopathology in a rabbit model of endophthalmitis. Ophthalmology 1991;
     98(9):1370–1375.
54. Das T, Jalali S, Gothwal VK, Sharma S, Naduvilath TJ. Intravitreal dexamethasone in
     exogenous bacterial endophthalmitis: results of a prospective randomised study. Br J
     Ophthalmol 1999; 83(9):1050–1055.
55. Shah GK, Stein JD, Sharma S, et al. Visual outcomes following the use of intravitreal
     steroids in the treatment of postoperative endophthalmitis. Ophthalmology 2000;
     107:486–489.
56. Mao LK, Flynn HW Jr, Miller D, Pflugfelder SC. Endophthalmitis causes by streptococ-
     cal species. Arch Ophthalmol 1992; 110:798–801.
57. Mandelbaum S, Forster RK, Gelender H, Culbertson W. Late onset endophthalmitis
     associated with filtering blebs. Ophthalmology 1985; 92:964–972.

Index